IMMOBILIZED ENZYMES, ANTIGENS, ANTIBODIES, AND PEPTIDES

PREPARATION AND CHARACTERIZATION

ENZYMOLOGY

A Series of Textbooks and Monographs

Volume 1. Immobilized Enzymes, Antigens, Antibodies, and Peptides: Preparation and Characterization, *edited by Howard H. Weetall*

In Preparation

Enzyme Structure and Function, *by S. Blackburn*
Subunit Enzymes: Biochemistry and Functions, *edited by Kurt Ebner*
Enzyme Reactions and Enzyme Systems, *By Charles Walter*

Additional Volumes in Preparation

IMMOBILIZED ENZYMES, ANTIGENS, ANTIBODIES, AND PEPTIDES

PREPARATION AND CHARACTERIZATION

edited by Howard H. Weetall

Research and Development Laboratories
Corning Glass Works
Corning, New York

MARCEL DEKKER, INC. New York

MARCEL DEKKER, INC.

270 Madison Avenue, New York, New York 10016

LIBRARY OF CONGRESS CATALOG CARD NUMBER: 75-11417

ISBN: 0-8247-6285-1

Current printing (last digit):
10 9 8 7 6 5 4 3 2 1

PRINTED IN THE UNITED STATES OF AMERICA

PREFACE

The study of immobilized biologically active materials has, within recent years, become a major area of endeavor. Many symposia have been held in the U.S. and abroad describing the many methods of immobilization and applications of immobilized enzymes, antigens, antibodies and other compounds. In addition to the symposia, several books on the subject have been published.

The published books have, for the most part, represented collected papers from symposia, or have generally been reviews of the literature. None of these volumes have been directly useful to the researcher trying to prepare or use a biologically active material.

The purpose of this book is to make available to the research community a document from which the reader can prepare and use an immobilized derivative of some biologically active compound. The contributors of this volume have attempted to present enough methodology to permit the reader, without further reference, to actually prepare and utilize an immobilized biologically active compound.

Of course, not all coupling methods or applications are covered. However, an attempt has been made to cover the major areas of immobilization technology and the major applications of this technology to enzymes, antigens, antibodies, and peptides.

Howard H. Weetall

CONTRIBUTORS

George Baum, Corning Glass Works, Corning, New York

Milton J. Becker, Metrix Division, Armour Pharmaceutical, Chicago, Illinois

Thomas Ming Swi Chang, McGill University, Montreal, Quebec, Canada

D. Dinelli, Laboratory for Microbiological Processes, SNAM Progetti, Monterotondo, Rome, Italy

George G. Guilbault, Louisiana State University, New Orleans, Louisiana

N. B. Havewala, Corning Glass Works, Corning, New York

W. E. Hornby, Department of Biochemistry, University of St. Andrews, St. Andres, Fife, United Kingdom

Richard A. Laursen, Department of Chemistry, Boston University, Boston, Massachusetts

William F. Line, Metrix Division, Armour Pharmaceutical, Chicago, Illinois

Merrill Lynn, Corning Glass Works, Corning, New York

W. Marconi, Laboratory for Microbiological Processes, SNAM Progetti, Monterotondo, Rome, Italy

F. Morisi, Laboratory for Microbiological Processes, SNAM Progetti, Monterotondo, Rome, Italy

D. L. Morris, Department of Biochemistry, University of St. Andrews, St. Andres, Fife, United Kingdom

W. H. Pitcher, Jr., Corning Glass Works, Corning, New York

Garfield P. Royer, Department of Biochemistry, The Ohio State University, Columbus, Ohio

Stanley J. Wrobel, Corning Glass Works, Corning, New York

CONTENTS

Chapter 1

INORGANIC SUPPORT INTERMEDIATES: COVALENT COUPLING OF ENZYMES ON INORGANIC SUPPORTS

Dr. Merrill Lynn

Corning Glass Works
Corning, N. Y.

I. INTRODUCTION

Chemical coupling of enzymes with reactive supports is currently the most frequently used method of enzyme immobilization. Chemical coupling techniques, which involve a reaction between an enzyme and a reactive support, have been discussed in a number of papers in recent years (1-7). Most of these reviews have been concerned primarily with organic supports, but many of the coupling reactions are also applicable to inorganic supports.

In general, the covalent bonding procedure involves the formation of an activated support, followed by the reaction of the activated support with an enzyme to form a composite. A single reaction may be involved, or several steps may be required in preparing the activated support and coupling with an enzyme. For simplicity, the coupling techniques can be separated into three categories:

1. Reactions in which an enzyme is coupled directly to a stable reactive intermediate, as in coupling to a reactive aldehyde intermediate (8);
2. Reactions involving the formation of an activated intermediate before coupling, as in the diazotization of an aryl amine support, followed by reaction with an enzyme (8); or,
3. Reactions involving the use of a third reactive component for the coupling step, as in the carbodiimide coupling of an enzyme to an alkyl amine intermediate (9).

It should be readily recognized that no single immobilization technique and no single support system are ideal for all enzymes. The activity of an enzyme on immobilization is affected by the coupling procedure, the enzyme site involved in coupling, the orientation of the enzyme on the support surface, and the support material. The minimal requirements for a support are that it be insoluble, have available reactive groups, and be durable under the conditions in which it will be used. The advantages of using inorganic supports for immobilizing enzymes are discussed elsewhere (8,10,11).

A variety of inorganic support materials has been used for immobilizing enzymes. These include alumina, bentonite, charcoal, glass, hydroxyapatite, nickel oxide, silica, titania, and zirconia. The inorganic materials may be porous or non-porous. Generally, the non-porous materials have small surface areas, limiting the reactive groups that can be formed on the particles. This can be overcome by using very small particles, as is done with charcoal and silica gel. However, the small particles are difficult to manipulate, cause problems in column operations, and require special handling on removal from batch operations. Another technique for increasing surface area is to apply a porous, inorganic coating to the particles, but only a few such materials are presently available.

Porous inorganic materials have large surface areas, most of which are internal to the particle. The disadvantage of this is that the enzyme and its substrate must be able to readily penetrate the pores (12,13). This is not a severe limitation, however, because a wide range of pore sizes are available with a number of inorganic materials.

Perhaps the best, and simplest, procedure for placing reactive organic groups on the surface of inorganic materials is through reaction with silane coupling agents. A silane coupling agent is a monomeric silane which has an organofunctional group at one end and groups that will react with the inorganic surface at the other end (14,15). In some cases, the silanized supports can be reacted directly with enzymes, but in most cases, the supports are further modified with organic intermediates before coupling with an enzyme.

It is well recognized, of course, that these chemical reactions must be carried out with those reactive groups on the enzyme which are not essential to the enzyme's biological activity. This presents no severe limitation, however, as there are a large number of groups available for reaction. Silman and Katchalski (1) described the following functional groups of proteins as available for covalent binding:

1. α- and ε- amino groups

2. α-, β- and γ- carboxyl groups

3. sulfhydryl groups of cysteine and hydroxyl group of serine.

4. imidazole group of histidine

5. phenol ring of tyrosine

The amine groups react readily with acylating and alkylating groups, with aldehydes, and with isocyanates and isothiocyanates. The hydroxyl groups react with acylating agents, and the sulfhydryl groups react with alkylating agents and organomercurial compounds. The imidazole and phenol groups can be coupled with diazonium salts. In short, most of the standard organic reactions can be used for coupling.

A point that must be stressed again is that there is no ideal or universal coupling procedure. The preferred method of attachment is dictated by the enzyme and the end use of the immobilized enzyme system. A number of reasons can be cited for lack of success in preparing an active immobilized enzyme. In some cases, an enzyme may react poorly (or it may not react at all) with the particular intermediate being used (10). In some cases, the immobilized enzyme may be inactive because reaction occurred at the active site. In either instance, different coupling procedures must be used.

In some cases, an enzyme may be denatured by being too tightly bound; i.e., too many sites on the enzyme have been reacted (16). In these cases, a support with fewer reactive groups should be used. In some cases, the bond formed in coupling may not be durable under long term reaction conditions, as in column operations (17). In these cases, the bonding group may be modified, or another coupling procedure must be used.

In all cases, judicious selection of reactive support and coupling procedures should produce a useful immobilized active enzyme.

II. METHODS FOR ATTACHING REACTIVE ORGANIC GROUPS TO INORGANIC SUPPORTS

Chemically coupling enzymes to inorganic supports requires the presence of reactive groups that can be readily manipulated. Most of the inorganic materials used for supports have on their surfaces hydroxyl and oxide groups which are reactive under relatively mild conditions, but the number of reactions available is limited. The most widely used technique for placing reactive organic groups on inorganic surfaces is through reaction with silane coupling agents. In more recent developments, inorganic surfaces have been activated with transition metal salts (18,19) and with cyanogen bromide (20). The mechanisms for the latter reactions have not been well characterized, so they will not be discussed here. However, these reactions offer additional choices for modifying inorganic surfaces, and future developments in these areas should be valuable to the immobilized enzyme researcher.

A. Silanization of Inorganic Supports

The major approach to chemically bonding reactive organic groups to inorganic surfaces is through reaction with silane coupling agents. Silane coupling agents are commonly used in several areas, including reinforced plastics and chromatography. In reinforced plastics, they are used to improve the adhesion of resins to inorganic fillers. In chromatography, they are used to prepare chemically bonded packings. Among the inorganic materials that have been silanized successfully are alumina, aluminum silicate, clays, glass, nickel oxide, quartz, silica, talc (magnesium silicate), titania, and wollastonite (calcium silicate) (21,22,23).

Silane coupling agents have dual functionality, with inorganic functionality at one end and organic functional groups at the other. The inorganic functional groups, which can be esters, halides, or silanols, condense with hydroxyl groups on inorganic surfaces. If there are no reactive groups on the surface, the coupling agent cannot chemically couple to the inorganic particle. As a result, only mechanical adhesion can occur. For example, calcium carbonate does not appear to react with silane coupling agents (21).

The mechanism of silane reaction with an inorganic support surface involves condensation of the ester, halide or silanol with hydroxyls on the support surface. In the absence of water, the halide or ester groups condense with the hydroxyls on the surface. In the presence of water, the ester and halide groups hydrolyze to form hydroxyls which react with the hydroxyls on the inorganic surface. A typical reaction for silane esters is:

$$(RO)_3\,Si\text{-}(CH_2)_3\text{-}X \xrightarrow{H_2O} (HO)_3\,Si\text{-}(CH_2)_3\text{-}X\;; \qquad (1)$$

$$I\text{-}OH + (HO)_3\,Si\text{-}(CH_2)\text{-}X_3 \longrightarrow I\text{-}O\text{-}\underset{\diagup O}{\overset{\diagdown O}{Si}}\text{-}(CH_2)_3\text{-}X\,. \qquad (2)$$

Silane coupling agents can have one, two or three hydrolyzable groups attached to the silicon atom. Theoretically, one hydroxyl group should be sufficient for bonding to inorganic surfaces, but a more stable product is produced with coupling agents that have three hydrolyzable groups. Though ideally, all of the hydrolyzable groups on the silane should react with surface hydroxyls, this does not occur. The condensation of hydroxyl groups between silane molecules is a competitive reaction which prevents complete reaction with the surface. In addition, steric factors prevent complete condensation, which leaves some unreacted hydroxyls on the surface.

The silane layer on the inorganic particle surface is a multi-layer coating (24,25). The thickness of this layer is affected by silane concentration, reaction solvent, pH, reaction temperature, reaction time, and surface area of the inorganic particle (26). The concentration of silane on the surface is a somewhat critical parameter because a minimum number of reactive groups is needed for a stable immobilized enzyme preparation, but if too many reactive groups are present, the enzyme may be bound too tightly. The optimum concentration of reactive groups on the surface lies within a relatively narrow range.

A variety of organic groups is available in silane coupling agents. The more common commercially available silane coupling agents are listed in Table 1.

Following are three typical silanization procedures for inorganic materials. In each case, the support should be clean and free of organic contaminants. Inorganic materials can be cleaned by heating above 500°C, by heating in 5% nitric acid, by extraction with solvents, or by other commonly used procedures.

TABLE 1

Silane Coupling Agents

γ-aminopropyltriethoxy silane	$H_2N-(CH_2)_3-Si(OC_2H_5)_3$
N-β-(aminoethyl)-γ-aminopropyltrimethoxy silane	$H_2N-(CH_2)_2-NH-(CH_2)_3-Si(OCH_3)_3$
β-(chloromethylphenyl)-ethyltrimethoxy silane	$Cl-CH_2-C_6H_4-CH_2-CH_2-Si(OCH_3)_3$
γ-chloropropyltriethoxy silane	$Cl-(CH_2)_3-Si(OC_2H_5)_3$
γ-cyanopropyltrimethoxy silane	$NC-(CH_2)_3-Si(OCH_3)_3$
β-(3,4-epoxycyclohexyl)-ethyltrimethoxy silane	$O<(S)-(CH_2)_2-Si(OCH_3)_3$
γ-glycidoxypropyltrimethoxy silane	$CH_2(O)CH-CH_2-O-(CH_2)_3-Si(OCH_3)_3$
γ-mercaptopropyltrimethoxy silane	$HS-(CH_2)_3-Si(OCH_3)_3$
γ-methacryloxypropyltrimethoxy silane	$H_2C=C(CH_3)-C(=O)-O-(CH_2)_3-Si(OCH_3)_3$

1. Aqueous Silanization of Porous Glass with γ-Aminopropyltriethoxy silane

One gram of porous glass is added to five ml of a 10% aqueous solution of γ-aminopropyltriethoxy silane which has been adjusted to pH 5 with acetic acid. The reaction mixture is stirred and heated to 80°C. for one to two hours. The solution is decanted and the porous glass is washed several times with water. This silanized glass is then heated for approximately two hours at 120°C. The resulting product is an alkyl amine glass.

2. Organic Silanization of Alumina with γ-Aminopropyltriethoxy silane

One gram of alumina is added to five ml of a 5% solution of γ-aminopropyltriethoxy silane in toluene. The reaction mixture is stirred and heated to reflux for one to two hours. The toluene solution is decanted and the alumina is washed several times with toluene, followed by several washings with acetone. The product is then heated for one to two hours at 120°C. The resulting product is an alkyl amine derivative of alumina.

3. Aqueous Silanization of Porous Glass with γ-Mercaptopropyltrimethoxy silane

One gram of porous glass is added to five ml of a 10% aqueous solution of γ-mercaptopropyltrimethoxy silane which has been adjusted to pH 5 with 6 N HCl. The reaction mixture is stirred and heated to reflux for about two hours. The reaction solution is decanted and the porous glass is washed several times with pH 5 solution, followed by several washings with water. The product is then heated for two hours at 120°C. The resulting product is a sulfhydryl glass derivative.

The concentration of reactive groups on the support surface generally can be determined by modifications of standard chemical techniques used for functional group analysis. In addition, pyrolysis-Gas Chromatography is an excellent technique for determining functional groups on inorganic supports.

B. Preparation of Stable Reactive Intermediates

Inorganic supports with many different reactive groups can be prepared by silanization. In some cases, enzymes can be reacted directly with the silanized supports. However, greater versatility in coupling procedures is obtained by further modifying the supports to provide additional organofunctional groups. Some of the modifications are relatively simple reactions, such as hydrolysis of the cyanopropyl support to produce a carboxyl derivative, or hydrolysis of the epoxy silane supports to produce dihydroxy derivatives. Other modifications are fairly elaborate, such as the preparation of aryl amines or active esters from alkyl amine glass.

Organofunctional groups desirable for enzyme coupling include alkyl and aryl amines, carbonyl, carboxyl, azide, substituted triazines, hydrazine, isocyanate, isothiocyanate, hydroxyl and activated esters. With a little imagination, almost any functional silane support can be used as a starting point for preparing these derivatives. For example, the preparation of derivatives from alkyl amine glass has been discussed extensively in the literature (8,10,11). The descriptions for a number of these preparations follow. The main advantage of this approach is its use of a single silane derivative as the starting material; but again, a number of other silane derivatives can be used in similar fashion. The procedures described here primarily illustrate the versatility and ease in manipulating reactive inorganic intermediates.

1. Preparation of Reactive Intermediates from Alkyl Amine Derivatives

a. Carbonyl Derivative. A reactive aldehyde intermediate is readily prepared by reacting glutaraldehyde with an amine support:

$$\text{I-NH}_2 + \text{H-}\overset{\overset{\displaystyle O}{\|}}{C}\text{-CH}_2\text{-CH}_2\text{-}\overset{\overset{\displaystyle O}{\|}}{C}\text{-H} \longrightarrow \text{I-N=CH-CH}_2\text{-CH}_2\text{-}\overset{\overset{\displaystyle O}{\|}}{C}\text{-H.} \quad (3)$$

The reaction is carried out in a neutral or acidic solution at room temperature. Ten ml of a 2.5% glutaraldehyde solution in water or in 0.1 M phosphate buffer, pH 7.0, is added to one gram of alkyl amine support, and the reaction is allowed to continue for about one hour. The reaction mixture may be stirred or shaken during this period. The glutaraldehyde solution is decanted (or the mixture can be filtered) and the product is washed several times with water.

The product, which has an active aldehyde group, can be used immediately for coupling, or it can be dried and stored for later use.

b. Carboxyl Derivatives. Reactive carboxyl intermediates can be prepared by reacting succinic anhydride with amine supports, or by reacting dicarboxylic acids with amine supports in the presence of a carbodiimide.

(1) Succinic Anhydride Reaction.

$$\text{I-NH}_2 + \begin{array}{c} \text{H}_2\text{C}-\overset{\overset{\text{O}}{\|}}{\text{C}} \\ | \\ \text{H}_2\text{C}-\underset{\underset{\text{O}}{\|}}{\text{C}} \end{array}\!\!>\text{O} \longrightarrow \text{I-NH}-\overset{\overset{\text{O}}{\|}}{\text{C}}-\text{CH}_2-\text{CH}_2-\overset{\overset{\text{O}}{\|}}{\text{C}}-\text{OH.} \quad (4)$$

One gram of alkyl amine support is added to 10 ml of a 1% aqueous succinic anhydride solution which has been adjusted to pH 6 with sodium hydroxide. The pH of the reaction mixture is maintained between 6 and 7 until no further adjustment is needed. After the pH has stabilized, the reaction is continued for about one hour at room temperature. The carboxylated product is washed exhaustively with water, then dried at 120°C for two hours.

(2) Dicarboxylic Acid - Carbodiimide Reaction (27)

$$\text{I-NH}_2 + \text{R}'\text{-N=C=N-R}'' + \text{HO-}\overset{\overset{\text{O}}{\|}}{\text{C}}\text{-R-}\overset{\overset{\text{O}}{\|}}{\text{C}}\text{-OH} \longrightarrow \text{I-NH-}\overset{\overset{\text{O}}{\|}}{\text{C}}\text{-R-}\overset{\overset{\text{O}}{\|}}{\text{C}}\text{-OH.} \quad (5)$$

Carbodiimide reagents have been used extensively as dehydrating agents in organic reactions, and have been widely used in peptide synthesis (28,29). Either aqueous or organic solvents may be used for carbodiimide reactions, but only the aqueous system will be presented here.

One gram of amine support, 0.4 gram of azelaic acid, and 0.1 gram of 1-cyclohexyl-3-(2-morpholinoethyl)-carbodiimide metho-p-toluenesulfate are added to 10 ml of water. The mixture is placed in a shaker bath at 60°C and allowed to react overnight. The reaction mixture is decanted, and the product washed with water until the washings are neutral. The carboxylated product is dried at 120°C for two hours, then stored for further use.

The carboxyl derivatives can be used for coupling proteins, either through the carbodiimide reaction, through an aryl halide intermediate, or through active ester intermediates.

(3) Acyl Chloride Intermediate. The acyl chloride intermediate is prepared from the carboxyl support by reaction with thionyl chloride:

$$\text{I-COOH} + SOCl_2 \longrightarrow \text{I-}\overset{\overset{\displaystyle O}{\|}}{C}\text{-Cl.} \qquad (6)$$

One gram of carboxyl support is added to 10 ml of a 10% solution of thionyl chloride in dry chloroform, and the reaction mixture is refluxed for four hours. The reaction solution is decanted and the product is washed several times with chloroform. The acyl chloride product is dried for 30 minutes in a vacuum oven, and is generally used for coupling shortly after drying. However, the product can be stored for short periods in a desiccator before use.

(4) N-Hydroxysuccinimide Ester Intermediate. Active ester intermediates, such as the N-hydroxysuccinimide derivative, are stable, easy to use, and quite versatile. The N-hydroxysuccinimide supports are made by reacting a

carboxyl support with N-hydroxysuccinimide in the presence of a carbodiimide (30):

$$\text{I-COOH} + \text{HO-N}\!\left(\begin{smallmatrix}C=O\\ |\\ C=O\end{smallmatrix}\right) \xrightarrow{\text{R-N=C=N-R}'} \text{I-}\overset{\overset{\displaystyle O}{\|}}{C}\text{-O-N}\!\left(\begin{smallmatrix}C=O\\ |\\ C=O\end{smallmatrix}\right) \qquad (7)$$

In a typical preparation with inorganic supports, one gram of carboxyl support is added to 10 ml of dioxane and the mixture is stirred for a short time. To this mixture is added 0.2 g N-hydroxysuccinimide and 0.4 g N,N'-dicyclohexylcarbodiimide, and this mixture is stirred for about four hours. The reaction solution is decanted and the product is washed several times with dioxane, followed by several washings with methanol to remove the dicyclohexylurea which precipitates during the reaction, followed by several more washings with dioxane. The product can be used at this point, or it can be dried at 80 C in an evacuated oven for one hour. The N-hydroxysuccinimide product is stable for long periods. This active ester support is excellent for coupling through primary amine groups on proteins.

(5) p-Nitrophenyl Ester Intermediate. The p-nitrophenyl ester derivative is another stable and easy to use active ester intermediate. Somewhat like the ester preparation described previously, this derivative is prepared by reacting a carboxyl support with p-nitrophenol in the presence of a carbodiimide:

$$\text{I-COOH} + \text{HO-}\langle\bigcirc\rangle\text{-NO}_2 \xrightarrow{\text{R-N=C=N-R}'} \text{I-}\overset{\overset{\text{O}}{\|}}{\text{C}}\text{-O-}\langle\bigcirc\rangle\text{-NO}_2. \tag{8}$$

One gram of carboxyl support is added to 10 ml of dimethylformamide and the mixture is stirred for a short time. To this mixture is added 0.2 g of p-nitrophenol and 0.2 g of 1-cyclohexyl-3-(2-morpholinoethyl)-carbodiimide metho-p-toluenesulfonate, and the reaction mixture is stirred about four hours. The reaction solution is decanted and the product is washed thoroughly with water. Like the above active ester, the product can be used at this point or it can be dried at 120°C for two hours. This active ester is excellent for following protein coupling reactions because the formation of p-nitrophenol is readily detected.

c. Aryl Amine Derivative. There are a number of potential routes for the preparation of an aryl amine derivative, but the most commonly used route has been through the reaction of alkyl amine supports with p-nitrobenzoyl chloride, followed by the reduction of the nitro group:

$$\text{I-NH}_2 + \text{Cl-}\overset{\overset{\text{O}}{\|}}{\text{C}}\text{-}\langle\bigcirc\rangle\text{NO}_2 \longrightarrow \text{I-NH-}\overset{\overset{\text{O}}{\|}}{\text{C}}\text{-}\langle\bigcirc\rangle\text{NO}_2 \xrightarrow{[\text{H}]} \text{I-NH-}\overset{\overset{\text{O}}{\|}}{\text{C}}\text{-}\langle\bigcirc\rangle\text{NH}_2. \tag{9}$$

One gram of alkyl amine support is added to 10 ml of a chloroform solution containing 0.1 gram of p-nitrobenzoyl chloride, and the reaction mixture is refluxed for one hour. The reaction solution is decanted and the aryl nitro product is washed several times with chloroform. The product can be air dried or heated for 30 minutes at 80°C to remove the chloroform.

The aryl nitro product is reduced in a 1% aqueous dithionite solution. One gram of the nitro product is added to 10 ml of a 1% aqueous dithionite solution adjusted to pH 8.5 with sodium hydroxide. The reaction mixture is heated to reflux and the pH is maintained between 8-9. Refluxing is continued for one hour after the pH has stabilized. The reaction solution is decanted and the product is washed with water.

The aryl amine support can be used immediately or can be dried and stored for later use. The aryl amine can be treated as a primary amine for coupling, but is generally coupled with proteins through the diazonium salt.

d. Phenylhydrazine Derivative. The phenylhydrazine intermediate, which is readily prepared from the aryl amine (31), is excellent for coupling through carbonyl groups (32):

$$\text{I-C}_6\text{H}_4\text{-NH}_2 + \text{NaNO}_2 \xrightarrow{\text{HCl}} \text{I-C}_6\text{H}_4\text{-N}_2^+\text{Cl}^- \xrightarrow{\text{H}} \text{I-C}_6\text{H}_4\text{-NH-NH}_2 \quad (10)$$

One gram of aryl amine support is added to 10 ml of 2N HCl, and the reaction mixture is cooled in an ice bath. To this mixture is added 0.25 g of sodium nitrite, and the reaction is allowed to continue for 30 minutes. The reaction solution is decanted, and the product is washed with cold water, then with a 1% solution of cold sulfamic acid, and several more times with water.

The diazonium chloride is reduced in a 1% dithionite solution at pH 8.5. Ten ml of a 1% aqueous dithionite solution adjusted to pH 8.5 with sodium hydroxide is added to the diazonium chloride product and the reaction mixture is heated to reflux. The pH of the reaction solution is maintained between 8-9, and refluxing is continued for about one hour after the pH has stabilized. The phenylhydrazine product is washed several times with water, and can be used within a short time after preparation. For longer storage the hydrochloride derivative should be made, and should be stored at low temperatures.

e. Triazine Derivative. Triazine derivatives have not been widely used, but some fairly good results have been reported with organic supports. E.g., O'Neill immobilized amyloglucosidase on DEAE-cellulose through a triazine derivative (33), and Baum immobilized the same enzyme onto aminopolystyrene beads through a triazine derivative (34). The preparation of this derivative involves a reaction between an amine support and a substituted triazine:

$$\text{I-NH}_2 + \text{Cl-}(\text{C}_3\text{N}_3)(\text{Cl})(\text{R}) \longrightarrow \text{I-NH-}(\text{C}_3\text{N}_3)(\text{Cl})(\text{R}). \qquad (11)$$

On gram of amine support, 0.2 ml of triethylamine, and 0.3 g of 1,3-dichloro-5-methoxytriazine are added to 10 ml of benzene and the mixture is stirred for two hours at 50°C. The reaction mixture is decanted and the product is washed several times with toluene. The triazine product is dried in an evacuated oven at 100°C for one hour.

f. Isothiocyanate Derivative. The isothiocyanate is not widely used, either, but is an easy derivative to use in the coupling step. The isothiocyanate derivative is prepared from amine supports by reaction with thiophosgene:

$$\text{I-NH}_2 + \text{Cl-C(=S)-Cl} \longrightarrow \text{I-NCS}. \qquad (12)$$

One gram of amine support is added to 50 ml of a 10% solution of thiophosgene in chloroform, and the reaction mixture is refluxed for four hours. The reaction mixture is decanted and the product washed several times with chloroform. The isothiocyanate support is dried for 30 minutes in a vacuum oven, and is generally used for coupling shortly after drying. However, the product can be stored for short periods in a desiccator.

The above preparations by no means represent an exhaustive list of derivatives for immobilizing proteins. These derivatives are among those that have been used by a number of researchers. A few of the less publicized derivatives that can be made from amine supports are isocyanate derivatives prepared by reaction with phosgene,

$$\text{I-NH}_2 + \text{Cl-}\overset{\overset{\displaystyle O}{\|}}{C}\text{-Cl} \longrightarrow \text{I-NCO;} \qquad (13)$$

bromoacetyl derivatives prepared by reaction with O-bromoacetyl-N-hydroxysuccinimide,

$$\text{I-NH}_2 + \text{Br-CH}_2\text{-C-O-N}\langle\text{succinimide}\rangle \longrightarrow \text{I-NH-}\overset{\overset{\displaystyle O}{\|}}{C}\text{-CH}_2\text{-Br;} \qquad (14)$$

and thiol derivatives prepared by reaction with a thiol lactone,

$$\text{I-NH}_2 + \text{(N-acetylhomocysteine thiolactone: S, C=O, N-}\overset{\overset{\displaystyle O}{\|}}{C}\text{-CH}_3) \longrightarrow \text{I-NH-C-}\underset{\underset{\displaystyle \text{NH-}\underset{\underset{\displaystyle O}{\|}}{C}\text{-CH}_3}{|}}{CH}\text{-CH}_2\text{-CH}_2\text{-SH.} \qquad (15)$$

Wachter, Machleidt, Hofner and Otto (35) recently discussed the preparation and use of an interesting isothiocyanate derivative. They prepared the derivative by reacting p-phenylene diisothiocyanate with alkyl amine glass:

$$\text{I-NH}_2 + \text{SCN-C}_6\text{H}_4\text{-NCS} \xrightarrow{\text{DMF}} \text{I-NH-}\overset{\overset{\displaystyle S}{\|}}{C}\text{-NH-C}_6\text{H}_4\text{-NCS.} \qquad (16)$$

The group had excellent results in coupling peptides to this derivative.

The list of derivatives available for coupling will be continually expanding. The virtues of the various coupling derivatives will have to be examined by the researchers for their own systems.

III. METHODS FOR THE COVALENT ATTACHMENT OF ENZYMES TO INORGANIC SUPPORTS.

As previously mentioned, the covalent bonding procedure is simply the reaction of an enzyme with an activated support to form the immobilized enzyme composite. In the general reaction procedure, a solution of the enzyme is added to the reactive support and the reaction is allowed to proceed for some period of time ranging from 30 minutes to overnight. The reaction product is washed to remove unbound enzyme. Adsorbed or ionically bound protein is removed by washing with solutions of sodium chloride, urea, guanidine hydrochloride, etc., depending upon the preference of the researcher.

Immobilized enzymes have been stored under buffer, as a wet cake, partially dried, or lyophilized; at ambient temperatures, refrigerator temperatures, or freezer temperatures, depending upon the enzyme and the user's preference. For estimating the activity of immobilized enzymes, most of the assay techniques applicable to soluble enzyme may be used. Total protein loadings can be determined by a ninhydrin procedure, or can be estimated by determining the difference between the initial enzyme in solution and the total enzyme recovered after coupling and washing.

Generally, 50 to 100 mg of protein per gram of support is sufficient for coupling, but if the enzyme is expensive or in short supply, quantities less than 10 mg can be used per gram of support. When proteases are coupled, the reaction times should be as short as feasible, because the soluble protease undergoes autodigestion in solution. Soluble protease will also cleave the immobilized enzyme, but when the immobilized product is isolated and washed free of unbound protease, there is no further problem. The immobilized proteases will not undergo autodigestion.

A. Coupling of Enzymes to Supports with Stable Reactive Intermediates

The most convenient procedure for preparing immobilized enzymes is the reaction of an enzyme with stable reactive intermediates. Stable reactive intermediates that have been used include aldehyde, ester and triazine derivatives. As described earlier, these derivatives can be prepared and stored for extended periods, and a number of the reactive inorganic supports are available commercially. Following are procedures for immobilizing enzymes on some of these derivatives.

1. Coupling to a Carbonyl Intermediate-Schiff Base Coupling.

Aldehyde groups on an inorganic support react with primary amines to form an imine coupling:

$$I-CHO + H_2N{\sim}E \longrightarrow I-CH{=}N-E. \tag{17}$$

The reaction, which should be run in acidic or neutral solutions, is quite simple and gives a fairly stable linkage. The imine linkage is somewhat unstable in basic solutions, but the stability can be improved by reducing the imine with sodium borohydride (if the enzyme is not affected).

In a typical coupling procedure, enough enzyme solution in buffer (usually 0.1 M phosphate, pH 5-7) is added to the carbonyl derivative to just cover the support. The coupling reaction can be carried out at any temperature at which the enzyme is stable, but is conveniently run at room temperature, in an ice bath, or in a refrigerator. The mixture is slurried and the reaction is allowed to continue for approximately one hour. The product is washed several times with water or buffer and is stored under buffer or as a wet cake in the refrigerator.

2. Coupling to a Carboxyl Intermediate - Amide Coupling

Carboxyl groups on supports are activated for coupling by forming esters or aryl halides, as described in the previous section. The active carbonyl supports react with primary amines to form an amide linkage (the activated supports can also react with hydroxyl groups under similar conditions):

$$\mathrm{I{-}\overset{\overset{\displaystyle O}{\|}}{C}{-}X + H_2N{\sim}E \longrightarrow I{-}\overset{\overset{\displaystyle O}{\|}}{C}{-}NH{\sim}E}, \tag{18}$$

where X is a halide or ester group.

The reaction is run under very mild conditions, and gives a very stable linkage. The reactions can be run in slightly acidic or basic solutions, but neutral to basic solutions are preferred.

a. Acyl Chloride Intermediate. Enzymes are coupled to the acyl chloride support by adding to the support a solution containing 10 to 100 mg active protein at pH 8-9, adjusted with NaOH. The pH of the reaction solution is maintained at 8-9 by the addition of 2N NaOH. The reaction is continued at room temperature for approximately one hour, or overnight at refrigerator temperatures. The reaction solution is decanted, and the product is washed several times with water or buffer.

b. N-Hydroxysuccinimide Active Ester. Enzymes are coupled to the N-hydroxysuccinimide support in a reaction similar to that for the acyl chloride. In a typical procedure, one gram of N-hydroxysuccinimide support is added to 10 ml of an enzyme solution in 0.1 M phosphate buffer, pH 5 to 9, in an ice bath. Cuatrecasas (30) found with agarose that the active ester reacted optimally with the α-amino group of L-alanine at about pH 8.5, but coupling decreased at higher pH values because of hydrolysis of the active ester.

Reaction is continued at ice bath temperatures for several hours, then the reaction solution is decanted and the product is washed several times with buffer. To ensure removal of all active ester groups, the immobilized enzyme product can be shaken with a 1 M glycine (30) or ethanolamine solution for several hours at room temperature. The immobilized enzyme is then washed several more times with 0.1 M phosphate buffer before use or storage.

c. p-Nitrophenol Active Ester. The reaction of enzymes with p-nitrophenol active ester supports is very similar to the reaction of the N-hydroxysuccinimide derivative. One gram of p-nitrophenol support is added to 10 ml of an enzyme solution in 0.1 M phosphate buffer, pH 8, in an ice bath. The reaction is continued at ice bath temperatures for several hours, then the reaction solution is decanted and the product is washed several times with buffer. Again, as above, the product can be reacted with glycine or ethanolamine to ensure complete removal of all the active ester groups.

One advantage to using the p-nitrophenol active ester is that the enzyme coupling reaction can be followed by monitoring the appearance of p-nitrophenol at 318 nm.

3. Coupling to an Isothiocyanate Derivative - Thiourea Coupling.

The isothiocyanate derivatives have not been widely used because of the obvious odor associated with the intermediate. However, the derivative is easy to make, couples under fairly mild conditions, and produces a relatively stable bond. It has been reported that the reaction predominantly involves the ϵ-amino groups of lysyl residues. This is most likely the result of the high pH at which the reaction is run:

$$\text{I-NCS} + \text{H}_2\text{N}\sim\text{E} \longrightarrow \text{I-NH-}\overset{\overset{\displaystyle S}{\|}}{\text{C}}\text{-NH-E.} \qquad (19)$$

A solution of enzyme in 0.05 $NaHCO_3$ buffer, pH 9-10, is added to the isothiocyanate support. The mixture is stirred and the reaction is allowed to continue for approximately two hours. The solution is then decanted and the immobilized enzyme product is washed several times with water, and several times with the buffer solution in which the enzyme will be used.

4. Coupling to an Isocyanate Derivative - Urea Coupling

Isocyanate derivatives have not been used to any great extent for coupling, perhaps because of the use of phosgene in preparing the derivatives. However, like the isothiocyanate derivative, the isocyanate is relatively easy to prepare and produces a stable bond on coupling. Also like the isothiocyanate, the reaction is reported to predominately involve the ε-amino groups of lysyl residues:

$$I\text{-}NCO + H_2N\text{-}E \longrightarrow I\text{-}NH\text{-}\overset{\overset{\displaystyle O}{\|}}{C}\text{-}NH\text{-}E \qquad (20)$$

The isocyanate derivative is prepared by adding amine support to toluene, heating to reflux, saturating the toluene with dry HCl, then passing phosgene into the solution for 30 to 60 minutes (36,37,38). The reaction solution is decanted, and the support is washed several times with water. The product is dried for 30 minutes in a vacuum oven and is used for coupling shortly after drying, or it can be stored for a short period in a desiccator.

The reaction of the isocyanate support with enzymes is similar to that for the isothiocyanate derivative. A solution of enzyme in 0.05 Na HCO_3 buffer, pH 9-10, is added to the isocyanate support. The mixture is stirred, and the reaction is allowed to proceed for approximately two hours. The solution is then decanted, and the product is washed several times with buffer solution or water.

5. Coupling to a Triazine Derivative - Triazine Coupling.

As pointed out in the section on preparation of triazine supports, triazine derivatives have not been used to any great extent with inorganic supports. However, these derivatives should get further attention because of the environmental changes that can be made. Only two of the reactive groups on the triazine ring are involved in the coupling procedure:

$$\text{I-NH-}C_3N_3(\text{Cl})(\text{R}) + H_2N\text{-E} \longrightarrow \text{I-NH-}C_3N_3(\text{NH-E})(\text{R}) \quad (21)$$

The third group can be a reactive chlorine, or the chlorine can be replaced by other groups which can change the environment of the immobilized enzyme. The procedure described here uses a methoxy group; O'Neill, Dunnill and Lilly (33) used an amino derivative to study the coupling of amyloglucosidase; Kay and Lilly (39) also used the amino derivative for coupling to DEAE-Cellulose.

In a typical coupling procedure, one gram of the chloromethoxy-triazine support described earlier is added to several ml of enzyme solution in 0.1 M phosphate buffer, pH 8, at 5°C, and the reaction is allowed to proceed overnight at 5°C. The reaction solution is decanted, and the product is washed several times with buffer and water.

B. Coupling of Enzymes to Supports with Activated Intermediates

Some coupling reactions involve the formation of reactive intermediates which have limited stability and must be used in situ. These are described in a separate section because the procedures are different from those described earlier. A number of the stable reactive supports can be purchased commercially and used without further preparation. Others, like the carboxyl, isothiocyanate, cyanate, and triazine derivatives, while not available commercially, can be made and stored for some time before use. The following descriptions deal with activated derivatives which are generally made by the researcher and are used immediately after preparation because of their limited storage stability. These preparations include the azide, cyanogen bromide, and diazonium chloride reactions.

1. Coupling Through a Diazonium Salt - Azo Coupling

Diazonium salts are most commonly prepared by reacting nitrous acid with aromatic amines in the presence of hydrochloric acid. The salts are not usually isolated because they are unstable, and are explosive in solid form. Aryldiazonium fluoroborate salts are stable, and should be examined for enzyme coupling. To date, however, we have found no reference for their use in enzyme immobilization. Diazonium salts are extremely reactive, and will react with a large number of groups. The desired reaction here is to couple with the phenolic ring of tyrosine or the imidazole ring of histidine. Reaction can occur with the primary amines on lysine, arginine, and cysteine, but the resulting diazoamine linkage is not stable. The diazonium coupling reaction can be run in alkaline, neutral, or weakly acidic solutions, but is generally run in alkaline solutions:

$$I-C_6H_4-NH_2 \xrightarrow{NaNO_2} I-C_6H_4-N_2^+Cl^- \xrightarrow{HO-C_6H_4-E} I-C_6H_4-N=N-C_6H_3(OH)-E \quad (22)$$

The aryl amine support is diazotized by first adding 10 ml of 2N HCl to a gram of the aryl amine derivatibe. The reaction mixture is cooled in an ice bath, then 0.25 g of sodium nitrite is added. (An aqueous solution of sodium nitrite can also be used.) The reaction is allowed to proceed for about 30 minutes, then the reaction solution is decanted, and the product is washed first with cold water, then with a 1% solution of cold sulfamic acid (if desired), and several more times with cold water.

The diazotized product is coupled with enzymes by adding a solution of active protein in 0.05 M $NaHCO_3$ buffer, pH 8, to the support. The reaction is continued for one hour or longer at $0^{\circ}C$, then the solution is decanted and the product is washed with water or buffer.

2. Coupling Through an Azide Intermediate

Azide coupling has been carried out primarily with carboxymethylcellulose as the support (40,41,42), but the reaction can be carried out with inorganic carboxyl supports. The activated azide support reacts with primary amines under mild alkaline conditions. The first step in the procedure is the preparation of an azide derivative, which can be carried out in several ways.

The preparation of azides is well documented for the Curtius reaction in which an acid is converted into an amine through azide and isocyanate intermediates. Two of the more common preparations involve the reaction of an acyl chloride derivative with sodium azide (43),

$$\mathrm{I{-}\overset{O}{\overset{\|}{C}}{-}Cl + NaN_3 \longrightarrow I{-}\overset{O}{\overset{\|}{C}}{-}N_3}, \tag{23}$$

and the reaction of a methyl ester derivative with hydrazine to form a hydrazide which is then converted to the azide by reaction with nitrous acid (40):

$$\mathrm{I{-}\overset{O}{\overset{\|}{C}}{-}OCH_3 + H_2N{-}NH_2 \longrightarrow I{-}\overset{O}{\overset{\|}{C}}{-}NH{-}NH_2 \xrightarrow[NaNO_2]{HCl} I{-}\overset{O}{\overset{\|}{C}}{-}N_3}. \tag{24}$$

Enzymes are coupled to the azide support in an alkaline solution, such as 0.05 M borate buffer at pH 8.5, at $0^{\circ}C$ for 30 to 60 minutes:

$$\mathrm{I{-}\overset{O}{\overset{\|}{C}}{-}N_3 + H_2N{-}E \longrightarrow I{-}\overset{O}{\overset{\|}{C}}{-}NH{-}E} \tag{25}$$

The immobilized enzyme product is washed with a weak HCl solution and with water.

3. Coupling Through a Cyanogen Bromide Intermediate

Coupling of enzymes to agarose and dextran supports through the cyanogen bromide reaction has been used extensively, and there are many references in the literature (44,45,46). This technique has also been used with inorganic supports such as dextran-coated glass, which is essentially a combination of the organic and inorganic matrices. In addition, the reaction can be carried out with monohydroxy, dihydroxy, monoamine and diamine supports.

The usual mechanism suggested for cyanogen bromide reactions involves reaction with vicinal hydroxyls to produce an imidocarbonate which reacts with amines to form a carbamate (urethane) linkage:

$$I \begin{cases} CH{-}OH \\ CH{-}OH \end{cases} + CNBr \longrightarrow I \begin{cases} CH{-}O \\ CH{-}O \end{cases} C{=}NH \xrightarrow{H_2N-E} I \begin{cases} CH{-}O{-}\overset{O}{\overset{\|}{C}}{-}NH{-}E \\ CH{-}OH \end{cases} \qquad (26)$$

Presumably, a urea linkage would be formed if a vicinal amine were used instead of the hydroxyl support.

Cyanogen bromide reacts with monohydroxy supports to produce a cyanate intermediate (46), which is unstable and very reactive:

$$I{-}CH_2OH + CNBr \longrightarrow I{-}CH_2{-}O{-}C{\equiv}N. \qquad (27)$$

Reactive amines should add across the -C≡N bond to produce an

$$-O-\overset{\overset{NH}{\|}}{C}-NH-E$$

linkage, which should hydrolyze rapidly to a carbamate.

To activate supports such as dextran-coated glass, amines, or dihydroxy derivatives, procedures similar to those used for agarose can be used. One gram of dextran-coated support is added to 10 ml of solution containing 0.25 g of cyanogen bromide. (Extreme caution should be used in handling cyanogen bromide, and these reactions should be carried out in a hood.) The pH of the reaction mixture is immediately adjusted to approximately 11.0 by the addition of 2-4N NaOH. The reaction temperature should be maintained at 20°C by the addition of ice as needed. The pH of the reaction mixture is maintained around 11.0 by the addition of NaOH. The reaction is continued until the pH stabilizes, which usually requires about 10 minutes. The reaction mixture is decanted, and the activated support is washed exhaustively with cold water.

Enzymes are coupled to the activated support by adding a solution of active protein at pH 8-9 to the support. The pH may be adjusted with 2M NaOH or 0.05 $NaHCO_3$, and the pH should be maintained between 8-9 throughout the reaction. The reaction should be complete within two hours, but may be continued overnight to ensure complete reaction. The reaction solution is decanted, and the product is washed several times with water.

C. Coupling of Enzymes to Supports in the Presence of Other Reactive Components

Some coupling reactions are conveniently carried out by combining the enzyme, the support, and other reactive agents in a single step. The most common reaction of this type uses carbodiimides as coupling reagents. Another reaction that falls in this category is the Passerini reaction, which has been modified by Ugi and utilized by Axen, Vretblad and Porath for coupling chymotrypsin (47). Since the latter reaction has not yet been used with inorganic supports, it will be discussed only briefly.

Carbodiimide coupling can be carried out with either an amine support or a carboxyl support. The reactions can also be carried out in either aqueous or organic solutions, but the aqueous system should be used when active proteins are coupled. Procedures for carbodiimide reactions in organic systems are presented in an earlier section on esterification of carboxyl supports and will not be discussed here. Carbodiimide reactions can also be carried out stepwise, in which case the reaction might have been included in the previous section. However, the reaction is more conveniently run in a single step.

1. Carbodiimide Coupling of Enzymes to Amine Supports

The overall reaction involved in this coupling is a condensation between the amine on the support and an acid group on the protein to produce an amide linkage. A substituted urea is produced as a by-product:

$$\text{I-NH}_2 + \text{HO-}\overset{\text{O}}{\overset{\|}{\text{C}}}\text{-E} \xrightarrow[\text{H}^+]{\text{R'N=C=N-R''}} \text{I-NH-}\overset{\text{O}}{\overset{\|}{\text{C}}}\text{-E}. \qquad (28)$$

One gram of amine support is added to 10 ml of a protein solution containing approximately 50 mg of protein. To this mixture is added 50 mg of 1-cyclohexyl-3-(2-morpholinoethyl)-carbodiimide metho-p-toluenesulfate, and the pH is adjusted to 4.0 with 6 M HCl. The reaction is continued for two to three hours at room temperature, or overnight at 4°C. The reaction solution is then decanted, and the product is washed extensively with water.

2. Carbodiimide Coupling of Enzymes to Carboxyl Supports

If a carboxyl derivative is used for coupling, the reaction can be run in a single step as described earlier or it can be run in two steps. In the first step, excess carbodiimide is added to the carboxyl support to form a psuedourea. The excess carbodiimide is then removed from the reaction and a solution of the enzyme is added to complete the coupling:

$$\text{I-}\overset{\text{O}}{\overset{\|}{\text{C}}}\text{-OH} + \underset{\underset{\text{R}'}{|}}{\underset{\text{N}}{\underset{\|}{\overset{\overset{\text{R}}{|}}{\overset{\text{N}}{\overset{\|}{\text{C}}}}}}} \xrightarrow{H^+} \text{I-}\overset{\text{O}}{\overset{\|}{\text{C}}}\text{-O-}\underset{\underset{\text{R}'}{|}}{\underset{\text{NH}^+}{\underset{\|}{\overset{\overset{\text{R}}{|}}{\overset{\text{NH}}{\overset{|}{\text{C}}}}}}} \xrightarrow{H_2N\text{-}E} \text{I-}\overset{\text{O}}{\overset{\|}{\text{C}}}\text{-NH-E.} \qquad (29)$$

To one gram of carbonyl support is added a solution of 50 mg of 1-cyclohexyl-3-(2-morpholinoethyl)-carbodiimide metho-p-toluenesulfate in 10 ml of water. The reaction mixture is adjusted to pH 4-5 with 6M HCl, and the reaction is allowed to continue at room temperatue for two to three hours. The support is then washed several times with water to remove the excess carbodiimide.

Enzymes are coupled to the activated support by adding an enzyme solution to the support, adjusting the solution to pH 4-5, and allowing the reaction to continue for two to three hours at room temperature, or overnight at 4°C. The reaction solution is then decanted and the product is washed with water or buffer.

3. Use of Isocyanides for Coupling Enzymes to Amine and Carboxyl Supports

The formation of arylamides from the reaction of aryl isocyanides, carbonyl compounds, and carboxylic acids is known as the Passerini Reaction (48). This reaction has been studied and modified by Ugi (49), who extended the reaction to amines and carbonyls with isocyanides and carboxylic acids (among other modifications). The latter reaction involves the formation of an immonium ion from the reaction of an amine with a carbonyl compound. The immonium ion reacts with an isocyanide to produce an intermediate which reacts with nucleophilic reagents such as carboxylic acids. The final product has an amide linkage which is formed by a rearrangement:

$$\text{I-NH}_2 + \text{H-}\overset{\overset{\text{O}}{\|}}{\text{C}}\text{-H} \longrightarrow \text{I-}\overset{+}{\text{N}}\text{H=CH}_2\text{, AND} \qquad (30)$$

$$\text{I-}\overset{+}{\text{N}}\text{H=CH}_2 + \text{R-N=C} + \text{R'-}\overset{\overset{\text{O}}{\|}}{\text{C}}\text{-OH} \longrightarrow \text{I-}\underset{\underset{\text{R'}}{|}}{\underset{\text{C=O}}{\underset{|}{\text{N}}}}\text{-CH}_2\text{-}\overset{\overset{\text{O}}{\|}}{\text{C}}\text{-NH-R.} \qquad (31)$$

Axen, Vretblad and Porath (47) extended this reaction to the attachment of proteins, peptides, and amino acids to amino and carboxyl polymers. The supports they used included carboxymethyl agarose, agarose-lysine ethyl ester containing free amino groups, polyacrylamide with amine groups and keratin. To our knowledge, this coupling procedure has not yet been tried with inorganic supports, but this is certainly feasible and should prove to be an interesting approach.

IV. SUMMARY

The choice of reactions for coupling enzymes to supports is fairly extensive, and the number is steadily increasing. Many of the derivatized inorganic supports are available commercially, and many more will be made available in the future.

At the risk of appearing repetitive, we again stress that no single immobilization technique or single support is ideal for all enzymes. Although a few basic coupling procedures will be used for the majority of enzyme immobilizations, the researcher should look at several procedures before locking-in on a particular technique.

Most of the immobilization procedures are relatively easy to carry out, and can be performed under mild conditions. Although the reaction mechanisms appear complicated, the procedures are easy to follow.

A brief summary of reactive support derivatives and coupling procedures is given in Table II. While they are not listed in this table, it should be recognized that many of the reactions involving amine groups on the enzymes can also be carried out with hydroxyl and sulfhydryl groups.

Table III lists enzymes that have been coupled to inorganic supports, and the procedures used. This list is not intended to be complete, but is intended simply to give those who have not used inorganic supports a chance to see what can be done with inorganic materials.

TABLE II

Coupling to Inorganic Supports

Derivative	Intermediate	Enzyme Group Involved	Coupling Linkage
$-R-NH_2$	Carbodiimide	$HO-C(=O)-$	$-NH-C(=O)-$
$-C_6H_4-NH_2$	$-C_6H_4-N_2^+Cl^-$	$HO-C_6H_4-$	$-C_6H_4-N=N-C_6H_3(OH)-$
$-C_6H_4-NH-NH_2$		$-C(=O)-$	$-C_6H_4-NH-N=C$
$-C(=O)-OH$	$-C(=O)-Cl$	H_2N-	$-C(=O)-NH-$
$-C(=O)-OH$	$-C(=O)-O-C_6H_4-NO_2$	H_2N-	$-C(=O)-NH-$
$-C(=O)-OH$	$-C(=O)-O-N$ (succinimide)	H_2N-	$-C(=O)-NH-$
$-C(=O)-OH$	$-C(=O)-N_3$	H_2N-	$-C(=O)-NH-$
$-C(=O)-OH$	Carbodiimide	H_2N-	$-C(=O)-NH-$
$-C(=O)-H$		H_2N-	$-C=N-$
$-NCO$		H_2N-	$-NH-C(=O)-NH-$
$-NCS$		H_2N-	$-NH-C(=S)-NH-$
$-SH$	Carbodiimide	$HO-C(=O)-$	$-S-C(=O)-$
$-C(=O)-CH_2Br$		H_2N-	$-C(=O)-CH_2-NH-$
vicinal diol (-OH, -OH)	CNBr	H_2N-	$-O-C(=O)-NH-$ / $-OH$
chlorotriazine (Cl, R)		H_2N-	triazine (NH-, R)

TABLE III

Enzymes Covalently Coupled to Inorganic Supports

Enzyme	Support	Coupling Procedure	References
Acetylcholinesterase	Glass frit	Carbodiimide	9
	Glass frit	Diazo	9
Alkaline phosphatase	Porous glass	Diazo	10,50,51,52
Alkaline protease	Silica	Thiourea	50
L-Aminoacid oxidase	Porous glass	Diazo	53
Aminopeptidase M	Porous glass	Diazo	54
B-Amylase	Bentonite	Triazine	55
Arylsulphatase	Porous glass	Diazo	56
Carbamyl phosphokinase	Porous glass	Schiff base	57
Catalase	Nickel oxide on silica-aluminum	Thiourea	58
Cellulase	Bentonite	Triazine	55

TABLE III (continued)

Enzyme	Support	Coupling Procedure	Reference
α-Chymotrypsin	Iron oxide	Schiff base	59
	Iron oxide-cellulose	Cyanogen bromide	59
	Porous glass	Diazo	60
	Porous glass	Schiff base	61
Deoxyribonuclease I	Porous glass	Diazo	62
Ficin	Porous glass	Diazo	17,50
	Porous glass	Thiourea	17,50
β-Galactosidase	Iron oxide	Schiff base	59
	Iron oxide-cellulose	Cyanogen bromide	59
	Porous glass	Diazo	63,64
	Porous glass	Schiff base	61
	Zirconia-coated/ Porous glass	Schiff base	65
Glucoamylase	Porous glass	Diazo	13,66
	Porous glass	Schiff base	13,66
	Zirconia-coated/ Porous glass	Diazo	13
	Zirconia-coated/ Porous glass	Schiff base	13
Glucose isomerase	Porous glass	Diazo	67

TABLE III (continued)

Enzyme	Support	Coupling Procedure	Reference
Glucose oxidase	Alumina	Thiourea	17,50
	Hydroxyapatite	Thiourea	17,50
	Nickel oxide	Thiourea	50,68
	Porous glass	Diazo	17,50,69,70
	Porous glass	Thiourea	17,50,71
	Silica	Thiourea	50
Hexokinase	Porous glass	Diazo	72
Invertase	Bentonite	Triazine	55
	Glass	Triazine	73
	Porous glass	Diazo	74
Lactate dehydrogenase	Glass beads	Diazo	75
	Porous glass	Carbodiimide	76
	Porous glass	Diazo	77
	Porous glass	Schiff base	77,78
Leucine aminopeptidase	Porous glass	Diazo	54
Papain	Porous glass	Amide	79
	Porous glass	Diazo	17,50,79,80
	Porous glass	Schiff base	79
	Porous glass	Thiourea	17,80
	Silica	Diazo	17
	Silica	Thiourea	17
	Zirconia-coated/ Porous glass	Amide	79

TABLE III (continued)

Enzyme	Support	Coupling Procedure	Reference
Papain	Zirconia-coated/ Porous	Schiff base	79
Pepsin	Porous glass	Carbodiimide	81
Peroxidase	Nickel oxide	Thiourea	10,50
	Porous glass	Diazo	10,50
Pronase	Porous glass	Diazo	82
Pyruvate kinase	Glass beads	Diazo	75
Ribonuclease A	Bentonite	Triazine	55
Steroid esterase	Porous glass	Diazo	83
Trypsin	Bentonite	Triazine	55
	Porous glass	Carbodiimide	50
	Porous glass	Diazo	17,50,78,80 84,85,86
	Porous glass	Thiourea	17,50,80
	Silica	Diazo	17
	Silica	Thiourea	17
Urease	Betonite	Triazine	55
	Porous glass	Diazo	50,87

REFERENCES

(1). I. H. Silman, and E. Katchalski, Annual Rev. Biochemis. 35, 873 (1966).
(2). E. M. Crook, K. Brocklehurst and C. W. Wharton, Methods in Enzymology 19, 963 (1970).
(3). L. Goldstein, Methods in Enzymology 19, 935 (1970).
(4). K. Mosbach, Sci. Amer. 226 (3), 26 (1971).
(5). G.J.H. Melrose, Rev. Pure Appl. Chem. 21, 83 (1971).
(6). H.D. Orth and W. Brummer, Angew. Chem. Internat. Edit. 11, 249 (1972).
(7). O.R. Zaborsky, Immobilized Enzymes, CRC Press, Cleveland, Ohio, 1973.
(8). M. Lynn, Chemie Et Industrie-Genie Chemique 106, 601 (1973).
(9). G. Baum, F. B. Ward and H. H. Weetall, Biochim. Biophys. Acta 268, 411 (1972).
(10). H. H. Weetall and R. A. Messing in The Chemistry of Biosurfaces, ed. M. L. Hair, Chapter 12, Marcel Dekker, N.Y., 1972.
(11). H. H. Weetall, Research/Development, Dec. 1971, 18.
(12). R. A. Messing, 166th Amer. Chem. Soc. Nat. Meet., Chicago, Ill., Aug. 1973.
(13). H. H.Weetall and N. B. Havewala, Enzyme Engineering, ed. L. B. Wingard, Jr., 241, Interscience Publishers, N.Y., 1972.
(14). S. Sterman and J. G. Marsden, Bulletin F-41920, Union Carbide Corporation, 1968.
(15) Silane Coupling Agents, Bulletin, Dow Corning Corporation, 1970.
(16). D. F. Ollis and R. Datta, 25th Southeastern Reg. Meet. Amer. Chem. Soc., Charleston, S. C., Nov. 7-9, 1973.
(17). H. H. Weetall, Biochim. Biophys. Acta 212, 1 (1970).
(18). S. A. Barker, A. N. Emery and J. M. Novais, Process Biochem. 5, 11 (1971).
(19). A. N. Emery, J. S. Hough, J. M. Novais and T. P. Lyons, Chemical Eng., Feb. 1972, 71.
(20). H. H. Weetall, et. al, Manuscript submitted to Science.
(21). B. M. Vanderbilt and J. J. Jaruzelski, I & EC Product Res. Develop. 1, 188 (1962).
(22). M. W. Ranney, S. E. Bergen and J. G. Marsden, Proceedings SPI Ann. Tech. Conf. 1972, Sec. 21-D.
(23). W. D. Bascom, Macromolecules 5, 792 (1972).
(24). S. Sterman and H. B. Bradley, SPE Trans. 1, 224 (1961).
(25). M. E. Schrader, I. Lerner and F. J. D'Oria, Modern Plastics 45, 195 (1967).
(26). M. Lynn and A. M. Filbert, 166th Amer. Chem. Soc. Nat. Meeting, Chicago, Ill., Aug. 1973.

(27). G. Baum, Private Communication.
(28). L. F. Fieser and M. Fieser, Reagents for Organic Synthesis, Vol. I, 231, John Wiley and Sons, N.Y., 1967.
(29). M. Goodman and G. W. Kenner, Adv. In Protein Chem. 12, 488 (1957).
(30). P. Cuatrecasas and I. Parikh, Biochemistry 11, 2291 (1972).
(31). A. I. Vogel, Practical Organic Chemistry, 3rd Edition, 636, Longman, London, 1946.
(32). K. Dean, W. H. Scouten, and C. C. Sweeley, Manuscript in Preparation.
(33). S. P.O'Neill, P. Dunnill and M. D. Lilly, Biotech. Bioeng. 13, 337 (1971).
(34). G. Baum, 166th Amer. Chem. Soc. National Meeting, Chicago, Ill., August 1973.
(35). E. Wachter, W. Machleidt, H. Hofner and J. Otto, FEBS Lett. 35, 97 (1973).
(36). L. F. Fieser and M. Fieser, Reagents for Organic Synthesis, Vol. I, 842, John Wiley and Sons, New York, 1967.
(37). H. Brandenberger, Angew. Chem. 67, 661 (1955).
(38). H. Brandenberger, Helv. Chim. Acta 40, 61 (1957).
(39). G. Kay and M. D. Lilly, Biochim. Biophys. Acta 198, 276 (1970).
(40). F. Micheel and J. Ewers, Makromol. Chem. 3, 200 (1949).
(41). M. A. Mitz and J. Summaria, Nature 189, 576 (1961).
(42). W. E. Hornby, M. D. Lilly, and E. M. Crook, Biochem. J. 98, 420 (1966).
(43). L. Fieser and M. Fieser, Reagents for Organic Synthesis, Vol. I, 1041, John Wiley and Sons, N.Y., 1967.
(44). R. Axen, J. Porath and S. Ernback, Nature 214, 1302 (1967).
(45). R. Axen and S. Ernback, Eur. J. Biochem. 18, 351 (1971).
(46). G. J. Bartling, et al, Biotech. Bioeng. 14, 1039 (1972).
(47). R. Axen, P. Vretblad and J. Porath, Acta Chem. Scand. 25, 1129 (1971).
(48). M. Passerini, Gazz. Chim. Ital. 51, 126 (1921).
(49). I. Ugi, Angew. Chem. Internat. Edit. 1, 8 (1962).
(50). R. A. Messing and H. H. Weetall, US Patent 3,519,538 (1970).
(51). H. H. Weetall, Nature 223, 959 (1969)
(52). R. A. Zingaro and M. Uziel, Biochim. Biophys. Acta 213, 371 (1970).
(53). H. H. Weetall and G. Baum, Biotech. Bioeng. 12, 399 (1970).
(54). G. P. Royer and J. P. Andrews, J. Biol. Chem. 248, 1807 (1973).
(55). P. Monsan and G. Durand, FEBS Lett. 16, 39 (1971).

(56). H. H. Weetall, Nature 232, 473 (1971).
(57). D. L. Marshall, Biotech. Bioeng. 15, 447 (1973).
(58). A. D. Traher and J. R. Kittrell, Biotech. Bioeng. 16, 413 (1974).
(59). P. J. Robinson, P. Dunnill and M. D. Lilly, Biotech. Bioeng. 15, 603 (1973).
(60). K. Tanizawa and M. L. Bender, J. Biol. Chem. 249, 2130 (1974).
(61). P. J. Robinson, P. Dunnill and M. D. Lilly, Biochim. Biophys. Acta 242, 659 (1972).
(62). A. R. Neurath and H. H. Weetall, FEBS Lett. 8, 253 (1970).
(63). L. E. Wierzbicki, V. H. Edwards and F. V. Kosikowski, Biotech. Bioeng. 16, 397 (1974).
(64). J. H. Woychik and M. V. Wondolowski, Biochim. Biophys. Acta 289, 347 (1972).
(65). H. H. Weetall, et. al., Biotech. Bioeng. 16, 295 (1974).
(66). D. R. Marsh, Y. Y. Lee and G. T. Tsao, Biotech. Bioeng. 15, 483 (1973).
(67). G. W. Strandberg and K. L. Smiley, Biotech. Bioeng. 14, 509 (1972).
(68). H. H. Weetall and L. S. Hersh, Biochim. Biophys. Acta 206, 54 (1970).
(69). M. K. Weibel, W. Dritschilo, H. J. Bright and A. E. Humphrey, Anal. Biochem. 52, 402 (1973).
(70). M. K. Weibel and H. J. Bright, Biochem. J. 124, 801 (1971).
(71). B. J. Rovito and J. R. Kittrell, Biotech. Bioeng. 15, 143 (1973).
(72). J. Shapira, J. Lecocq, A. Furst and H. H. Weetall, Experientia 28, 1261 (1972).
(73). P. Monson and G. Durand, C. R. Acad. Sci., Ser. C. 273, 33 (1971).
(74). R. D. Mason and H. H. Weetall, Biotech. Bioeng. 14, 637 (1972).
(75). T. L. Newirth, M. A. Diegelman, E. K. Pye and R. G. Kallen, Biotech. Bioeng. 15, 1089 (1973).
(76). I. C. Cho and H. Swaisgood, Biochim. Biophys. Acta 334, 243 (1974).
(77). J. E. Dixon, F. E. Stolzenbach, J. A. Berenson, and N. O. Kaplan, Biochem. Biophys. Res. Comm. 52, 905 (1973).
(78). F. Widmer, J. E. Dixon and N. O. Kaplan, Anal. Biochem. 55, 282 (1973).
(79). H. H. Weetall and R. D. Mason, Biotech. Bioeng. 15, 455 (1973).
(80). H. H. Weetall, Science 166, 615 (1969).
(81). W. F. Line, A. Kwong and H. H. Weetall, Biochim. Biophys. Acta, 242, 194 (1971).

(82). G. P. Royer and G. M. Green, Biochem. Biophys. Res. Comm. 44, 426 (1971).
(83). M. J. Grove, G. W. Strandberg and K. L. Smiley, Biotech. Bioeng. 13, 709 (1971).
(84). P. J. DeJong and P. L. Kumler, J. Chem. Ed. 51, 200 (1974).
(85). J. R. Ford, A. H. Lambert, W. Cohen and R. P. Chambers, Enzyme Engineering, ed. L. B. Wingard, Jr., 267, Interscience Publishers, N. Y., 1972.
(86). W. F. Shipe, G. Senyk and H. H. Weetall, J. Dairy Sci. 55, 647 (1972).
(87). H. H. Weetall and L. S. Hersh, Biochim. Biophys. Acta, 185, 464 (1969.

Chapter 2

THE KINETICS OF IMMOBILIZED ENZYMES

Garfield P. Royer

Department of Biochemistry
The Ohio State University
Columbus, Ohio

I. INTRODUCTION

The rates of chemical reactions catalyzed by soluble enzymes are dependent on substrate concentration, hydrogen ion concentration, temperature, solvent composition, and the presence of inhibitors. (In this context, the term "substrate" is defined as a reactant of the enzyme catalyzed reaction; it is not the support to which the catalyst is bound.) Catalysis by fixed enzymes depends on these variables and others resulting from immobilization. The kinetic behavior of an enzyme which is insolubilized by binding to a solid support may be altered as a result of perturbation of the three-dimensional structure or by chemical modification of the amino acid side chains.

In addition, the environment of a given enzyme in the bound and free states may be considerably different. The kinetic behavior of an enzyme fixed on an apolar membrane, for instance, can be dramatically altered from that of the same enzyme dissolved in a dilute buffer. Studies of enzymes bound to charged supports have demonstrated striking changes in the pH-dependence and binding of charged substrates.

The following depicts the effect of substrate and product diffusion in a heterogeneous enzyme system.

Soluble Enzyme

$$E + S \underset{K_m}{\rightleftarrows} ES \xrightarrow{k_{cat}} E + P$$

Fixed Enzyme

$$\begin{array}{ccc} S_b & & P_b \\ k_{ds} \downarrow & & \downarrow k_{dp} \\ E + S_o \underset{K_m}{\rightleftarrows} ES & \xrightarrow{k_{cat}} & E + P_o \end{array} \qquad (1)$$

The subscripts b and o represent bulk solution and the enzyme vicinity, respectively. The rate constant k_{ds} governs the transport of substrate from bulk solution to the vicinity of the enzyme, and k_{dp} governs the diffusion of product into the bulk phase. The constants K_m and k_{cat} occur in the Michaelis-Menten equation.

$$v_o = \frac{V_m S_o}{K_m + S_o}$$

$$V_m = k_{cat} E_o \qquad (2)$$

The terms are defined as: v_o, initial velocity; V_m, velocity at S_o K_m; S_o, initial substrate concentration; E_o, total enzyme concentration; k_{cat} is the rate constant which governs the decomposition of the ES complex; and K_m is the Michaelis constant which, in the simplest case, is a dissociation constant. It is evident from Eq. 1 that the reaction rate catalyzed by a fixed enzyme can be affected by the rate of transport of substrate to the enzyme, and by the rate of transport of product to the bulk solution. In fixed multienzyme systems, the diffusion of substrates and products is of special importance, since the effects may be amplified along the reaction sequence. The reduction of lag time in the appearance of the ultimate product is the principal effect of fixing the enzymes of a reaction sequence in close proximity. This effect is important both in the consideration of the intracellular action of enzymes and in the practical applications of bound enzymes.

The purpose of this chapter is to consider and illustrate the kinetic effects of immobilization, and not to review the literature. Some practical aspects of collection and treatment of rate data are also included. Excellent books which deal with kinetics of soluble enzymes (e.g. 1-3) are available. Although, to my knowledge, there are no books or reviews of immobilized enzyme kinetics, general review articles which deal with the subject have been published (e.g., 4-7).

II. CHEMICAL AND CONFORMATIONAL MODIFICATION

The activity of bound enzymes is generally lower than the corresponding soluble enzymes (8). Reasons for this include: 1) The reaction of amino acid side chains necessary for catalysis or substrate binding; 2) steric hindrance of the approach of substrate; 3) disruption of the three-dimensional structure of the protein; and 4) diffusional limitations. The decrease in activity may result from an increase in K_m or a decrease in V_m(k_{cat} or E_o). Alteration of the binding site or steric hindrance would produce an increase in K_m. The decrease in V_m can result from total destruction of part of the bound enzyme, or from partial reduction of activity of all the bound enzyme, or a combination of both. In two laboratories, the actual fractions of active sites was determined for bound trypsin derivatives (Table 1).

TABLE 1

Fraction of Active Trypsin Covalently Linked To Porous Glass[a]

Pore Diameter	Particle Size	Retention of Active Sites[b]
2000 Å	35 μ	0.39[c]
2000 Å	150 μ	0.33[c]
550 Å	177-840 μ	0.42[d]

[a]Bound protein was determined by amino acid analysis according Spackman, et al. (9).

[b]Corrected for inactive material in the soluble enzyme preparation (usually 20-30%).

[c]By active site titration with p-nitrophenyl p'-guanidino-benzoate (10).

[d]By active site labeling with a spin-label 1 oxyl-2,2,6,6-tetramethyl-4-piperidinyl methylphosphorofluoridate (11).

A considerable fraction of the bound enzyme remains active. The activation energy for the 550-A derivative is identical to that of the soluble enzyme (12), which is consistent with an all-or-none inactivation process. This observation was also made with insoluble derivatives of pronase (13), leucine aminopeptidase (14), glycerate mutase (15), and L-asparaginase (16). It should be stressed that similar activation energies of bound and soluble enzyme forms is by no means proof that the lowering of activity is an all-or-none destruction process, but is merely consistent with this interpretation.

The three-dimensional conformation of an enzyme would be susceptible to change when the enzyme is adsorbed or covalently bonded to a solid support. Denaturation may result from non-covalent interaction of the protein with the support; e.g. inactivation at hydrophobic surfaces (17). Strain resulting from the introduction of covalent bonds could also result in disruption of the conformation. Conversely, binding may have a beneficial effect on the stability of the enzyme in the presence of denaturants and at elevated temperatures. Gabel, et al (18) demonstrated that trypsin bound to sephadex retained activity in 8 M urea. This discovery is significant in that proteins insoluble in dilute buffer but soluble in urea could be digested with bound trypsin for structure and sequence studies. Thermal stability of bound enzymes in several cases (4) is better than the free enzyme form. However, improvement of thermal stability by immobilization is not a general rule. In fact, in a number of instances, the bound enzyme is actually less stable than the free enzyme at elevated temperature (4,13,19).

The information and kinetic behavior of bound enzymes may be affected by the presence of a specific substrate during the attachment process. Trypsin was linked to porous glass (20) and agarose (21) in the presence and in the absence of benzoyl-L-arginine ethyl ester at the pH optimum of the enzyme. The two immobilized trypsin derivatives differed significantly with respect to the points of attachment to the support, reactivity of histidine-57 with iodoacetamide, Michaelis constant, pH-dependence, and thermal stability. These results support the "induced fit" theory (22) which states that the binding of a specific substrate brings about a conformational change accompanied by the movement of catalytic groups to a more reactive arrangement.

Trapping of the "induced" enzyme conformer is possible by attachment to a support in a flow system. The substrate and enzyme are mixed just prior to exposure to a column of activated support. During the residence of the enzyme in the column, substrate saturation is maintained. If the geometry of some areas of the matrix conforms to the shape of the "induced ES form" of the enzyme, the conformer should be preserved as a result of covalent attachment.

Expansion of specificity of enzymes by this technique of immobilization should be possible. The "induced" conformer should catalyze the hydrolysis of non-specific substrates better than the native enzyme. Practical importance may be attached to this in applications where narrow enzyme specificity is a hindrance.

III. POLARITY OF THE SUPPORT

A. Electrostatic Effects

The dramatic effects of ionic support materials on enzyme action were demonstrated 10 years ago (23,24). Figure 1 illustrates the effect of shifting the pH optimum of the enzyme. This may be rationalized as a microenvironment effect. In the immediate vicinity of the enzyme active site, the "local concentration" of hydrogen ion is either less than that in the bulk solution (positively charged matrix) or greater than that in the bulk solution (negatively charged matrix). To achieve maximal activity, the required alkalinity of the bulk solution is less for the enzyme bound to a positively charged support. More alkalinity is required for the enzyme on a negatively charged support, since hydrogen ion is found in the vicinity of the enzyme active site.

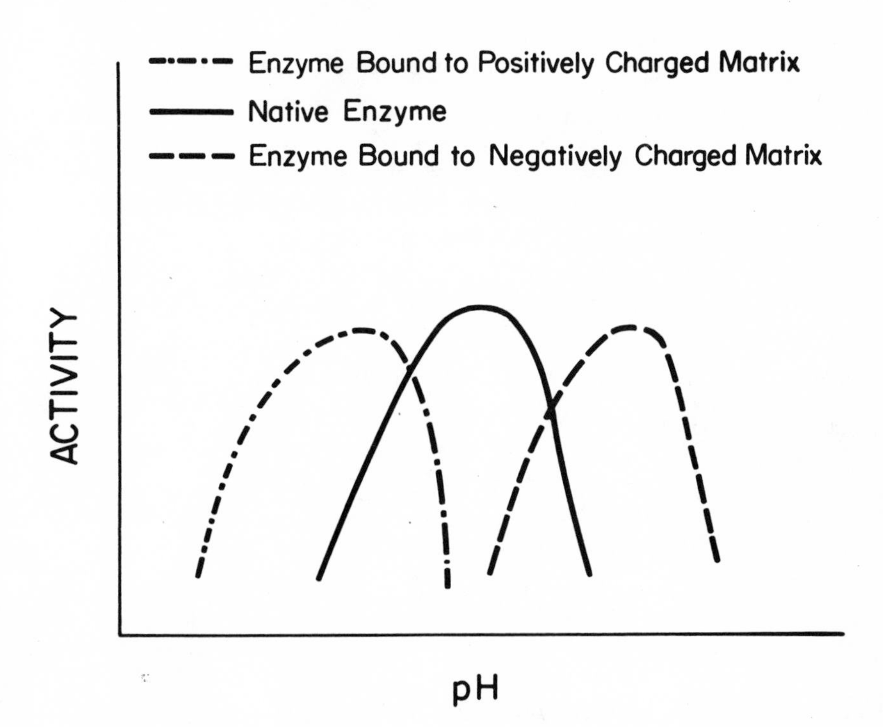

Fig. 1. Shift of the pH optimum by charged supports. A positively charged carrier will lower the pH optimum. A negatively charged carrier will increase the pH optimum. By proper selection of support material, such shifts permit the use of enzymes in environments which would normally inactivate them.

A rigorous thermodynamic explanation of the electrostatic shifts of pH optima is possible. The chemical potential (25) of a substance in a charged matrix is given by:

$$\mu' = \mu^{\circ} + RT \ln a' + ZF\psi, \tag{3}$$

where Z is the electrovalency of the substance, F is the Faraday constant, and ψ is the electrostatic potential. The chemical potential of the substance in the neutral, bulk solution is:

$$\mu = \mu^{\circ} + RT\ln a. \tag{4}$$

Subtraction of Eq. 4 from Eq. 3 gives:

$$\mu' - \mu = RT\ln(a'/a) + ZF\psi. \tag{5}$$

Division of Eq. 5 by Avogadro's number and imposing equilibrium yields:

$$kT\ln(a'/a) = -Z\varepsilon\psi,$$

or:

$$\ln a' - \ln a = -Z\varepsilon\psi/kT, \tag{6}$$

where ε is the charge on the electron. When the substance under consideration is the hydrogen ion, Eq. 6 becomes:

$$pH' - pH = \Delta pH = 0.43\varepsilon\psi/kT. \tag{7}$$

When the potential on a support is 100 mV, for example, ΔpH, which corresponds to the change in the pH optimum, would be 1.7 at room temperature. The shift is to higher pH for a negatively charged carrier ($\psi > 0$), and to lower pH for a positively charged carrier ($\psi < 0$).

There are a number of interesting academic and practical considerations of shifts in pH optima of enzymes bound to charged supports. The comparison of the pH-dependence of a free enzyme with the pH-dependence of the same enzyme bound to a membrane or other subcellular structure could provide valuable information on the natural environment of the enzyme. Of considerable practical importance is the stabilization of enzymes by fixing them to charged supports. When an enzyme is unstable or inactive in the pH range of the given process application, attachment of the enzyme to a charged support can be used to extend the active range of the enzyme. Shifts of pH optima may also be useful in discouraging certain types of microbial growth. Enzymes which are optimally active at neutral pH can be used in acid or basic media where microbial growth is retarded.

The attraction or repulsion of charged substrates and inhibitory products has considerable effect on the Michaelic constant and enzyme inhibitor dissociation constant, K_I. These constants would be increased (less binding) when the charge on the support is the same as the charge on the substrate or inhibitor. Equation 6 may be rearranged to

$$a' = ae^{-Z\epsilon\psi/kT}, \qquad (7)$$

where a' is the activity of the substrate or inhibitor in the vicinity of the enzyme and a is the activity of the substrate or inhibitor in the bulk solution. Substitution of concentrations for activities yields

$$[S]' = [S]e^{-Z\epsilon\psi/kT}. \qquad (8)$$

When K_m represents a dissociation constant,

$$K_m = \frac{[E][S]}{[ES]}; \quad K_m' = \frac{[E][S]'}{[ES]}. \tag{9}$$

The combination of Eqs. 8 and 9 gives

$$K_m' = K_m e^{-Z\epsilon\psi/kT}. \tag{10}$$

When Z is positive and ψ is positive (negatively charged carrier), $K_m' < K_m$; the substrate is attracted by the support; and the bulk concentration required to reach half-saturation is lowered. When Z is positive and ψ is negative (positively charged carrier), $K_m' > K_m$; and the apparent binding is decreased by repulsion of the positively charged substrate by the positively charged carrier. In applications, the minimization of K_m would be possible when a charged substrate is employed. If the reaction mixture were composed of neutral and charged species, both of which could react, the apparent specificity of the enzyme for the charged species would be enhanced.

An equation analogous to Eq. 10 may be derived for competitive inhibition:

$$K_I' = K_I e^{-Z\epsilon\psi/kT}, \tag{11}$$

where K_I' and K_I are the enzyme-inhibitor dissociation constants for the bound and free enzyme forms, respectively. The constant K_m/K_I is often used as a parameter for the effectiveness of a competitive inhibitor. Division of Eq. 10 by Eq. 11 yields

$$K_m'/K_I' = K_m/K_I(\exp[(Z_I-Z_S)\varepsilon\psi/kt]), \qquad (12)$$

in which Z_I and Z_S represent the charge on the inhibitor and substrate, respectively. Let us consider the following system: An enzyme bound to a negatively charged support (ψ= 100 mv), a positively charged substrate (Z_S = +1), and a negatively charged competitive inhibitor (Z_I = -1). The parameter K_m/K_I will be reduced by a factor of 4.4×10^{-4}. The effects of other possible combinations of charges of substrate and inhibitor are illustrated in Table 2. Like charges on substrate and inhibitor will result in no change of inhibitor potency, since the binding effect is the same for both species. A very large increase in the inhibitor effectiveness is apparent when the substrate carries the same charge as the support and the inhibitor carries the opposite charge.

When extended to macromolecular substrates, the previously discussed electrostatic effects would become quite large. Even when a support which exerts a relatively small potential is used, the charge effect is still significant. It is not uncommon for a protein to carry a charge of $\pm$ 10 or greater. β-lactoglobulin, for example, carries a charge of about -20 at pH 8.0.

Table 2

Effects of Competitive Inhibitors[a]

$\frac{K_m'/K_I'}{K_m/K_I}$	Z_I	Z_S	(Z_I-Z_S)
4.4×10^{-4}	-1	+1	-2
2.1×10^{-2}	-1	0	-1
1	-1	-1	0
1	+1	+1	0
4.8×10^{1}	+1	0	+1
2.3×10^{3}	+1	-1	+2

[a]The enzyme is a negatively charged support. Values of t = 300 K and ψ = 100 mV are used in Eq. 12.

The influences of charged supports may, of course, be minimized by the presence of salts. For the hydrolysis of benzoyl-L-arginine ethyl ester (positively charged) by bromelain bound to carboxymethyl cellulose (negatively charged) Wharton, Crook, and Brocklehurst (26) found that K_m increased in a hyperbolic fashion with increasing ionic strength. The experimental data were described satisfactorily by

$$K_m' = \frac{\gamma K_m \mu}{\gamma Z m_c/2 + \mu}, \qquad (13)$$

where μ is the ionic strength, Zm_C is the effective concentration of fixed charges on the support, and λ is the ratio of mean ion activity coefficients for the matrix and the bulk solution.

B. Hydrophobic Effects

In addition to electrostatic interactions between the support and substrates or inhibitors, hydrophobic interactions may also be important. A non-polar molecule would be partitioned between the support and the bulk solution (27). Brockman, Law, and Kezdy recently demonstrated this using porcine pancreatic Lipase B (28). These workers coated glass beads with an apolar silicone. The hydrolysis of tripropionine by lipase was faster in the presence of the hydrophobic surface of the glass. The explanation given was that the lipase adsorbed to the glass along with the non-polar substrate tripropionine. Thus, in analogy to the earlier discussed charged supports, the local concentration of substrate is increased as a result of an attraction of the substrate by the enzyme matrix. It is not always possible to take advantage of this attraction as a consequence of protein denaturation on apolar surfaces. However, for enzymes which naturally occur in hydrophobic areas and which use substrates with some apolar character, it is only logical to select a hydrophobic carrier. We have looked at a hydrophobic support which could well be of use for such enzymes (29).

A crosslinked, sulfonated polystyrene called Amberlyst is available commercially (Mallinkrodt). The resin has a macroporous, rigid structure which is very desirable for large-scale applications. The material may be conveniently transformed to the sulfonyl chloride by reaction with thionyl chloride in dimethylformamide. The sulfonyl chloride may be used directly in reaction with amino groups of proteins to form the sulfonamide. Alternatively, the support may be reacted with a diamine such as hexamethylenediamine. Although this material is not a good support for the enzyme trypsin, it should be ideal for enzymes such as lipases.

IV. DIFFUSIONAL EFFECTS

A. External Resistance

As Eq. 1 illustrates, the rate of transport of substrate to the enzyme or product from the enzyme can bring about considerable change in the kinetic behavior of an enzyme. The model of Fig. 2 will be used for much of the following discussion. The substrate concentration of the bulk solution, S_b, will be considered a constant. Use of initial velocities justifies this assumption. In addition, chemostat devices could be used. The following abbreviations and symbols are used:

$\underline{l}$	thickness of the unstirred layer (cm)
D	diffusion coefficient (cm^2/sec)

V	volume of the solution (cm^3)
A	surface area of support material (cm^2)
K_m	normal Michaelis constant moles/cm^3)
K_m'	Michaelis constant for bound enzyme (moles/cm^3)
k_{cat}	first order rate constant for the decomposition of the ES complex to product (sec^{-1})
k_d,	$AD/V\underline{l}$ (sec^{-1})
E_o'	enzyme loading (moles/cm^2)
E_o	enzyme concentration (moles/cm^3;AE_o'/V)
V_m'	maximal velocity (moles/cm^2/sec;$k_{cat}E_o'$)
V_m	maximal velocity (moles/cm^3/sec;$k_{cat}E_o$ or AV_m'/V)
S_b	substrate concentration in the bulk solution (moles/cm^3)
S_o	substrate concentration at support surface (moles/cm^3)
P_b	product concentration in the bulk solution (moles/cm^3)
P_o	product concentration at the support surface (moles/cm^3)
v'	reaction rate (moles/cm^2/sec)
v	reaction rate (moles/cm^3/sec)

The diffusional effects on the K_m of particle-bound enzymes have been considered by a number of workers (30-32). Of fundamental importance throughout the discussion is the consideration of units. It is hoped that the system mentioned earlier will be helpful. Notice the distinction between V_m and V_m'. In some published expressions for K_m', this distinction is either neglected or assumed to be obvious.

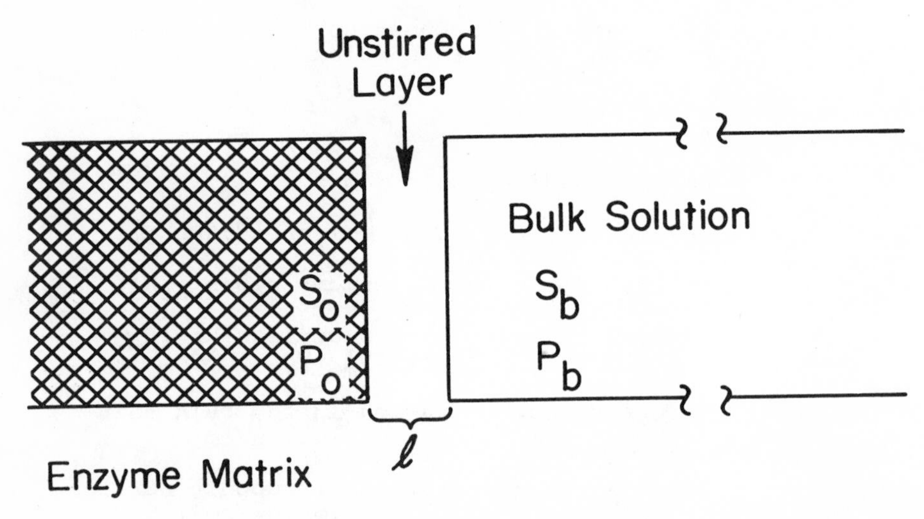

Fig. 2. Illustration of diffusional limitation by an unstirred layer. The substrate supply to the enzyme at the matrix surface may be controlled by the rate of transport across the unstirred layer. The thickness of the layer, l, can be reduced by increased stirring or flow rate.

The enzyme particle may be covered with an unstirred (Nernst) layer of solvent. The diffusion rates of substrates and products through this film may be limiting. Assuming a steady state in S_o we may write:

$$v' = \frac{D}{l}(S_b - S_o) = \frac{V_m' S_o}{K_m + S_o} = v'. \qquad (14)$$

The term on the left represents the diffusion rate of

substrate through the unstirred layer; this is the rate of substrate consumption as catalyzed by the enzyme. Elimination of S_o from Eq. 14 gives a quadratic in v':

$$v'^2 + v'[-(K_m D/\underline{l} + S_b D/\underline{l} + V_m)] + V_m' S_b D/\underline{l} = 0 \quad (15)$$

The solution of Eq. 15 may be approximated by -c/b, where $c = V_m' S_b D/\underline{l}$ and $b = -(K_m D/\underline{l} + S_b D/\underline{l} + V_m')$. This is shown as

$$v' = -\frac{c}{b} = \frac{V_m' S_b}{K_m' + S_b} \quad ; \; K'_m = K_m + V'_m \underline{l}/D. \quad (16)$$

The normal rate will be given by

$$v = dP_b/dt = \frac{A}{V} v' = \frac{V_m S_b}{K_m' + S_b}. \quad (17)$$

Notice that V_m in the numerator of Eq. 16 is the normal maximal velocity. The V_m' in the K_m' term remains in units of moles/cm^2/sec. The multiplication of V_m' by $\underline{l}$/D Eq. 16 provides units consistent with the conventional K_m. In accordance with Eq. 16 the K_m' should vary with enzyme loading and the width of the unstirred layer. The variation of K_m' as a function of flow rate or stirring rate is shown in Fig. 3. As the stirring or flow rate increases, the width of the unstirred layer decreases. It is important, therefore, to consider flow rates and enzyme loading when comparing Michaelis constants.

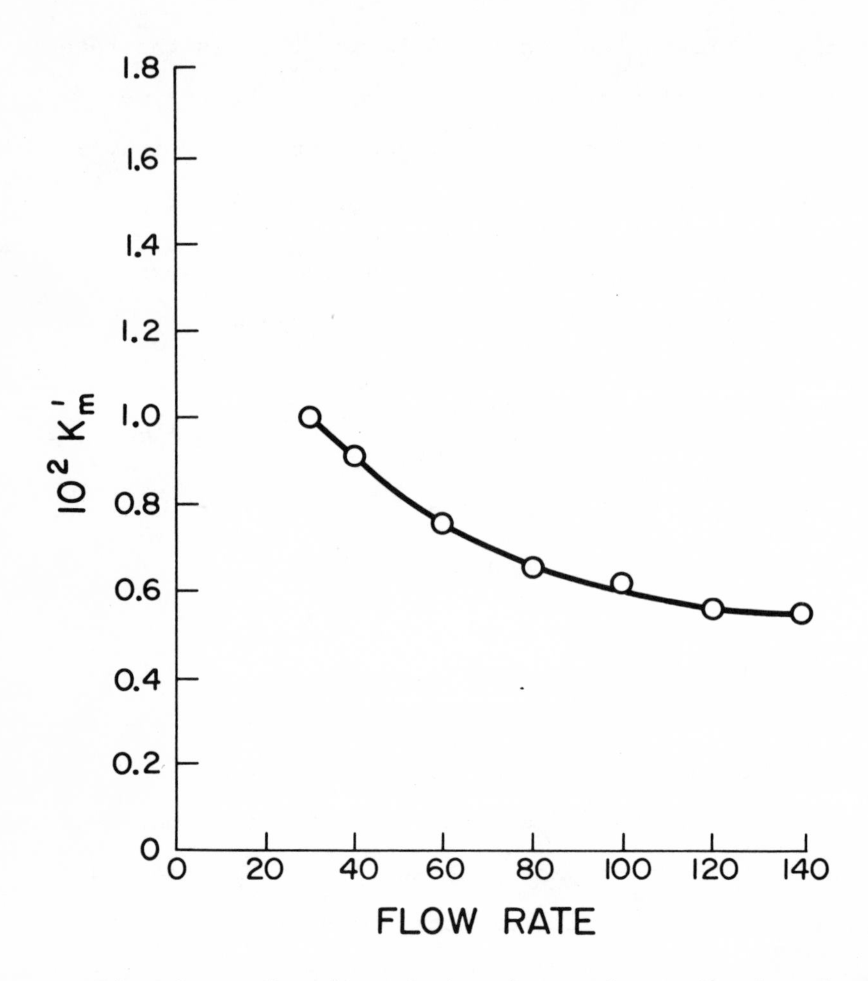

Fig. 3. Reduction of the observed K_m of a bound enzyme with increasing flow rate. Presumably, the effect is a result of the reduction of the thickness of the unstirred layer. (Redrawn from Ref. 28)

Hornby, et al. (30-31) also considered the effect of an electrostatic potential on the diffusion rate of substrate. In this case, the transport rate has an additional term:

$$v' = \frac{D}{\underline{l}}(S_b - S_o) - \frac{DzS_b}{RT} F \text{ grad } \psi$$

$$= \frac{V_m' S_o}{K_m + S_o}. \qquad (18)$$

Proceeding as before obtains the expression

$$K_m' = (K_m + V_m'\underline{l}/D)(1-\lambda)^{-1}$$

$$\lambda = \underline{l}z \text{ F grad } \psi/RT. \qquad (19)$$

In a more rigorous approach, Shuler, et al took into account the variation of substrate concentration through the electrostatic field; they used the potential distribution of the Gouey diffuse double-layer (32). Their expression for K_m' is

$$K_m' = K_m e^{\lambda} \left[1 + \frac{V_m'}{MD/\underline{l}(S_b + K_m \underline{l}\lambda)} \right]$$

$$M = 1/\left[1 + (1/k\underline{l}) \sum^{\infty} \lambda^n/n.n! \right], \qquad (20)$$

in which $1/k$ is the thickness of the diffuse double-layer. When $K_m e^{\lambda} >> S_b$, $M = e^{-\lambda}$, and $e^{-\lambda} = 1-\lambda$, Eq. 20 reduces to Eq. 19. In the absence of a potential ($\lambda=0$), both equations reduce correctly to $K_m' = (K_m + V_m'l/D)$.

We have observed the type of curve illustrated in Fig. 4. One possible explanation for this curve is the rate-limiting diffusion of product out of the matrix. This model would require that the diffusion coefficient for product be different than that for the substrate, which does not seem unreasonable.

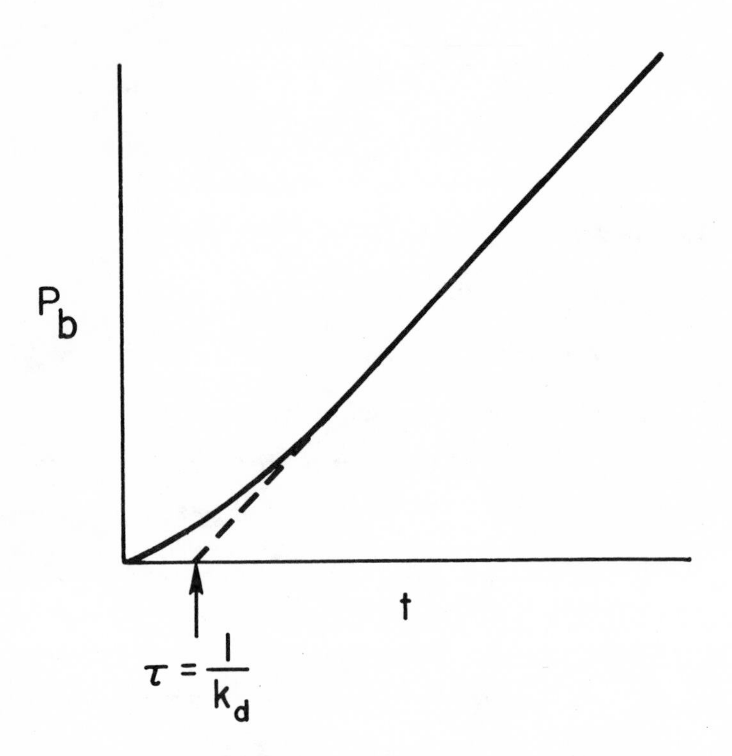

Fig. 4. Appearance of product in the bulk solution observed with several support materials. One explanation for the lag period is the rate of product diffusion out of the matrix.

The overall rate would be given by

$$\frac{dP_b}{dt} = k_d \ (P_o - P)_o - k_d \ P_o, \qquad (21)$$

in which $k_d = AD/V\underline{l}$. When the enzyme is saturated, the change in concentration of product at the matrix surface will be given by

$$\frac{dP_o}{dt} = V_m - k_d P_o . \qquad (22)$$

Integration of Eq. 22 gives

$$P_o = \frac{V_m}{k_d}\,(1-e^{k_d t}). \qquad (23)$$

Substitution of Eq. 23 into Eq. 21 results in

$$\frac{dP_b}{dt} = V_m(1-e^{-k_d t}). \qquad (24)$$

Integration gives

$$P_b = V_m t + \frac{V_m}{k_d}\,(e^{-k_d t}-1), \qquad (25)$$

which would describe the curve in Fig. 4. The linear part would be represented by

$$P_b = V_m t - V_m/k_d. \qquad (26)$$

The value of $1/k_d$ may be obtained by extrapolating the linear part of the curve to $P_b = 0$; the intercept on the time axis, τ, will equal $1/k_d$. An alternate way of obtaining k_d is based on

$$\Delta = \frac{V_m}{k_d}\;e^{-k_d t} \qquad (27)$$

which is obtained by subtracting Eq. 26 from Eq. 25. The value of Δ are taken as the differences at various points between the curved and extrapolated lines of Fig. 4. It is significant that k_d increases as the stirring rate increases (1 decreases), and τ thus goes to zero.

B. Enzyme Membranes

Considerable theoretical and experimental work has been done on enzyme membranes (4, 27, 34-39). It would require an entire book to describe in detail all of the models and boundary conditions employed. One example will be illustrated to help understand the approach taken in these systems. An enzyme attached to a membrane is exposed to identical concentrations of substrate and product on either side: $S_b = P_b = S_o = P_o$ and $D_s = D_p = D$.

Assumption of a steady state and $S \ll K_m$ allows

$$\frac{Dd^2S}{dx^2} = k_{cat}E_oS/K_m, \qquad (28)$$

or, for convenience,

$$\frac{d^2S}{dx^2} = (\alpha)^2S; \quad \alpha = \frac{k_{cat}E_o}{DK_m}^{1/2}. \qquad (29)$$

Integration of Eq. 29 yields

$$S = A\,e^{\alpha x} + B\,e^{-\alpha x}. \qquad (30)$$

The constants A and B may be evaluated using the boundary conditions: at $x = o$ and $x = l$, $S = S_b$. Insertion of these values of A and B into Eq. 30 gives

$$S = S_b\frac{\text{Sinh}\,\alpha x + \text{Sinh}\,\alpha(l-x)}{\text{Sinh}\,\alpha l}. \qquad (31)$$

Differentiation of Eq. 31 gives

$$\frac{dS}{dx} = \frac{\alpha S_b[\text{Cosh}(\alpha x) - \cosh\alpha(l-x)]}{\text{Sinh}\,\alpha l}. \qquad (32)$$

The rate of substrate entry will be given by

$$v_t = AJ_S^o + AJ_S^l$$

$$= -AD\frac{ds}{dx}_{x=o} + AD\frac{ds}{dx}_{x-l}$$

$$= 2\alpha ADS_b\frac{\cosh\alpha l - 1}{\text{Sinh}\,\alpha l}. \qquad (33)$$

In the steady state, v_t also equals the rate of product efflux. The rate per unit volume of supported enzyme will be given by

$$v' = v_t/A\underline{l} = 2\alpha DS_b \frac{\text{Cosh}\,\alpha\underline{l} - 1}{\underline{l}\,\text{Sin}\,\alpha\underline{l}} \qquad (34)$$

With Eqs. 29 and 34, it may be shown that

$$v' = \frac{2k_{cat}E_oS_b}{K_m} \; \frac{\text{Cosh}\,\alpha\underline{l} - 1}{\alpha\underline{l}\,\text{Sinh}\,\alpha\underline{l}} \qquad (35)$$

Since $v = k_{cat}E_oS_b/K_m$,

$$\frac{v'}{v} = F = \frac{2(\text{Cosh}\,\alpha\underline{l} - 1)}{\alpha\underline{l}\,\text{Sinh}\,\alpha\underline{l}}. \qquad (36)$$

The function F is an effectiveness factor expressing the ratio of the diffusion-limited rate to the diffusion-free rate.

Bunting and Laidler (34) expressed F as

$$F = \frac{\tanh \gamma\underline{l}}{\gamma\underline{l}}$$

$$\gamma = \alpha/2 = \left(\frac{k_{cat}\,E_o}{4DK_m}\right)^{1/2}. \qquad (37)$$

These workers tested Eq. 37 by investigating β-galactosidase entrapped in polyacrylamide gel. Gels containing varying amounts of enzyme were sliced into discs of widths ranging from 25 to 1000μ. They found that diffusion control was negligible for thin slices containing little enzyme. Figure 5 shows the excellent agreement between the theoretical and experimental values of F.

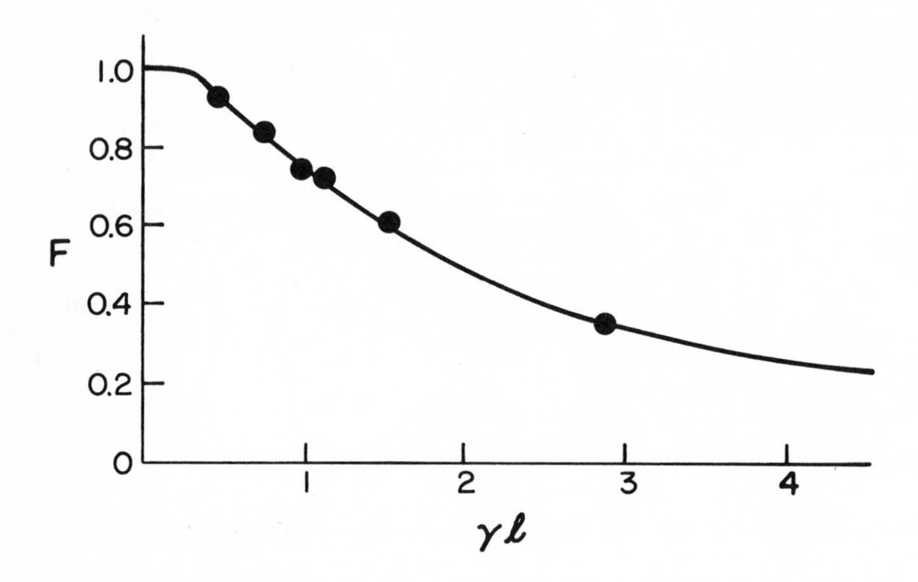

Fig. 5. Enzyme membrane effectiveness factor plotted against $\gamma \underline{l}$. $F = \tanh\gamma\underline{l}/\gamma\underline{l}$, $\gamma = (k_{cat}E_o/4DK_m)1/2$, and $\underline{l}$ = thickness of the membranes.

V. MULTIENZYME SYSTEMS

Simultaneous fixation of a number of enzymes which catalyze consecutive reactions is of interest for a number of reasons. In nature, enzyme groups associated with a given sequence of reactions are often organized in multienzyme complexes (40) or are associated on membranes. Moreover, enzymes appear to be important in biomembranes for facilitation of transport (38). Studies on simple model membranes which contain enzymes are important for getting precise information on well-defined systems.

Mosbach and Mattiasson bound hexokinase (HK) and glucose-6-phosphate dehydrogenase (G-6-PDH) to agarose and a matrix of poly (acrylamide/acrylic acid). These enzymes catalyze the reactions:

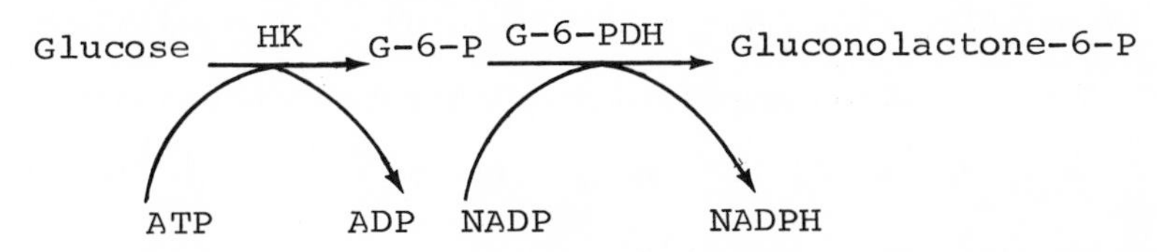

The lag time in the appearance of the final product (NADPH) decreases when the enzymes are immobilized. The activities of the enzymes were determined separately. Bound hexokinase activity was measured by incubation in the presence of glucose and soluble G-6-PDH. The reaction was followed by observing the reduction of NADP. The activity of the second enzyme, G-6-PDH, was assayed directly with excess G-6-P and NADP. Amounts of enzyme which correspond to the activities of the bound enzymes were then assayed free in solution. Curves of the type shown in Fig. 6 were obtained.

An effort to explain these results has been made by Goldman and Katchalski (39,42). The model shown in Fig. 2 is also applicable here. For the two-reaction sequence, Eq. 1 becomes

$$\begin{array}{ccccc} S_b & & P_{1b} & & P_{2b} \\ k_d \Big\downarrow & \xrightarrow{\text{Enzyme 1}} & \Big\uparrow k_d & \xrightarrow{\text{Enzyme 2}} & \Big\uparrow k_d \\ S_o & & P_{1o} & & P_{2o} \end{array} \qquad (38)$$

Several simplifying assumptions were made: 1) S_b is constant; 2) at t=0, $P_{1b} = P_{2o} = 0$; 3) both reactions are first order in S; 4) the activity of enzyme 1 is constant; 5) there is no inhibition by either product; 6) substrate and product flows are perpendicular to the matrix surface; 7) the diffusion coefficients of substrates and products are all equal; 8) the pH is constant; and 9) the matrix is electrically neutral.

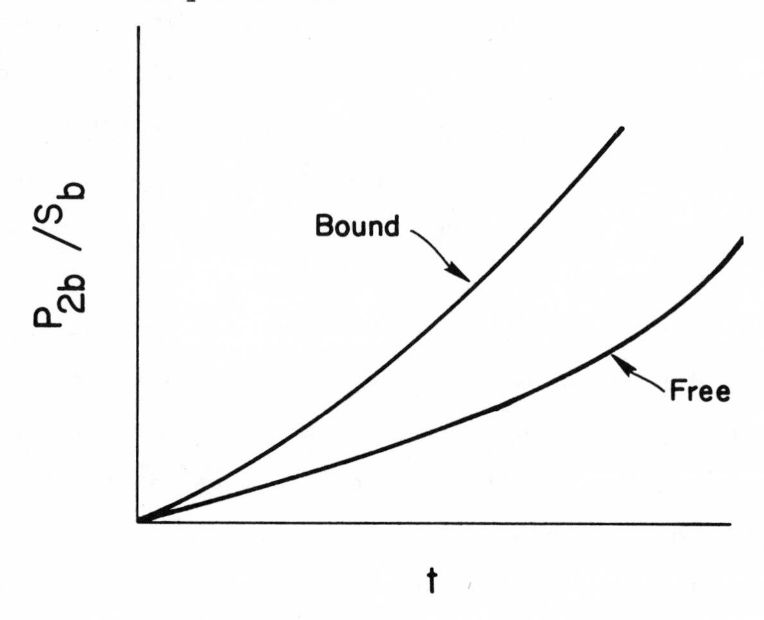

Fig. 6. Time course for the appearance of the final product of two consecutive enzyme-catalyzed reactions. The individual activities of the enzymes are the same in each case.

At x=0,

$$-\frac{D(S_b - S_o)}{\underline{l}} = -k'_1 S_o \; ; \; k'_1 = \frac{V'_{m1}}{K_{m1}}$$

$$S_o = \frac{DS_b}{k'_1 \underline{l} + D} \, . \tag{39}$$

Assuming that

$$\frac{-D(P_{1b} - P_{1o})}{\underline{1}} = k'_1 S_o - k'_2 P_{1o};\ k'_2 = \frac{V'_{m2}}{K_{m2}}. \tag{40}$$

Substitution for S_o gives

$$P_{1o} = P_{1b} + \left(\frac{k'_1 \underline{1}\, S_b}{k'_1 \underline{1} + D}\right)\left(\frac{D}{k'_2 \underline{1} + D}\right). \tag{41}$$

An expression for P_{1b} must now be obtained. It follows from Eq. 40 that the overall rate of P_{1b} production is

$$\frac{dP_{1b}}{dt} = \frac{A}{V}\,(k'_1 S_o - k'_2 P_{1o}). \tag{42}$$

Substitution for S_o and P_{1o} and integration give

$$P = \frac{k'_1 D S_b}{k'_2 (k'_1 \underline{1} + D)}\ 1 - \exp_{-} - \frac{Ak'_2 Dt}{V(k'_2\ \underline{1} + D)}\ . \tag{43}$$

P_{1o} may now be eliminated from Eq. 41 to give

$$P_{1o} = \left\{ D + k'_2 \underline{1} - D \exp \left[\frac{Ak'_2 Dt}{V(k'_2 \underline{1} + D'}\right]\right\} \left[\frac{Dk'_1 S_b}{k'_2 (k'_2 \underline{1} + D)\ (k'_1 \underline{1} + D)}\right]. \tag{44}$$

The overall rate of P production will be given by

$$\frac{dP_{2b}}{dt} = \frac{A}{V} k'_2 P_{10}. \tag{45}$$

Substitution of Eq. 44 into Eq. 45 and integration gives

$$P_{2b} = \frac{Dk'_1 S_b}{k_2'(k_1'\ \underline{1} + D)} \left\{ \frac{A}{V} k'_2 t\ \exp\left[- \frac{Ak'_2 Dt}{V(k'_2\ \underline{1} + D)} - 1\right]\right\}. \tag{46}$$

This equation is of the form

$$P_{2b} = At + B \left[e^{-Ct} - 1 \right]$$

$$A = \frac{ADk'_1 S_b}{V(k'_1 \underline{1} + D)} ; \quad B = \frac{Dk'_1 S_b}{k'_2 (k'_1 \underline{1} + D)} ;$$

$$C = \frac{Ak'_2 D}{V(k'_2 \underline{1} + D)} . \qquad (47)$$

The linear part of the curve (compare to Fig. 4) will be given by

$$P_{2b} = At - B. \qquad (48)$$

The t intercept, τ, will be given by

$$\tau = B/A = 1/k'_2 . \qquad (49)$$

The corresponding equation for the homogeneous two-enzyme system can be derived in a similar manner. The rate equation for P_{1b} is given by

$$\frac{dP_{1b}}{dt} = k_1 S_b - k_2 P_{1b} . \qquad (50)$$

Integration of Eq. 50 gives

$$P_{1b} = k_1 S_b \left[1 - e^{-k_2 t} \right] . \qquad (51)$$

The rate equation for P_{2b} is given by

$$\frac{dP_{2b}}{dt} = k_2 P_{1b} . \qquad (52)$$

Substituting Eq. 51 into Eq. 52 and integration yields

$$P_{2b} = k_1 S_b t + \frac{k_1 S}{k_2} (e^{-k_2 t} - 1) \qquad (53)$$

Equation 53 is identical in form to Eq. 47 with $A = k_1$, $B = k_1 S_b/k_2$, $C = k_2$. Notice that C in Eq. 47 reduces to k_2 when $\underline{1} = 0$. Also, the rate constant k_2 appears in the numerator of C, indicating the hyperbolic dependence of C on enzyme loading, E_o'. It would seem justified to look at the variation of C with a system containing varying amounts of enzyme 2 at a fixed flow rate. Goldman and Katchalski looked at the variation C as calculated with k_{cat} and K_m values from the literature.

Notice that the linear part of Eq. 53 is

$$P_{2b} = k_1 S_b t - \frac{k_1 S_b}{k_2}. \tag{54}$$

In this case, τ is also given by $1/k_2$. The intercepts of the curves for the bound and free enzymes should be the same when the activities of the bound and free forms of enzyme 2 are equal. As mentioned earlier, Mosbach and Mattiasson did their experiments at equal activities of enzymes for the heterogeneous and homogeneous systems. Inspection of their data shows that the values of τ are never the same for the bound and free forms of the enzymes. This does not mean the theory does not hold, however, because Mosbach and Mattiasson determined their activities at saturating levels of substrate ($v = k_{cat}E_o$ or $v' = A\, k_{cat}E_o/V$). However, the rate constant k_2 is given by

$$k_2 = \frac{k_{cat}E_o}{K_m} \text{ or } k_2 = \frac{A\, k_{cat}E_o'}{K_m V}. \tag{55}$$

Clearly, a change in K_m by immobilization could explain the differences in the τ values.

Mattiasson and Mosbach (43) have reported on a three-enzyme system in which the reactions mentioned are preceded by the hydrolysis of lactose by β-galactosidase. There appears to be a cumutative effect when the third enzyme is added. In another three-enzyme system, Srere, Mattiasson, and Mosbach (44) showed considerable rate enhancement was brought about by immobilization. The enzymes used were malate dehydrogenase, citrate synthetase and lactate dehydrogenase. The rate enhancements observed in these cases are important in the explanation of the low rates observed for soluble enzymes assayed as physiological concentration of substrate.

VI. PRACTICAL CONSIDERATION FOR KINETIC ANALYSIS

A. Assays

There are many carriers and coupling methods for enzyme immobilization. The assay conditions often depend on the type of support and linkage used. One must bear in mind that the pH, stirring rate, and substrate concentration appropriate for a given soluble enzyme need not apply to the bound enzyme. This section will point up some practical differences in the manipulation and assay of bound and free enzymes.

Immobilized enzymes are assayed in a sitrred suspension or in a packed bed. Bound enzymes can be assayed titrimetrically or spectrophotometrically in suspension. The advantage of the titrimetric (pH-stat) method is that the experimental set up is virtually identical to that used for soluble enzymes. It uses radiometer pH-stats which have overhead mechanical stirrers. This type of stirrer is particularly desirable for assay of enzymes on friable materials such as porous glass, because normal magnetic stirrers pulverize porous glass. The enzyme may be weighed or pipetted from a stock suspension in which the enzyme is evenly disbursed. Generally, it is undesirable to allow porous particles to dry, because trapped air will prevent rehydration of the interior of the particle. Though applying a vacuum will allow hydration, it could also denature the bound enzyme. Some gels may be dried with retention of enzyme activity, but, for most work, a stock suspension may be satisfactorily used.

Conditions for a typical titrimetric assay of immobilized trypsin are:

- 1-10mM substrate, benzoyl-L-arginine ethyl ester in 10^{-4} M Tris · HCl, pH8.00, 25 mM $CaCl_2$
- 10^{-2} N standard base, protected by an Ascarite trap
- 20 ml reaction volume
- 10-100 mg of bound enzyme
- chart speed of 2cm/min

-temperature bath at 25°C constant

-N_2 over reaction vessel

The pH-stat system is checked occassionally by determining the specific activity of a soluble enzyme of known purity. The standard base is retitrated periodically. The base strength is conveniently checked by allowing total hydrolysis of a known amount of substrate and checking the the observed displacement on the chart against the calculated displacement.

The rate of base addition is equal to the rate of ester hydrolysis. This is not true for amides, where part of the protons produced on hydrolysis are taken up by the amine formed. If the pK of amine is known, an absolute rate may be calculated (45); no proton production will be observed if the pK of the amine is considerably higher than the pH of the assay. In the case of alkyl esters, the rate may be calculated by the formula

$$\text{rate} = \frac{\text{Slope(ml/min) x base concentration (moles/ml)}}{\text{reaction volume(liters)}}$$

Spectrophotometric assay of bound enzymes in suspension is sometime complicated by light scattering. For routine assays of active preparations, the enzyme may be stirred in the cuvette by a small magnetic stirring bar (44). For fragile materials which settle rapidly, a gas stirrer can be used. The gas flow can be stopped periodically for absorgance measurements. For most materials, a continuous assay using a flow cell in the spectrophotometer is effective. The reaction mixture is pumped through the cell with a peristaltic pump and returned to the reaction vessel. The outside volume is kept to a minimum, and the sampling tube, which is covered with nylon net, is placed near the stirring bar to prevent plugging. For enzymes bound to porous glass, mechanical stirring is used. Widmer, Dixon, and Kaplan (47) have described an apparatus which entails the use of a vortex mixer (Fig. 7). The principal advantage of this reactor is long-term use of the same enzyme preparation because no abrasive forces are used.

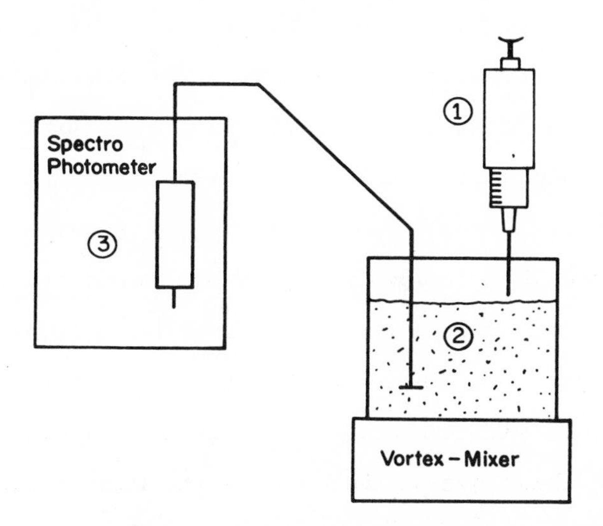

Fig. 7. Assay set-up for enzymes bound to fragile supports. The thermostated reaction vessel (2) is mounted in a vortex mixer. The bound enzyme can be suspended without abrasive forces. Activation of the syringe (1) forces the reaction mixture to a flow cell (3). The absorbance is read, and the contents of the cell are returned to the reaction vessel. (Redrawn from Ref. 45)

B. Kinetic Constants

For suspension assays, conventional rate equations are applicable. The valuation of the apparent Michaelis constant is done with a Lineweaver-Burk plot or an Eadie plot. For these plots, one determines the velocity as a function of initial substrate concentration. Use of the integrated rate equation is also possible (1). Regardless of the plotting form employed, some attempt at error estimation is necessary. We use a program based on the method of Wilkinson (48). Data should be collected over at least two orders of magnitude of initial substrate concentration, which should encompass K_m.

Long-term assays of immobilized enzymes in packed beds are often necessary. Problems which sometimes arise include: end effects, channeling, blockage, microbial growth, and changes in column dimensions which alter the residence time. The problem of end effects can be solved by diluting the bound enzyme with inert support. Channeling can be reduced by carefully pouring columns and by avoiding temperature changes. Preliminary screening of the feed stream will reduce the change of plugging the column. Microbial growth is perhaps the most serious of the problems mentioned here. Selection of temperature and pH which discourage growth is often helpful; periodic flushing with an antimicrobial agent is also possible in some cases. For the determination of kinetic parameters of laboratory column, microbial growth may be eliminated by storing the column in an antimicrobial agent.

A version of the integrated form of the Michaelis-Menten equation is used to evaluate the kinetics parameters of the column method. Without imposing initial velocity conditions, the Michaelis-Menten or Henri equation is

$$\frac{dP}{dt} = \frac{V_m S}{K_m + S} \quad , \tag{56}$$

where S is the free substrate concentration at any given time. Replacing S in Eq. 56 with $(S_o - P)$ and integrating gives

$$\frac{P}{t} = \frac{-K_m \ln}{t} \left(\frac{S_o}{S_o - P} \right) + V_m \tag{57}$$

For soluble enzymes, the time of the reaction is determined experimentally, and a plot of P/t vs $\ln \left(\frac{S_o}{S_o - P} \right) / t$ is made. The slope is $-K_m$, and intercept on the P/t-axis is V_m. For bound enzymes in columns, the approach is somewhat different. Equation 57 may be rearranged to

$$P = K_m \ln \left(\frac{S_o - P}{S_o} \right) + V_m t \ . \tag{58}$$

The residence time, t, may be replaced by V_v/Q where V_v is the void volume of the column and Q is the flow rate. For a given value of V_v/Q, P is determined as a function of initial substrate concentration (the feed concentration). A plot of P vs $\ln \left(\frac{S_o - P}{P} \right)$ is made, and the resulting slope is K_m (Fig. 8). The void volume of the column is determined by monitoring the column stream after applying a colored substance. Knowing the flow rate and measuring the time it takes for the compound to elute permits the direct calculation of the volume. The ordinary recording column monitors are useful for this measurement. A series of experiments are done at various flow rates. Notice the variation in the slopes of the plots of Fig. 8. After the apparent K_m values are plotted against flow rate, limiting value of K_m can be determined (Fig. 3).

Keyes and Semershy have proposed another method (49). They measured the rate at various flow rate at one initial change in substrate concentration. The change in product concentration is plotted against the residence time or the reciporcal of the flow and is extrapolated to zero time. These rates are determined at the various initial substrate concentrations and are plotted in the normal Lineweaver-Burk fashion. For agarose derivatives of RNAase and uricase, the K_m values found for the bound enzyme forms were identical to the K_m's for the soluble enzymes.

Regardless of the approach, meaningful kinetic properties may only be obtained on a stable enzyme preparation. The usual causes of activity loss are support erosion and release of weakly bound enzymes. Stability should be demonstrated before the kinetic study is done.

Checks during and after the study are also advisable. In the titrimetric assay, a part of the supernatant can be assayed. In the spectrophotometric assay, the sample in the cuvette, which is devoid of insoluble enzyme, should have an absorbance which is constant with time.

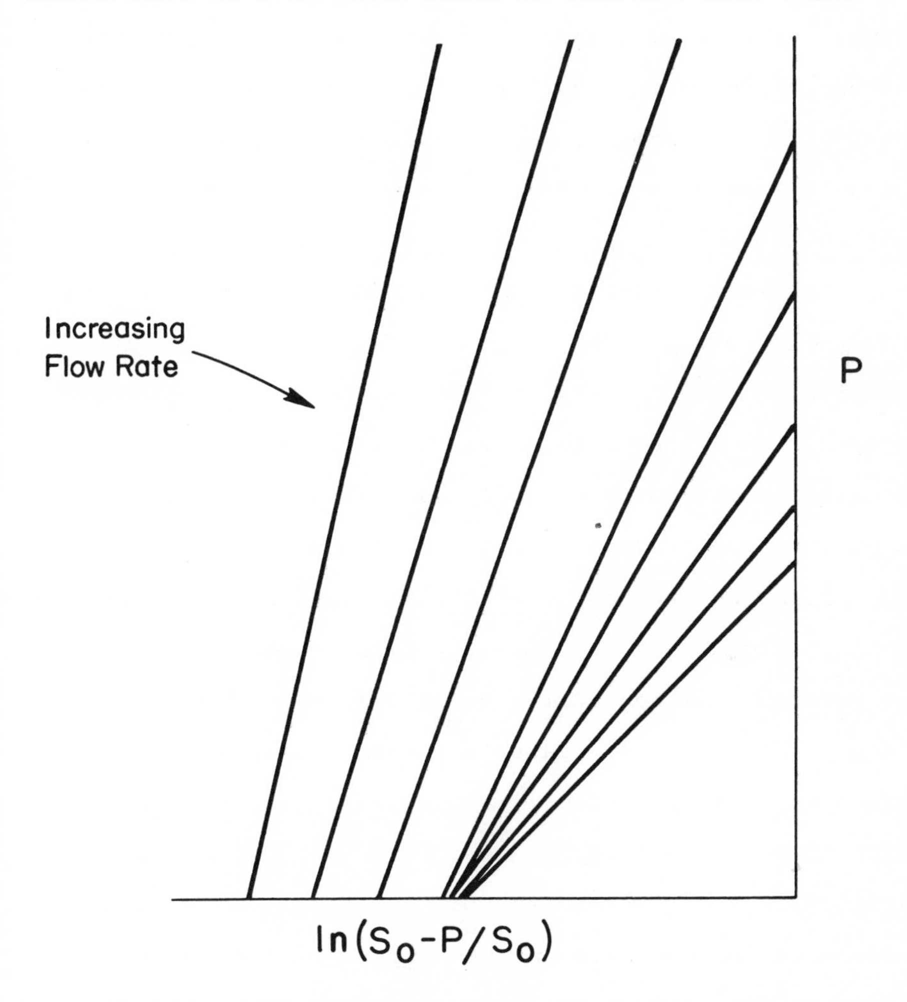

Fig. 8. Determination of the Michaelis constant of an immobilized enzyme packed in a column. The slope(K_m) decreases as the flow rate increases.

In column experiments, the enzyme is washed to constant activity. In some cases, however, the activity will remain constant for a long period before showing a sudden drop. Though some of the problems cited above could explain this, another explanation is possible. Suppose the enzyme bond is of two types: covalent, and strongly adsorbed. The adsorbed enzymes could be slowly eluted in the normal sense of column chromatography to produce loss of activity as the enzyme comes off the column.

Absolute values of k_{cat} are difficult to obtain for many soluble and bound enzymes. The value is calculated by dividing the V_m value by the concentration of active enzyme. When a soluble enzyme is shown by several physical criteria to be pure, it is reasonable to use the protein concentration as the active enzyme concentration. The preferred methods, however, are to titrate the enzyme active sites or to specifically label them.

The transition from soluble enzymes to bound enzymes is not difficult. In fact, a stable immobilized enzyme preparation is preferable to a soluble enzyme, since the same enzyme sample may be used repeatedly. As shown by Erlanger et al., immobilization can be quite useful in investigating unstable enzymes (50). In the study of lipid solubilized or stabilized enzymes or both, immobilization seems quite promising. When the straightforward methods for coupling and assay become better known, many laboratory operations performed normally with soluble enzymes will eventually be done with bound enzymes.

ACKNOWLEDGEMENT

The author is grateful for the support provided by NIH grant GM 19507.

REFERENCES

(1). K. J. Laidler, The Chemical Kinetics of Enzyme Action, Oxford, London, 1968.
(2). H. N. Christensen and G. A. Palmer, Enzyme Kinetics, W. B. Saunders, Philadelphia, 1967.
(3). K. M. Plowman, Enzyme Kinetics, McGraw-Hill, New York 1972.
(4). R. Goldman, L. Goldstein, and E. Katchalski in Biochemical Aspects of Reactions on Solid Supports, (G. R. Stark, ed.) Academic Press, New York, 1971 pp. 1-78.
(5). G. J. H. Melrose, Rev. Pure and Applied Chem., 21, 83 (1971).
(6). G. P. Royer, Chem Tech (in Press)
(7). P. V. Sundarum and K. J. Laidler, Chemistry of the Cell Interface, Part A, Academic Press, New York, 1971, Chap. V.
(8). J. P. Andrews, R. Uy, and G. P. Royer, Enzyme Technol. Digest, 1 99 (1973).
(9). D. Spackman, W. Stein, and S. Moore, Anal. Chem. 30, 1190 (1958).
(10). R. J. Ford, R. P. Chambers, and W. Cohen, Biochem. Biophys. Acta., 309, 175 (1973).
(11). L. J. Berliner, S. T. Miller, R. Uy, and G. P. Royer, Biochem. Biophys. Acta., 315, 195 (1973).
(12). R. Uy, and G. P. Royer, (unpublished)
(13). G. P. Royer, and G. M. Green, Biochem. Biophys. Res. Commun., 44 426 (1971).
(14). G. P. Royer, and J. P. Andrews, J. Biol. Chem., 248, 1807 (1973).
(15). P. Bernfeld, R. E. Bieber, and D. M. Watson, Biochem. Biophys. Acta., 191, 570 (1969).
(16). A. Ya. Nikolaw, Biochimya (Engl. Trans.) 27, 713 (1962)
(17). S. Ghosh, and H. B. Bull, Arch. Biochem. Biophys. 99, 121 (1962).

(18). D. Gabel, P. Vretbald, R. Axen, and J. Porath, J. Biochem. Biophys. Acta 214, 561 (1970).
(19). G. P. Royer and Andrews, J. Macromol. Sci. - Chem., A7(5) 1167 (1973).
(20). G. P. Royer and R. Uy, J. Biol. Chem., 248, 2606 (1973).
(21). R. Uy, V. S. H. Liu, and G. P. Royer (unpublished)
(22). D. E. Koshland, Jr., Advan. Enzymol., 22, 45 (1960)
(23). Y. Levin, M. Pecht, L. Goldstein, and E. Katchalski, Biochemistry 3, 1905 (1964).
(24). L. Goldstein, Y. Levin, and E. Katchalski, Biochemistry, 3, 1913 (1964).
(25). E. A. Guggenheim, Thermodynamics, Interscience, New York 1949, p. 332.
(26). E. W. Wharton, E. M. Crook, and K. Brocklehurst, Eur. J. Biochem., 6, 572 (1968).
(27). P. V. Sundaram, A. Tweedale, and K. J. Laidler, Can. Jour. Chem. 48, 1498 (1970).
(28). H. L. Brockman, J. H. Law and F. J. Keydy, J. Biol. Chem., 248 4965 (1973).
(29). G. P. Royer and G. M. Green, Polymer Preprints, 15, 387 (1974).
(30). W. E. Hornby, M. D. Lilly, and E. M. Crook, Biochem. J., 98 420 (1966).
(31). M. D. Lelly, and A. K. Sharp, Chem. Eng. (London) CE12 (1968)
(32). M. L. Shuler, R. Aris, and H. M. Tsuchiya, J. Theor. Biol. 35 67 (1972).
(33). B. Sinha, G. M. Green, and G. P. Royer (unpublished).
(34). P. S. Bunting, and K. J. Laidler, Biochemistry, 11, 4477 (1972).
(35). T. Kobayoshi, and M. Moo-Young, Can. J. Chem. Engin., 50, 162 (1972).
(36). E. Selegny, G. Brown, and D. Thomas, Physiol. Veg. 9, 15 (1971).
(37). D. Thomas, M. C. Tran, G. Gellf, D. Domurado, B. Paillot, R. Jacobsen and G. Brown, Biotechnol. Bioeng. Symp. No. 3, 299 (1972).
(38). D. Thomas and G. Brown, Biochimie, 55, 975 (1973).
(39). R. Goldman, Biochimie, 55, 953 (1973).
(40). L. J. Reed, Accts. Chem. Res., 7 (2), 40 (1974).
(41). K. Mosbach, and B. Mattiason, Acta Chem. Scand., 24, 2093 (1970).
(42). R. Goldman and E. Katchalski, J. Theor. Biol. 32, 243 (1971).
(43). B. Mattiasson and K. Mosbach, Biochem. Biophys. Acta., 235, 253 (1971).
(44). P. A. Srere, B. Mattiasson, and K. Mosbach, Proc. Nat. Acad. Sci., USA, 70, 2534 (1973)

(45). G. F. Bryce, and B. R. Rabin, Biochem. J., 90, 509 (1964).
(46). D.L. Marshall (personal communication).
(47). F. Widmer, J. E. Dixon, and N. O. Kaplan, Anal. Biochem., 55 282 (1973)
(48). G. N. Wilkinson, Biochem. J., 80, 324 (1961).
(49). M. H. Keyes, and F. E. Semersky, Abstracts, 164th National Meeting of the American Chemical Society, New York, New York, August 1972, No. 13007.
(50). B. F. Erlanger, M. F. Isambert, and A. M. Michelson, Biochem. Biophys. Research Commun. 40, 70 (1970).

Chapter 3

ENGINEERING ASPECTS AND REACTOR DESIGN OF IMMOBILIZED ENZYME SYSTEMS

W. H. Pitcher, Jr. and N. B. Havewala

Corning Glass Works
Corning, New York

I. INTRODUCTION

By and large, immobilized enzyme-induced reactions fall into the category of solid catalyzed liquid phase reactions. Sources of information on general heterogeneous catalytic reactor design are available in literature (1,2). This chapter deals with obtaining and utilizing engineering and reaction kinetics information for designing immobilized enzyme reactors.

Reactors utilizing immobilized enzymes fall into several general categories including batch reactors, continuous flow stirred tank reactors, fixed-bed reactors, and fluidized bed reactors (Fig. 1). A recent literature review of the subject has been published by Zaborsky (3).

The kind of reactor selected for a given process depends to a certain extent upon the type of immobilized enzyme (IME) used. For example, only the most durable of immobilized enzymes could be used in a stirred tank reactor, while micron-size particles in a packed column would result in unacceptably high pressure drop and plugging problems. The more versatile types of immobilization which use rigid particulate support materials can be adapted to most kinds of reactor systems.

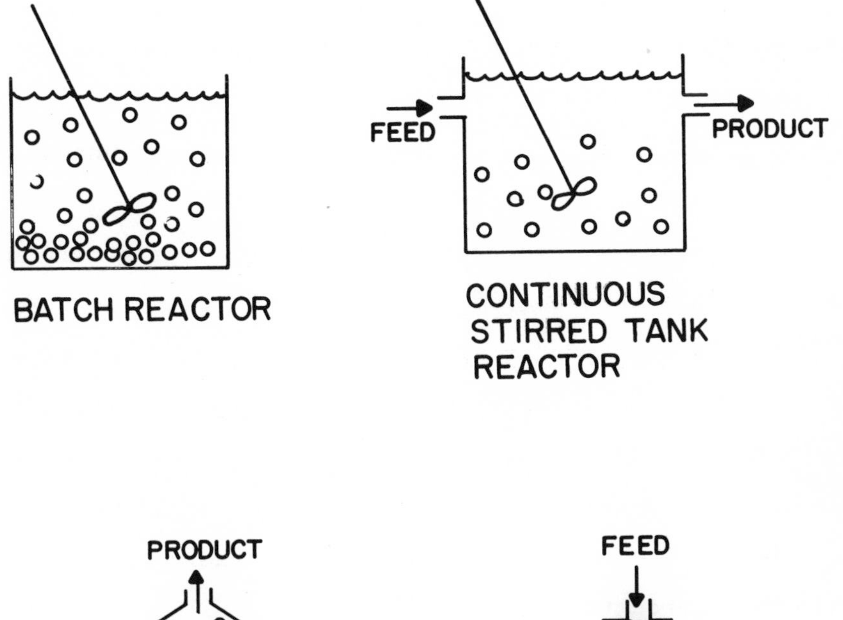

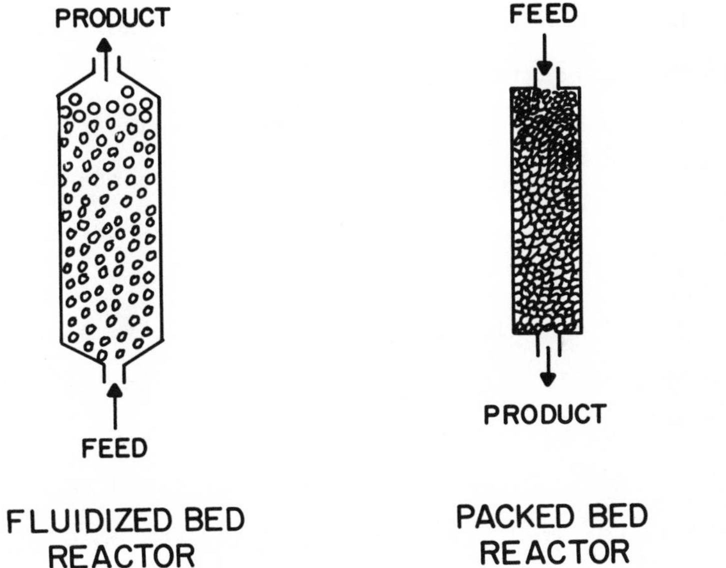

Fig. 1. Reactor Types

Even though a major advantage of immobilized enzymes is their use in continuous processes, they can also be utilized to advantage in batch-type stirred tank reactors from which the IME can be easily separated from the batch for subsequent re-use. Similarly, stirred tank reactors can be operated on a continuous basis with mesh across the inlet and outlet to prevent loss of IME.

The fluidized bed reactor is extremely versatile and can be used even when the reaction liquid contains fine suspensions. In practice, recirculation may be necessary to achieve the velocities required for fluidization and the total residence time necessary for the desired conversion. In this way, the reactor system could be operated in a manner approximating a batch reactor or a continuous stirred tank reactor.

O'Neill et al (4) have given a comparative study of immobilized glucoamylase using maltose substrate in a packed bed reactor and a continuous flow stirred tank reactor.

Using immobilized chymotrypsin on a carboxymethylcellulose, Lilly and Sharp (5) measured the rate of hydrolysis of acetyl tyrosine ethyl ester in a stirred reactor with different types of impellers and found the reaction rate to vary over a wide range of stirrer speeds.

Smiley (6) carried out conversion of starch to glucose with immobilized glucoamylase in a continuous flow stirred tank reactor. Havewala (7,8) has reported engineering and reactor design information for pilot plant-scale conversion of starch to glucose, using immobilized glucoamylase in a packed column. Some design parameters on this have been published elsewhere (9).

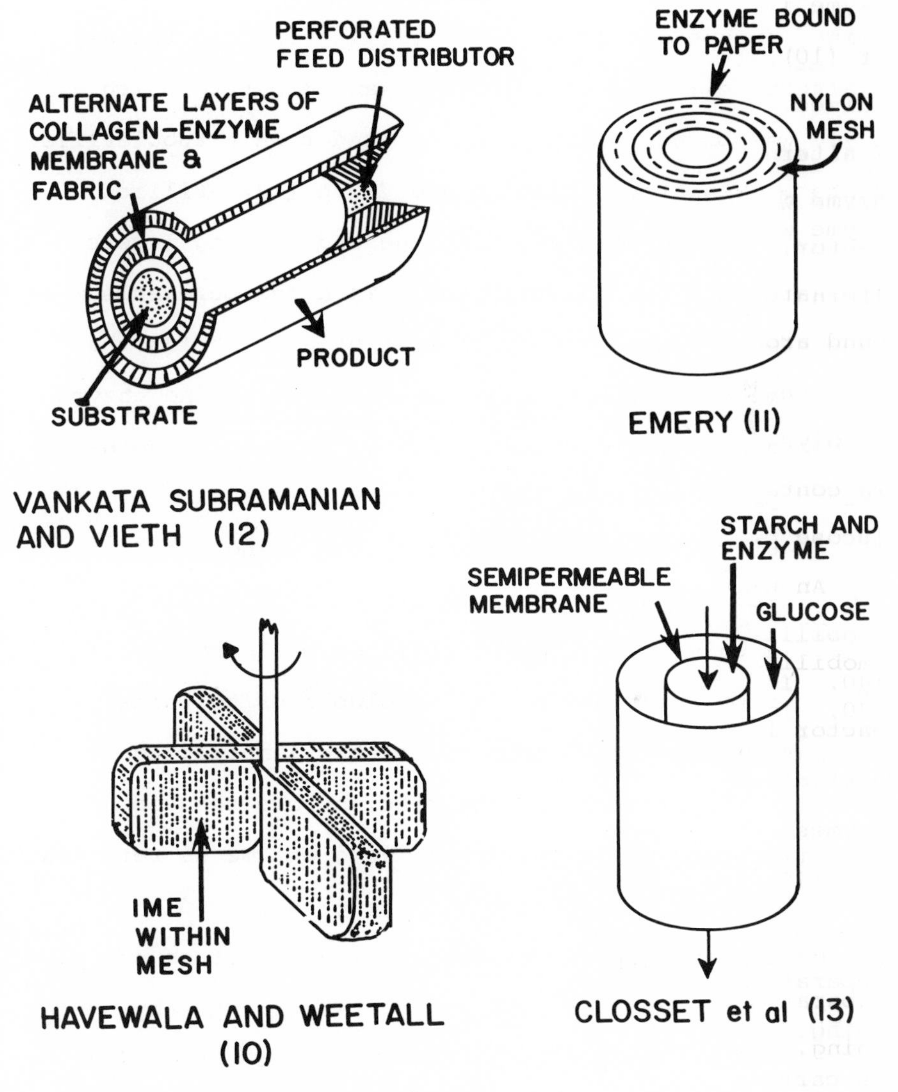

Fig. 2. Reactor Variants

Other variations of the basic reactor types (Fig. 2) include a stirred tank reactor in which the immobilized enzyme is enclosed in mesh containers attached to a stirrer (10), to give adequate agitation with minimal immobilized enzyme attrition. Emery (11) reported the use of alternate layers of nylon mesh and paper to which enzyme was bound rolled into a cylinder in a tubular flow reactor. Venkatasubramanian and Veith (12) utilized alternate collagen-enzyme membrane and backing layers wound around a feed distributor. In a semipermeable membrane reactor suggested by Closset et al (13) for the hydrolysis of starch by glucoamylase, starch and enzyme are contained by a membrane which is permeable to the glucose product.

An ultrafiltration-reactor system using a soluble immobilized enzyme has been described by O'Neill _et al_ (140. In this system, an ultrafilter cell provides a reactor into which may be fed insoluble or colloidal substrates to be acted upon by soluble immobilized enzymes. The products of the reaction are removed by ultrafiltration while the immobilized enzyme is retained.

Hornby and Filippusson (15) have described the preparation of trypsin attached to inner surface of nylon tubing. Hydrolysis of α-N-benzoyl-L-arginine ethyl ester was carried out by perfusing the substrate through the tube.

Chang (16,17,18) reported the use of semipermeable microcapsules. Enzymes incorporated during the micro-encapsulation process are prevented by their size from diffusing out of the microcapsules. Hollow fibers are attractive alternatives to microencapsulation. Such catalysts are prepared by drawing the enzyme solution by capillary action into a small-diameter hollow fiber and then sealing the two ends. Rony (19) has derived a theory of diffusion control within hollow fiber catalysts.

Robinson et al (20) have devised means for immobilizing enzymes or magnetizable particles. Using existing technology, these particles, even if very small, can be recovered magnetically.

Frequently, the most practical and efficient reactor for commercial applications is the fixed-bed or packed-column reactor. In large reactors of this kind, conditions should normally approach those of ideal plug flow. Plug flow and batch reactors are more efficient than continuous flow stirred tank reactors when reaction kinetics are higher than zero in order. Enzyme-promoted reactions frequently follow Michaelis-Menten kinetics which is between zero and first order. Figure 3 shows the ratio of ideal plug flow to continuous stirred tank reactor volume as a function of conversion for several values of S_o/K_m. Systems with product inhibition will benefit even more from plug flow versus continuous feed stirred tank reactor (CFSTR) design. On the other hand, the use of CFSTR may be advantageous where substrate inhibition is significant. Packed or fixed-bed reactors also have the advantage of avoiding the particle abrasion or fracturing common in stirred tank or fluidized bed reactors. Particle size becomes important in the packed bed reactor, since the pressure drop across the bed rises rapidly with decreasing particle size. However, diffusion limitations in porous particles decrease with decreasing particle size, possibly necessitating a compromise in the particle size chosen for a packed bed reactor.

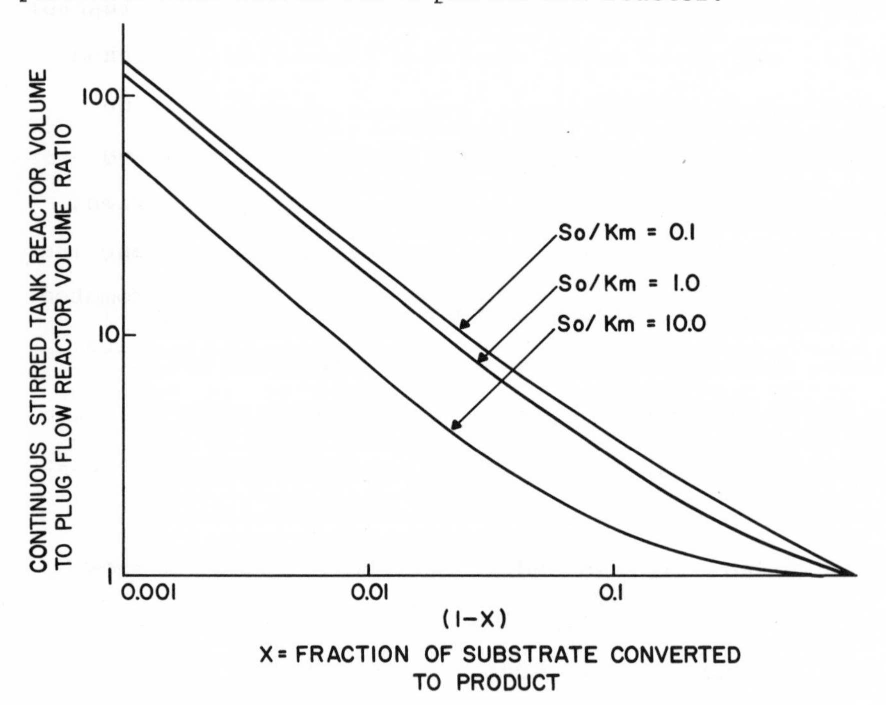

Fig. 3. Relative Reactor Volumes as a Function of Conversion.

II. ENZYME REACTOR BEHAVIOR

The behavior of ideal enzyme reactors can be predicted when the reaction kinetics and mode of reactor operation are know. Three ideal types of reactors considered here are batch reactor, continuous feed stirred tank reactor, and plug-flow reactor.

A. Irreversible Reactions following Michaelis-Menten Kinetics

For irreversible reaction, the rate of reaction given by the Michaelis-Menten kinetics is (21)

$$v = \frac{1}{V_s} \frac{dS}{dt} = \frac{kES}{K_m+S} \tag{1}$$

for the forward reaction. Equation 1 can be integrated to obtain the expression for a batch reactor

$$\frac{kEt}{V_s} = S_o X - K_m \ln (1 - X). \tag{2}$$

Similarly, for an ideal plug flow reactor in which each volume element of fluid proceeds through the reactor behaving as a batch reactor by not mixing with the adjacent fluid elements, Eq. 2 can be rewritten as

$$\frac{kE}{F} = S_o X - K_m \ln (1 - X). \tag{3}$$

For a CFSTR, the material balance is

$$v = F (S_o - S) \tag{4}$$

Thus, combining with Eq. 1,

$$F (S_o - S) = \frac{kES}{K_m+S}. \tag{5}$$

After substitution and rearrangement, the resulting equation is

$$\frac{kE}{F} = X \left[\frac{K_m}{1 - X} + S_o \right]. \tag{6}$$

B. Reversible Reactions Following Michaelis-Menten Kinetics

For reversible Michaelis-Menten kinetics (21),

$$v = \frac{kE \ (S - P/K)}{K_m + S + K_m P/K_p} ; \tag{7}$$

for batch or plug flow reactors,

$$\frac{kFt}{V_s} \text{ or } \frac{kE}{F} =$$

$$X_e S_t \left[(1 - K_m/K_p)(X_t - X_i) + \left(\frac{K_m}{S_t} + 1 - X_e + \frac{K_m X_e}{K_p} \right) \ln \frac{X_e - X_i}{X_e - X_t} \right] ; \tag{8}$$

and for a CFSTR,

$$\frac{kE}{F} = \frac{(X_t - X_i) \left[K_m + S_t - X_t S_t + \frac{K_m S_t X_t}{K_p} \right]}{1 - X - \frac{X_t}{K}}. \tag{9}$$

C. Reactions with Product Inhibition Effect

For competitive product inhibition (21),

$$v = \frac{kES}{S + K_m (1 + P/K_i)} ; \tag{10}$$

and for batch or plug flow reactors,

$$\frac{kET}{V_s} \text{ or } \frac{kE}{F} =$$

$$S_o\ (1 - K_m/K_i)\ \ (X_t - X_i) - (K_m + K_m S_t/K_i)\ \ln\left(\frac{1 - X_t}{1 - X_i}\right); \tag{11}$$

and for CFSTR,

$$\frac{kE}{F} = \frac{(X_t - X_i)\left[S_t\ (1 - X_t) + \dfrac{K_m + K_m X_t S_t}{K_i}\right]}{1 - X_t}. \tag{12}$$

D. Reactions with Substrate Inhibition Effect

For substrate inhibition (21),

$$v = \frac{kE}{1 + K_m/S + S/K'_m}. \tag{13}$$

For batch or plug flow reactors,

$$\frac{kEt}{V_s} \text{ or } \frac{kE}{F} =$$

$$S_o X - K_m \ln\ (1 - X) + \frac{S_o^2 X}{K'_m} - \frac{S^2 X^2}{2K'_m}. \tag{14}$$

For CFSTR,

$$\frac{kE}{F} = XS_o\ \left[1 + \frac{K_m}{S_o\ (1 - X)} + \frac{S_o\ (1 - X)}{K'_m}\right]. \tag{15}$$

Equations 2, 3, 6, 8, 11, 12, 14, and 15 can be expressed graphically as conversion vs. normalized residence time curves. Normalized residence time can be either.

$$\frac{\text{(wt. of IME)}\ \ \text{(elapsed time)}}{\text{(substrate volume)}}$$

or

$$\frac{\text{(total enzyme activity)}\ \ \text{(elapsed time)}}{\text{(substrate volume)}}$$

for batch reaction, or

$$\frac{\text{(wt. of IME)}}{\text{(substrate flow rate)}}$$

or

$$\frac{\text{(total enzyme activity)}}{\text{(substrate flow rate)}}$$

for continuous reactors. Havewala and Pitcher (22) have given such a graph (Fig. 4) illustrating immobilized glucose isomerase enzyme, a reversible reaction, for batch and column (packed bed) systems.

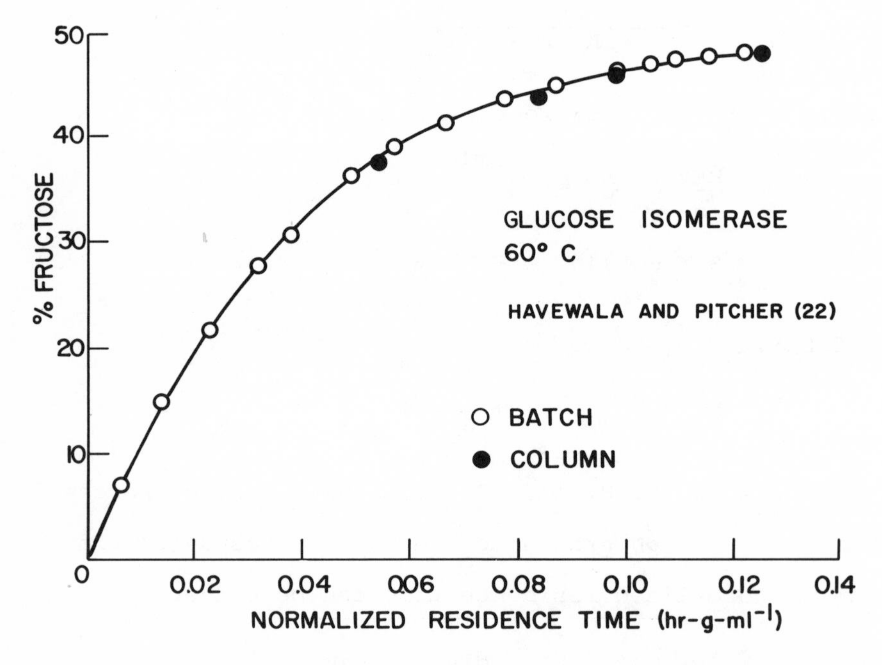

Fig. 4. Conversion as a Function of Normalized Residence Time.

For an ideal plug flow column and a batch reactor with no external mass transfer, the two curves should be identical.

This type of curve or an equation describing the curve can be determined experimentally and then used to calculate enzyme activity from flow rate and conversion measurements. If the conversion level is known, this type of curve can be used to find the normalized residence time. The enzyme activity can then be calculated using the values from substrate flow rate or substrate volume and elapsed time.

III. MASS TRANSFER

A. External Mass Transfer (23)

The rate of mass transfer from bulk solution to a surface can be written as

$$v = k_m a_m \quad (S_b - S_s). \tag{16}$$

This mass transfer rate can be equated with the reaction at the particle surface or the apparent reaction rate within the particle. For the steady-state case where internal diffusion is not significant and the reaction kinetics are simple Michaelis-Menten,

$$v = k_m a_m \quad (S_b - S_s) = \frac{kES_s}{K_m + S_s}. \tag{17}$$

Solving Eq. 16 for S_s yields

$$S_s = S_b - \frac{v}{k_m a_m}, \tag{18}$$

which can be substituted into Eq. 17 and rearranged in the Lineweaver-Burk plot form to give

$$\frac{1}{v} = \frac{1}{V_m} + \left(\frac{K_m}{V_m}\right)\left(\frac{1}{S_b}\right)\left(\frac{1}{1 - \dfrac{v}{S_b k_m a_m}}\right), \tag{19}$$

where, if S_b then v V_m and Eq. 19 approaches

$$\frac{1}{v} = \frac{1}{V_m} + \left(\frac{K_m}{V_m}\right)\left(\frac{1}{S_b}\right). \tag{20}$$

For S_b 0, Eq. 19 can be rewritten as

$$v^2 - (K_m a_m k_m + k_m a_m S_b + V_m)\ v + k_m a_m V_m S_b = 0, \tag{21}$$

which reduces to

$$\frac{1}{v} = \left(\frac{K_m}{V_m} + \frac{1}{k_m a_m}\right)\ \frac{1}{S_b}\ . \tag{22}$$

Figure 5 shows a Lineweaver-Burk plot for external mass transfer-limited reaction with the slope at large S_b approaching $\frac{K_m}{V_m}$ and at small S_b approaching $\frac{1}{k_m a_m} + \frac{K_m}{V_m}$ from Eq. 20 and 22, respectively. From these two slopes and the intercept (equal to $1/V_m$), the values of V_m, K_m, and $k_m a_m$ can be determined from experimental data on a Lineweaver-Burk plot of sufficient range.

Another approach to plotting is to rearrange Eq. 19 into the form

$$\frac{1}{V_m - v} = \frac{1}{K_m}\ \left(\frac{S_b}{v}\right)\ - \frac{1}{K_m k_m a_m}\ , \tag{23}$$

which results in a straight line plot (see Fig. 5). Equation 23 can be useful if, as is often the case, the slope of a Lineweaver-Burk plot at $1/S_b \to 0$ is difficult to determine.

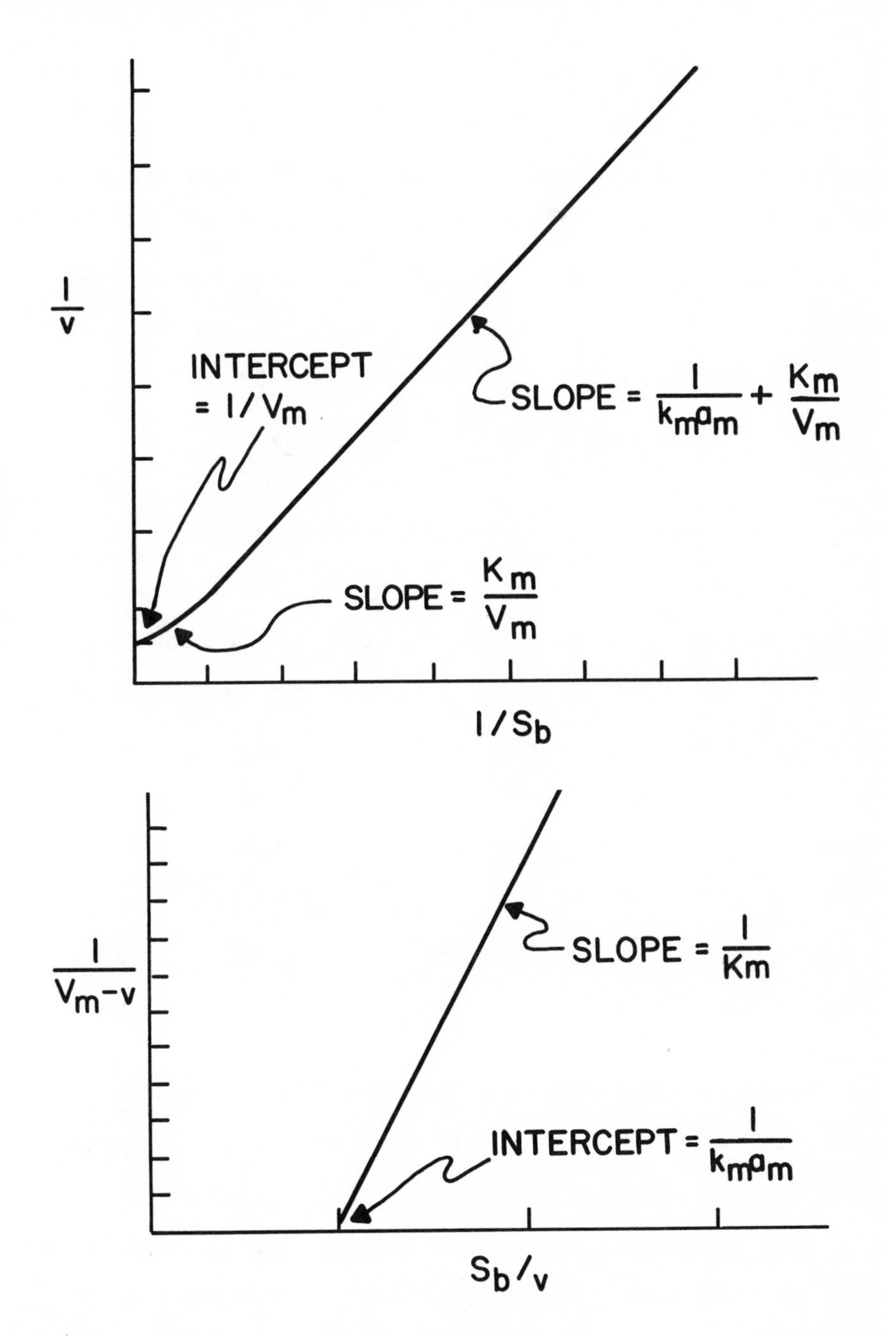

Fig. 5. Effect of External Mass Transfer Limitations.

The effect of film diffusion in a fixed-bed reactor can also be estimated from single-point reactor performance data by calculating the height of the bed required for the necessary mass transfer to occur, assuming film diffusion to be the controlling step. This height can be estimated from the correlation given by Satterfield (24):

$$z = \frac{\epsilon N_{Re}^{2/3} N_{Sc}^{2/3}}{1.09 a_v} \ln \frac{v_1}{v_2} . \qquad (24)$$

As an example, as reported by Havewala and Pitcher (22), consider an immobilized glucose isomerase packed bed reactor of 12.8 cm bed height operated at 100 ml/hr feed rate (50% glucose solution) with conversion of 45% of the glucose to fructose.

The mass velocity per unit superficial bed cross section G can be calculated from the flow rate and the column radius (0.75 cm).

$$G = \frac{(100 \text{ ml/hr})\ (1.23 \text{ g/ml feed density})}{\pi\ (.75 \text{ cm})^2\ \ 3600 \text{ sec/hr}}$$

$$= 0.01933 \text{ g cm}^{-2} \text{ sec}^{-1}.$$

The Reynolds number N_{Re} and Schmidt number N_{Sc} can be calculated from feed viscosity (about 3.6 centipoises at 60°C), G, particle diameter D_p (20/30 mesh or 0.0707 cm), feed density (1.23 g/cm^3), and diffusivity (approximated at 0.21 x 10^{-5} cm^2/sec at 60°C in 50% glucose solution).

$$N_{Re} = \frac{D_p G}{\mu} = \frac{(.0707\ \text{cm})\ (.01933\ \text{g cm}^{-2}\ \text{sec}^{-1})}{3.6 \times 10^{-2}\ \text{poises}} = 0.0380.$$

$$N_{Sc} = \frac{\mu}{pD} = \frac{3.6 \times 10^{-2}\ \text{poises}}{(1.23\ \text{g/cm}\ )\ (.21 \times 10^{-5}\ \text{cm/sec})} =$$

$$1.394 \times 10^4.$$

The surface area per unit volume of bed is equal to

$$\frac{(\text{surface area per particle } (1 - \epsilon)}{(\text{volume per particle})} =$$

$$\frac{4\pi \left(\frac{.0707\ \text{cm}}{2}\right)^2 (.65)}{4/3\pi \left(\frac{.0707\ \text{cm}}{2}\right)^3} = 55.16\ \text{cm}^2/\text{cm}^3.$$

Using Eq. 24, the bed height can then be estimated by

$$z = \frac{(.35)\ (.0380)^{2/3}\ (1.394 \times 10^4)^{2/3}}{(1.09)\ (55.16\ \text{cm}^2/\text{cm}^3)} \ln \frac{1}{1 - .45}$$

$$z = 0.23\ \text{cm}.$$

Thus, the calculated bed height, assuming film diffusion control in this case, was 0.23 cm. Since this is much less than the actual bed height, it indicates negligible film diffusion resistance.

Another practical method of checking for external mass transfer effects in packed bed reactors is to change the depth of the bed while keeping the residence time the same (1). If film diffusion effects are significant, the change in linear velocity will result in a change in conversion in the reactor. Changes in bed height by at least a factor of two or three are recommended to assure observable conversion differences. Weetall and Havewala (25) have applied this approach in studying the role of mass transfer in immobilized glucoamylase reactors.

The mass transfer coefficient k_m can be estimated for a fixed bed reactor by the following correlation of Wilson and Geankoplis (26):

For $0.0016\ N_{Re} < 55$, $0.35 < \varepsilon < 0.75$,

$$J = (k_m \rho / G) N_{Sc}^{2/3} = (1.09/\varepsilon) N_{Re}^{-2/3} \quad . \qquad (25)$$

For $55 < N_{Re} < 1500$,

$$J = (k_m \rho / G) N_{Sc}^{2/3} = (0.250/\varepsilon) N_{Re}^{-0.31} . \qquad (26)$$

Satterfield has discussed the problem of mass transfer in slurry reactors (stirred reactors or fluidized beds). He gives the correlations of Brian and Hales (27).

$$\left(\frac{k_m d_p}{D}\right)^2 = 4.0 + 1.21\ N_{Pe}^{\ 2/3}. \qquad (27)$$

He also gives the correlation

$$k_m'\ (N_{Sc})^{2/3} = 0.38\ \frac{g\mu\Lambda p}{p^2}^{1/3}. \qquad (28)$$

The problem of mass transfer in stirred batch reactors has also been discussed by O'Neill (28).

B. Internal Mass Transfer

When reaction promoted by an enzyme immobilized with a porous material occurs, a substrate concentration gradient is established in which concentration decreases as distance from the porous body surface increases. The general problem of simultaneous diffusion and reaction has been the subject of extensive literature. Probably the most comprehensive is a book by Satterfield (24).

Since much recent literature on the subject is available, treatment of this topic will be relatively brief. The ratio of observed or apparent reaction rate to the rate if no diffusion limitation exists is called the effectiveness factor n, which for immobilized enzyme systems will normally be less than unity. The differential

The differential equation describing diffusion in a sphere is

$$\frac{d^2S}{dr^2} + \frac{2}{r}\frac{dS}{dr} = \frac{v_i}{D_{eff}}\ . \qquad (29)$$

This equation can only be solved analytically for rate expression of the form

$$v_i = k\ S^m. \qquad (30)$$

Cases such as Michaelis-Menten kinetics must be solved analytically.

Figures 6 and 7 show generalized plots for Michaelis-Menten kinetics giving η as a function of a modulus for spherical and flat plate geometry, assuming D_{eff} is constant. The modulus is

$$\phi L = \frac{L^2}{D_{eff}} \left(\frac{1}{Vc}\frac{dn}{dt}\right) \frac{1}{S} = \frac{R^2}{9\ D_{eff}} \left(\frac{1}{Vc}\frac{dn}{dt}\right) \frac{1}{S} \qquad (31)$$

for flat plate and spherical geometry. The term $\frac{1}{Vc}\frac{dn}{dt}$ is the observed reaction rate per unit volume of porous carrier.

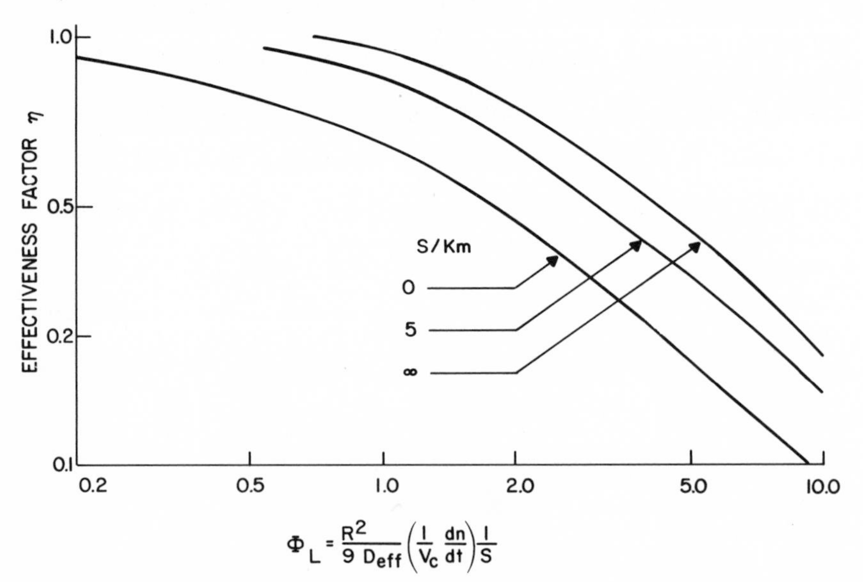

Fig. 6. Effectiveness Factors for Michaelis-Menten Kinetics in Spherical Geometry.

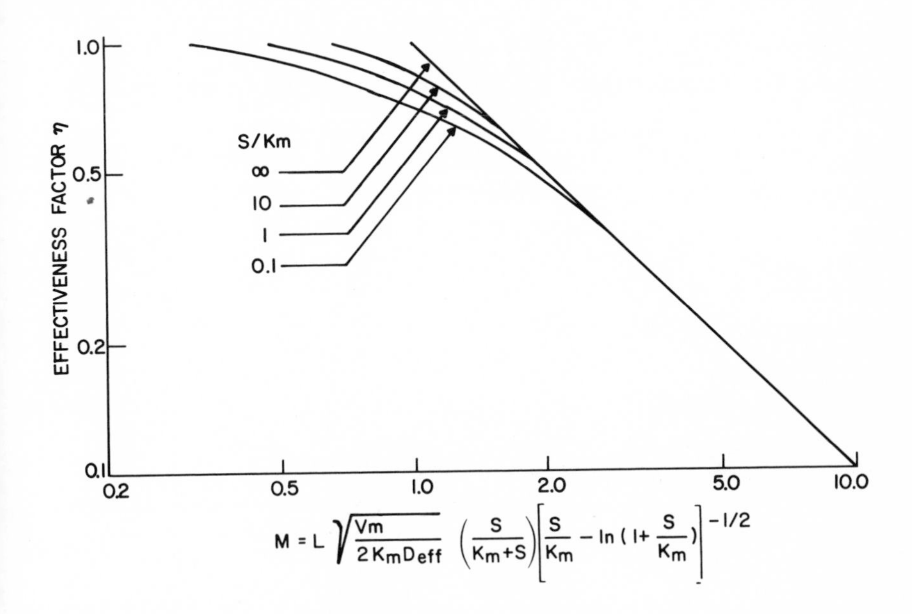

Fig. 7. Effectiveness Factors for Michaelis-Menten Kinetics in Flat Plate Geometry.

If the diameter of the diffusing substrate molecule is a significant fraction of the pore diameter, the effective diffusivity will be decreased. The following empirical correlation proposed by Pitcher (29) and by Satterfield et al (30) can be used to estimate the reduction diffusivity:

$$\log_{10} \frac{D_{eff}\tau}{D_o\ \theta} = -2.0\lambda,$$

where λ = the ratio of solute critical molecular diameter to pore diameter. Critical diameter is defined as the diameter of the smallest cylinder into which the molecule will fit without distortion. Bischoff (31) expressed n graphically (Fig. 8) as a function of a general modulus M for Michaelis-Menten kinetics, assuming D_{eff} is constant at various levels of S/K_m.

$$M = L\sqrt{\frac{V_m}{2\,K_m\,D_{eff}}}\left(\frac{S}{K_m + S}\right)\left[\frac{S}{K_m} - \ln\left(1 + \frac{S}{K_m}\right)\right]^{-1/2}. \qquad (33)$$

For M greater than 2, $n = \frac{1}{M}$.

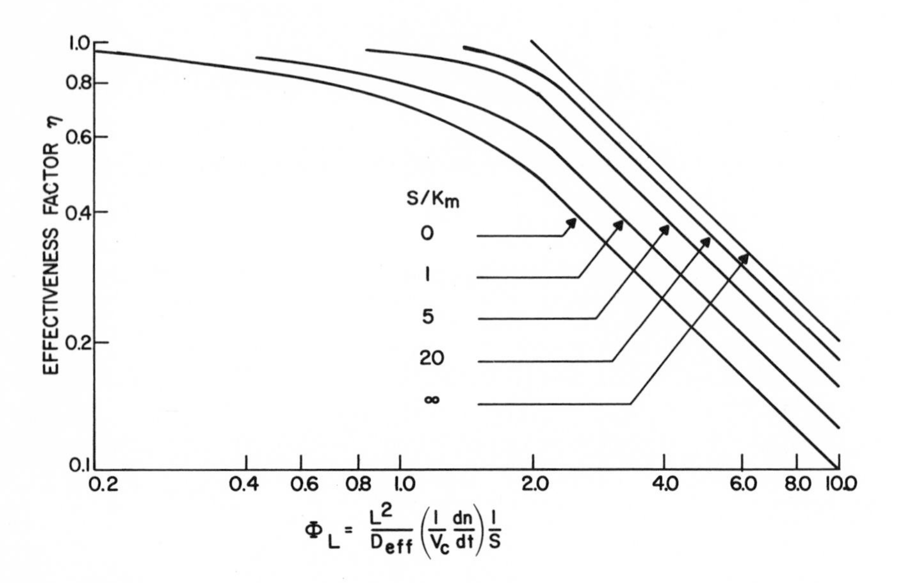

Fig. 8. Effectiveness Factors for Michaelis-Menten Kinetics.

More complex cases for substrate and product inhibition have been solved numerically by Moo-Young and Kobayaski (32).

Frequently, lack of knowledge of the value of D_{eff} lessens the value of the generalized plots. There are several practical methods of determining the effectivenss factor or extent of pore diffusion limitations. One method is to decrease particle size or membrane thickness until it results in no additional increase in observed reaction rate. The rate then observed is assumed to be the intrinsic rate, and n can be calculated for other cases by dividing the observed rate by the intrinsic rate. Another method of determining the effect of pore diffusion resistance is to determine the variation of reaction rate or activity with temperature for the immobilized as well as the soluble enzyme. If the Arrhenius plot (log of reaction rate vs. reciprocal absolute temperature) shows a straight line with a slope equal to that of an Arrhenius plot for the soluble enzyme, it usually implies that no pore diffusion problem exists. Usually, internal diffusion limitations will result in a curvature (decrease in slope) at higher temperatures. The slope of such a plot in the region of pore diffusion limitation will be less than the slope of the soluble enzyme plot at the same temperature. Figure 9 shows such a plot for immobilized glucoamylase (20/30-mesh porous glass particle carrier) where there is evidently some pore diffusion limitation at 60°C and above. Wheeler (33) has shown that as the effectivenss factor falls below unity the measured activation energy also falls, tending toward the arithmetic mean of activation energies for the diffusion process and the chemical reaction. Since the former energy is relatively small, the measured activation energy will approach half the true value as diffusion limitation becomes dominant.

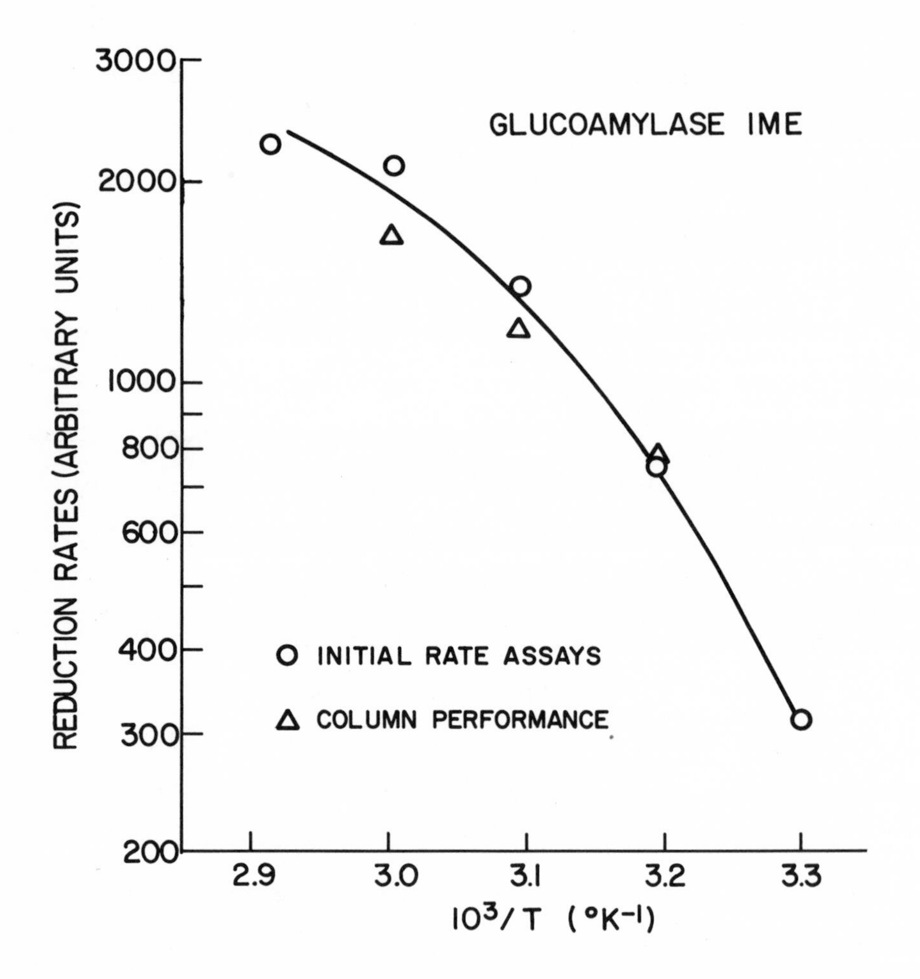

Fig. 9. Arrhenius Plot for Immobilized Glucoamylase.

Other specific examples of pore diffusion in the literature include Marsh et al (34) who reported the effectiveness factor for glucoamylase immobilized on glass. They determined the effectiveness factor experimentally by comparing reaction rates using particles of various sizes including some finely crushed. Figure 10 shows a plot of their data, n vs. a modulus assuming intrinsic zero-order kinetics which is justified because $S > K_m$. It can be seen that the experimental points agree fairly well with the predicted curve.

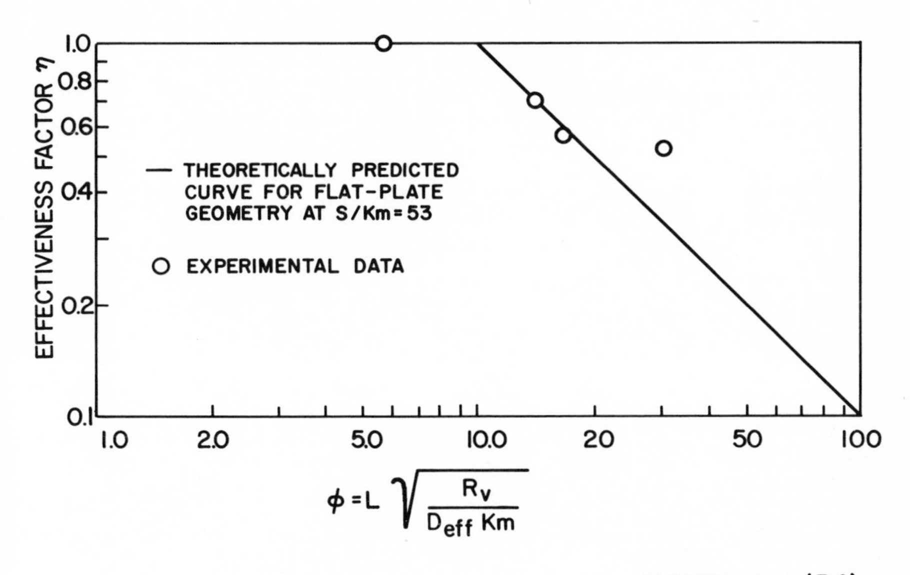

Fig. 10. Experimentally Determined Effectivenss Factors for Immobilized Glucoamylase.

Bunting and Laidler (35) have studied β-galactosidase trapped in polyacrylamide gel and obtained a typical Lineweaver-Burk plot (Fig. 11) showing the increase in slope with the increase in enzyme concentration. The slope of the straight-line portion of the curve at small S is $\frac{K_m}{\eta V_m}$, with all IME curves having the same intercept $(1/V_m)$. The soluble enzyme curve has a smaller slope, as expected. The different intercept is probably an indication that not all the immobilized enzyme is active.

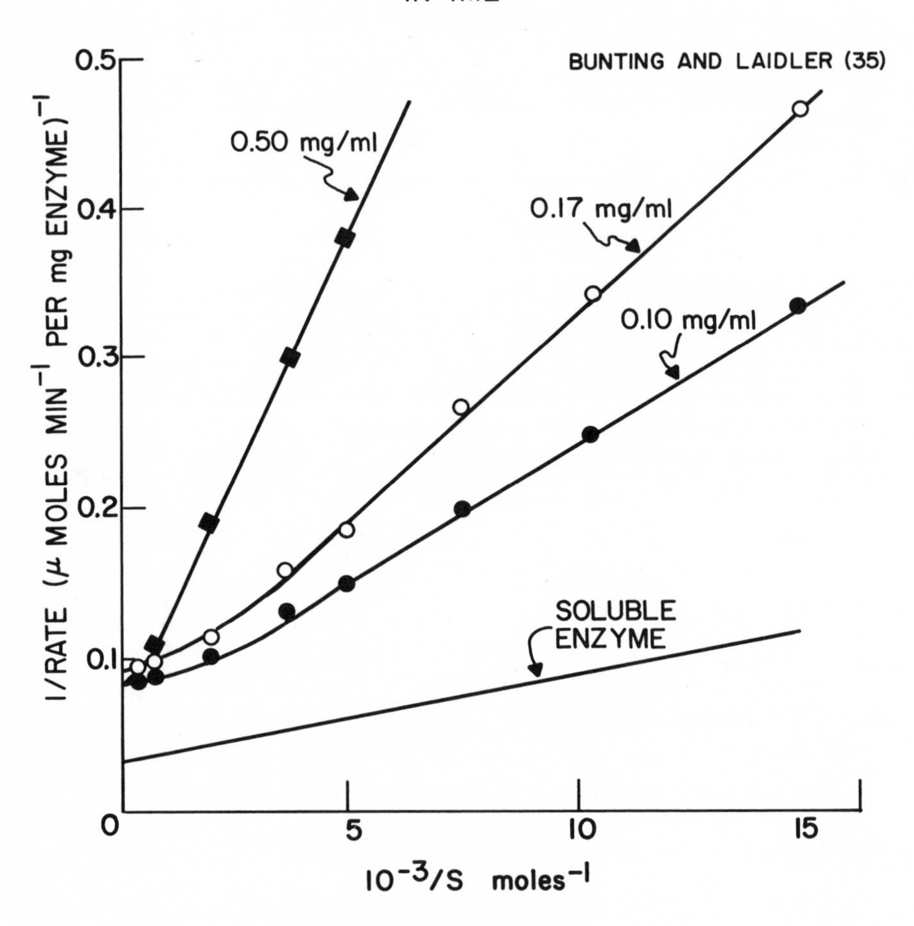

Fig. 11. Lineweaver-Burk Plot.

Goldman, Kedem and Katchalski (36) have reported extensive studies of the role of diffusion in the kinetics of a matrix entrapped enzyme.

IV. BACKMIXING

A possible source of column inefficiency is backmixing. The effect of backmixing can be estimated from a calculated dispersion coefficient as discussed by Levenspiel (1). The dispersion coefficient is defined as D/uL_c, where D = substrate diffusivity, u = fluid velocity, and L_c is bed height. For packed beds at low Reynolds numbers ($N_{Re} < 10$), the Peclet number, $N_{Pe} = \frac{ud_p}{D\varepsilon}$, where d_p= particle diameter, is approximately equal to 0.5 (McMaster (37) reported a value of about 0.6). The dispersion coefficient can be related to the Peclet number.

$$\frac{D}{uL_c} = (N_{Pe})^{-1} \frac{d_p}{\varepsilon L_c} \quad . \tag{34}$$

For high substrate concentrations ($S > K_m$), reactions following Michaelis-Menten kinetics become essentially zero-order and conversions are unaffected by backmixing. At low substrate concentrations, the same reaction is first order, a case considered by Levenspiel (1). For Michaelis-Menten kinetics at intermediate substrate levels, Kobayashi and Moo-Young (38) given generalized plots comparing reactor sizes against plug flow conditions in the same manner as developed by Levenspiel.

V. ELECTROSTATIC EFFECT

While the effect of diffusion on immobilized enzyme behavior has been the subject of much quantitative investigation, electrostatic effects usually have only been invoked as qualitative explanations for phenomena such as shifts in optimal pH for IME (39). Hornby, et al (40) considered combined electrostatic and diffusive effects by using a modified Michaelis constant. Shuler et al (41) treated quantitatively the effects of electrostatic fields and diffusion on IME-promoted reaction rates. Their treatment was based on a potential distribution for the electrical double layer valid for small surface potential. Hamilton et al (42) extended this treatment by employing the complete Gouy-Chapman solution, valid for higher surface potential, although the results of Shuler et al are sufficiently accurate for most practical cases. Hamilton et al show examples of the curvature of Lineweaver-Burk plots resulting from significant electrical diffusive effects. They also point out that oppositly charged substrate and surface can result in effectiveness factors greater than unity.

VI. TEMPERATURE EFFECT

The variation of reaction rate or deactivation rate with temperature for immobilized enzymes can generally be described by the Arrhenius equation. The activation energy is usually in the 5-20 k cal/g mole range, while deactivation energies may commonly range from about 10 to over 100 k cal/g mole. For example, the reported (7,8) value for immobilized glucoamylase is about 15 k cal/g mole, while the value for deactivation energy for the same enzyme is about 40-45 k cal/g mole.

VII. ENZYME DEACTIVATION

Enzyme activity frequently exhibits an exponential decay curve that appears as a straight line when the logarithm of activity (or reaction rate) is plotted vs. time, as shown in Fig. 12 (22). Another common shape of curve is seen in Figure 13, where two distinct slopes can be observed. One possible explanation for these two decay curves for the same system at different temperatures is that there are two activity loss mechanisms with the more rapid one affecting only a certain fraction of the enzyme. For example, perhaps there is a relatively rapid physical loss of certain weakly bonded enzyme molecules from the carrier and a slower denaturation of all the enzyme. Another possibility is that two enzyme species with different decay rates or bonding strengths are present. Both loss rates could be first order with respect to the enzyme involved (exponential decay), and would result in the type of curve shown in Fig. 13.

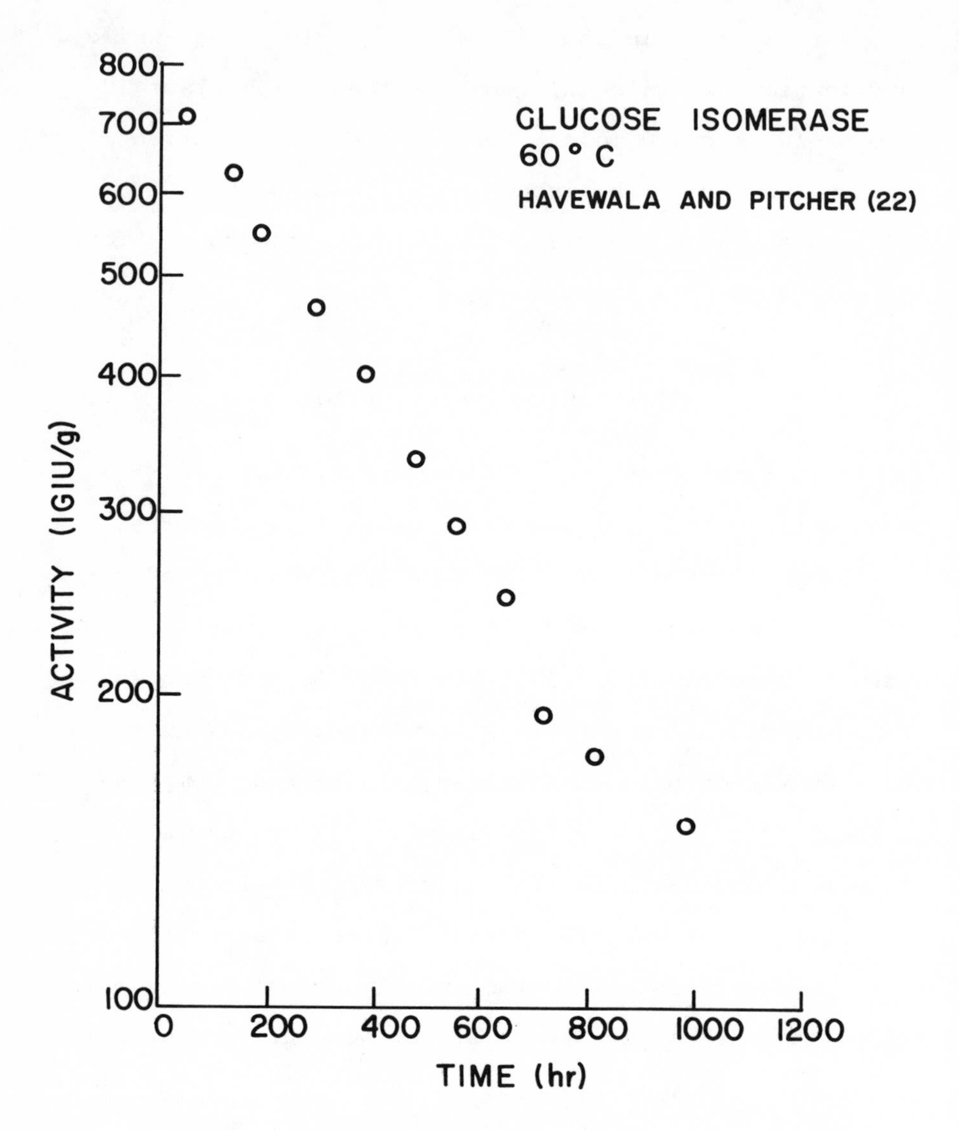

Fig. 12. Glucose Isomerase Activity versus Operating Time at 60°C.

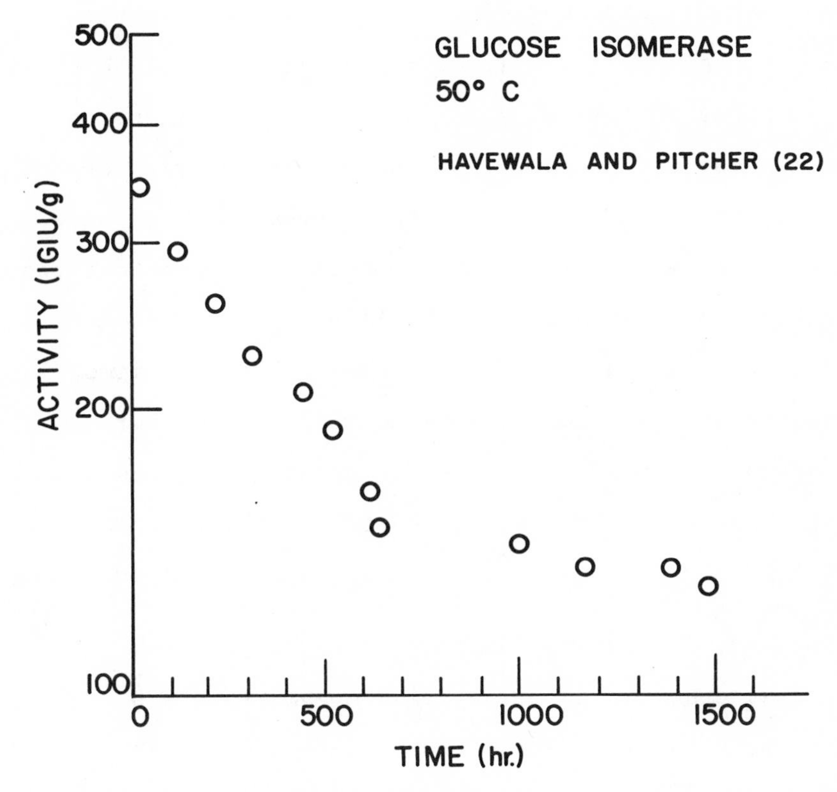

Fig. 13. Glucose Isomerase Apparent Activity versus Operating Time at 50°C.

Examples of downward concave semilog plots have also been observed. These perhaps result from gradual poisoning (heavy metals, etc.) that does not take effect until a threshold level is exceeded. It is evident that activity loss can result from enzyme denaturation, pore blockage (a reversible condition occasionally observed with porous carriers), and enzyme loss from the carrier. Enzyme loss from the carrier may occur as a result of carrier erosion or from the breaking of bonds between enzyme and carrier.

It should also be noted that immobilized enzymes whose apparent activity is diffusion-limited will exhibit slower apparent loss of activity than if no diffusion limitations were present. As the enzyme activity decreases, the effectiveness factor increases, thus slowing the apparent activity loss. This phenomenon has been discussed in greater detail by Ollis (43).

As pointed out previously, the activity of immobilized enzyme in a column can be monitored by measuring conversion and feed flow rate and using a standard conversion vs. normalized residence time curve to calculate activity.

VIII. HEAT TRANSFER

Heat transfer in immobilized enzyme reactors becomes important when close temperature control is desired to operate at optimum conditions. Relative to fixed bed reactors, heat transfer is generally good in fluidized bed and stirred tank reactors. Therefore only the case of a fixed bed reactor need be considered here. As an example, heat transfer on an immobilized lactase column reactor will be considered there.

To set an upper limit to the extent of heating or cooling possible in a fixed-bed reactor, the following assumptions are made. In order to calculate the maximum possible heat transfer rate:

1) The adiabatic temperature change occurs completely and instantaneously in the fluid as it enters the column. This gives the largest possible temperature gradients and the greatest heat transfer possible.

2) The temperature at the column wall is constant and equal to the desired column temperature, implying no external resistance to heat transfer.

3) A slug of liquid entering the column can be treated as an infinite cylinder (no axial heat transfer) cooled (or heated) externally for a period of time equal to the residence time in the IME column.

For small distances x from the column wall, the dimensionless temperature T* is the temperature difference between a given point and the wall which is expressed as a fraction of the difference between initial temperature T_i and wall temperature T_w, and which is given by (44)

$$T^* = \frac{T - T_w}{T_i - T_w} = \text{erf}\left(\frac{x}{2\gamma\alpha t}\right). \tag{36}$$

Thus, for a large column containing immobilized lactase using 20% lactose feed at 40°C,

$$\alpha = \frac{\left(0.72 \frac{\text{BTU}}{\text{hr ft}^{\circ}\text{F}}\right)}{(66.0 \text{ lb/ft}^3)\left(0.9 \frac{\text{BTU}}{\text{lb}^{\circ}\text{F}}\right)} = 0.0120 \text{ ft}^2/\text{hr}$$

from estimated values of k_t and c_p. The values of T* for a typical residence time of 10 minutes as a function of distance x from the column wall are:

x (inches)	T*
1.0	0.81
2.0	0.99

The value of T* at the center of a cylinder has been tabulated as a function of $\alpha t/r_o^2$, where r_o is the cylinder radius from the solution to the appropriate heat conduction equations. For the conditions specified, T* is at least 0.99 for a six-inch diameter column. It appears certain that as long as the wall temperature is not raised above 40°C (which, in this case, would be detrimental to the lactase enzyme) the adiabatic temperature change, resulting from heat of reaction, ΔH, will occur.

The adiabatic temperature change can be determined from the relationship

$$\Delta T_{adiabatic} = \frac{-(\Delta Hr)\ (\text{fractional conversion})\ (\text{reactant concentration})}{(\text{heat capacity})} \quad (37)$$

At 80% hydrolysis (20% lactose),

$$\Delta T = \frac{\frac{(-\gt 100\ \text{cal/g-mole})}{(342\ \text{g/g-mole})}\ (.8)\ (.2\ \text{g/g})}{0.9\ \text{cal/g}^\circ\text{C}} \quad 3.7^\circ\text{C}, \quad (38)$$

where ΔH_r was estimated from heats of combustion for lactose, glucose, and galactose with heats of solution and dilution neglected.

IX. PRESSURE DROP

One of the critical considerations in designing a fixed-bed IME reactor is the operational pressure drop. The problem of plugging is the extreme case. As result, small particle size desirable to reduce pore diffusion limitations increases pressure drop. Particles smaller than 30/40 mesh generally prove impractical. The correlation of Leva given by Perry (45) can be used to estimate pressure drop as

$$\Delta P' = \frac{2\, f_m\, G^2\, L_c\, (1-\epsilon)^{3-n}}{d_p \rho g_c\, \phi_s^{3-n}\, \epsilon^3} . \tag{39}$$

This correlation is sensitive to void fraction ϵ. At ϵ = 0.35, a 10% change in void fraction will change the pressure drop by a factor of 2. The implication, and it is a valid one, is that packing density can affect pressure drop. It should also be noted that a broad particle size distribution may reduce the value of ϵ and thus increase $\Delta P'$ due to small particles filling the interstices between larger particles.

Following is an example of a pressure drop calculation (22) for a 3-foot-high packed bed of 20/30 mesh particles operated at 60°C with a 50% glucose feed where:

$f_m = 100/N_{Re} = 100\,\mu/D_pG$ (for $N_{Re} < 10$);

$G = 0.356$ lb ft^2 sec^{-1};

$L = 3$ ft.;

$\varepsilon = .35$;

$n = 1$ (for $N_{Re} < 10$);

$d_p = 2.32 \times 10^{-3}$ ft (20/30 mesh);

$g_c = 32.17$ (lb) (ft)/(lb force) sec^2);

ϕ_s = shape factor = .65 (estimated for jagged particles, $\phi_s = 1$ for spherical particles).

$$\Delta P' = \frac{\dfrac{100\ (3.6\ \text{cp})\ (.000672\ \text{cp sec ft lb}^{-1})}{(2.32 \times 10^{-3}\ \text{ft})\ (.356\ \text{lb ft}^{-2}\ \text{sec}^{-1})}}{(2.32 \times 10^{-3}\text{ft})\ (32.17\ \text{lb ft/lb force}} \cdot \frac{(.356\ \text{lb ft}^{-2}\ \text{sec}^{-1})\ (3\ \text{ft})\ (.65)\ \dfrac{1\ \text{ft}^2}{144\ \text{in}^2}}{\text{sec}^2)\ (.65^2)\ (.35^3)\ (76.8\ \text{lb/ft}^3)}$$

= 6.3 psi.

X. IME REACTOR OPERATION

In this section, several aspects of reactor operation are considered without attempting comprehensive coverage of the subject.

The total production P of a reactor during a period of time t_p can be related to the feed rate F by

$$P = \int_0^{t_p} F\, dt. \tag{40}$$

For the case of exponential activity decay and operation at constant conversion,

$$P = \int_0^{t_p} F_i e^{-(\ln 2)\ t/t_{1/2}}\ dt, \tag{41}$$

where F_i = initial feed rate and $t_{1/2}$ = enzyme half life. This equation thus becomes

$$P = \frac{F_i t_{1/2}}{\text{in } 2} \left[1 - e^{\frac{-t_p \ln 2}{t_{1/2}}}\right], \tag{42}$$

after the integral is evaluated.

If decreasing flow rate to maintain conversion results in impractical variations in production rate, a system of multiple reactors with staggered start-up or reloading times may be utilized. By use of a sufficient number of reactors, the variation in production rate can be held to any desired level. The number of reactors required to maintain the production rate within given tolerance levels is a function of the immobilized enzyme utilization (number of half-lives reactor is operated before enzyme is replenished) as given by (22)

$$R_p = e^{\frac{-H \ln 2}{N}}. \tag{43}$$

As an example, the production rate is shown in Figure 14 as a function of time from start-up for a seven-reactor system, 40 day half-life, two half-life utilization.

Another strategy of column operation involves maintaining the initial reaction rate by raising the temperature as the IME loses activity. Determination of the optimal temperature vs. time profile for a reactor system is another, not necessarily so simple, problem. Other constraints such as maximum allowable residence time and equipment cost can become important as can the amount of product produced by a given amount of IME.

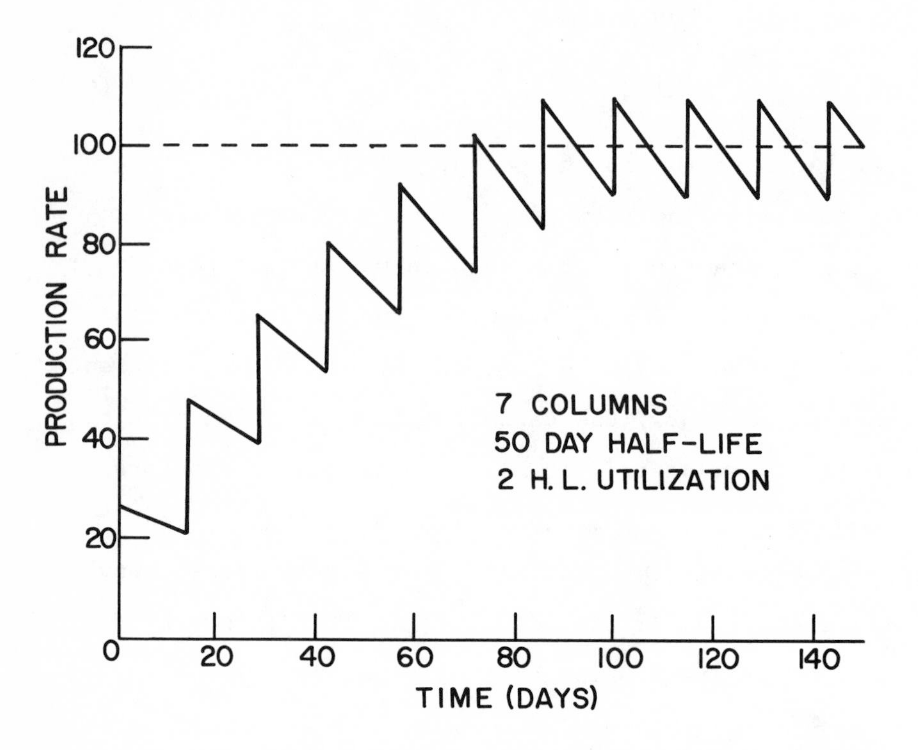

Fig. 14. Production Rate of Multiple Column System

XI. GENERAL IME SYSTEM DESIGN CONSIDERATIONS

Table 1 shows a partial list of design parameters and their dependent reactor system properties that go into determining the final cost or feasibility of processing a given product by an immobilized enzyme reactor system.

TABLE I

Design Parameters

Parameters	System Properties	Goal
Enzyme		
kinetics	Reaction Rate	
cost		
	Enzyme Loading	
Carrier		
cost	Coupling Efficiency	
particle size & shape		
surface area & pore size	Immobilized Enzyme Life	
composition		
surface treatment	Additional Processing Cost	Cost per pound product
Enzyme Immobilization	Equipment Cost	
technique		
temperature	Operating Cost	
concentration		
time	Heat Transfer	
pH		
	Pressue Drop (columns)	
Reactor		
type	Product Quality	
dimension		
Operating Conditions		
temperature		
feed rate		
feed composition		
pH		
pressure		

The first system property to be considered is the reaction rate or rate of product formation. It, in turn, depends on operating conditions, enzyme kinetics, and carrier shape and size (particle diameter, membrane thickness) because of pore diffusion limitations. Operating conditions include temperature, pH and feed composition (reactant, activator and inhibitor concentration).

Enzyme loading, or the amount of enzyme per pound of carrier, affects the ultimate cost as a function of carrier cost. Obviously, for a low-cost carrier, enzyme loading is less critical than for a high-cost carrier. Loading will also influence reactor size and equipment costs. Enzyme loading also depends on carrier surface area and pore size and may be affected by carrier composition and surface treatment. The enzyme attachment technique, whether covalent bonding, adsorption, or some other method, directly affects the amount of enzyme attached. Variables in this category include temperature, concentration, pH, and duration of coupling reaction.

Coupling efficiency is the percentage of enzyme activity, not recovered from the enzyme solution after attachment and which is observed as active immobilized enzyme.

$$\text{Coupling efficiency} = \frac{E_{IME}}{E_{initial} - E_{recovered}}, \qquad (44)$$

where E is activity. This depends on carrier shape and dimensions, since pore diffusion limitations will reduce the apparent coupling efficiency. The enzyme attachment technique and its variables are also important.

Enzyme stability or life depends on the enzyme itself, carrier durability, enzyme attachment method, and operating conditions, particularly temperature and pH. Storage stability of immobilized enzyme not in use is another problem of importance in enzyme use where operation is not continuous. During long-term continuous operation with immobilized enzyme, microbial growth or heavy metal contamination problem can severely affect the enzyme stability.

Depending on operating conditions, additional product processing may be required or may be eliminated by using an immobilized enzyme reactor system. For example, if electrolyte activators are introduced into the feed to increase enzyme activity or stability, ion exchange of the product may be required. Product discoloration may occur, a problem mainly associated with the manufacture of corn syrups using soluble enzyme, where the product is kept at elevated temperature for extended periods of time necessitating ion exchange or carbon treatment.

Enzyme attachment technique costs depend on the type of coupling. Adsorption is relatively inexpensive due to low labor and chemical cost, while covalent bonding may be more expensive, especially for processes requiring many steps.

Equipment costs depend on reactor design and size.

These, in turn, depend on reaction rate and other performance characteristics.

Similarly, operating costs depend on whether batch or continuous operation is used.

In reactor design, heat transfer problems must be considered. As discussed previously, most large packed columns under typical conditions (residence time of 15 minutes) must usually be operated adiabatically. Where this is impractical, internal coils, small diameter columns, or segmented columns with heat transfer between segments may be used. Other reactor types are easier to operate isothermally if desired.

Pressure drop is a problem unique to packed column reactors and depends upon feed rate, column height, feed composition and temperature as they affect viscosity, carrier particle size and packing. The trade-off between pressure drop and pore diffusion problems must frequently be considered.

These inputs from reaction rate through operating cost, including enzyme and carrier cost, can then be used in arriving at a reactor system design and its cost per pound of product.

NOMENCLATURE

a_m = surface area per unit volume

a_v = ratio of particle surface area to reactor volume

C_p = fluid heat capacity

d_p = particle diameter

D = substrate diffusivity

D_{eff} = effective diffusivity = $^{D\Theta}/_{\tau}$

D_o = bulk diffusivity

E = enzyme activity

f_m = friction factor = $^{100}/N_{Re}$ (for $N_{Re} < 10$)

F = flow rate

F_i = initial feed rate

g = acceleration due to gravity

G = mass velocity per unit superficial bed cross section

H = number of half-lives utilization of IME

H_r = heat of reaction

J = dimensionless group

k = turnover number

k_m = mass transfer coefficient

k_m' = value of k for a sphere settling at its terminal velocity

k_t = overall thermal conductivity in the packed bed

k_v = reaction velocity constant

K = equilibrium constant

K_i = product inhibition constant

K_m = Michaelis constant

K_m' = substrate inhibition constant

K_p = constant (Michaelis type for reverse reaction

L = flat plate thickness

L_c = bed height

m = order of reaction

M = general modulus

n = 1 (for $N_{Re} < 10$)

N = number of reactors

N_{Pe} = Peclet number = $\frac{d_p u}{D}$

N_{Re} - Renolds number

N_{Sc} - Schmidt number

P - product concentration

P' = pressure

r_o = cylinder radius

R = sphere radius

R_p = ratio of low to high production rate

S = substrate concentration

S_b = bulk substrate concentration

S_o = initial substrate concentration

S_s = substrate concentration at catalyst surface

S_t = total substrate concentration if all product converted to substrate

t = reaction time

$t_{1/2}$ = enzyme half life

t_p = total period of time of reactor operation

T = temperature

T_i = initial temperature

T_w = wall temperature

u = fluid velocity

v = reaction rate

v_i = intrinsic reaction rate

V_m = kE (maximum reaction velocity)

V_s = substrate volume

x = radial distant from column wall

X = $(S_o - S)/S_o$ = fractional conversion

X_e = X_t at equilibrium

X_i = $\frac{S_t - S_o}{S_t}$

X_t = $\frac{S_t - S}{S_t}$

Y_1 = mole fraction substrate in feed

Y_2 = mole fraction substrate in product

z = column height, assuming film diffusion to be rate controlling step

α = $\frac{k_t}{C_p}$

ϵ = void fraction

η = effectiveness factor

ϕ = modulus (Thiele-type)

ϕ_s = shape factor (0.65 for crushed glass)

ϕ_L = modulus

λ = ratio of diffusing solute critical molecular diameter to pore diameter

μ = fluid viscosity

ρ = fluid density

Θ = particle internal porosity

τ = tortuosity (ratio of actual diffusion path length to straight line distance)

REFERENCES

(1). O.Levenspiel, Chemical Reaction Engineering,John Wiley & Sons, Inc., New York, 1962.
(2). A. R. Cooper and G. V. Jeffreys, Chemical Kinetics and Reactor Design, Prentice - Hall, Inc., Englewood Cliffs, New Jersey, 1973.
(3). O. R. Zaborsky, Immobilized Enzymes, CRC Press, Cleveland, Ohio, 1973.
(4). S. P. O'Neill, P. Dunnill, and M. D. Lilly, Biotech. and Bioeng., 13, 1971, 337-352.
(5). M. D. Lilly and A. K. Sharp, The Chemical Engineer, Jan./Feb. 1968, 12-18.
(6). K. L. Smiley, Biotech. and Bioeng., 13, 1971, 309-317.
(7). N. B. Havewala, "Immobilized Enzyme Reaction Engineering", presented at the James B. Sumner Symposium, Cornell University, Ithaca, New York, 1972.
(8). N. B. Havewala, "Immobilized Enzyme Reactor Design for Commercial Scale Production of Dextrose", presented at the conference on "Immobilized Enzyme and Synthetic Enzyme Analogs", Battelle Columbus Laboratories, Columbus, Ohio, 1973.
(9). H. H. Weetall, Food Prod. Develop. 7, 94-100 (1973)
(10). N. B. Havewala and H. H. Weetall, "Method of Reacting an Insolubilized Enzyme in a Fluid Medium", U. S. Patent #3,767,535, 1973.
(11). A. Emery, "Annular Column Enzyme Reactors", presented at Enzyme Engineering Conference, Henniker, New Hampshire, 1973.
(12). K. Venkatasubramanian and W. R. Vieth, Biotech. and Bioeng. 15, 1973, 583-588.
(13). G. P. Closset, Y. T. Shah, and J. T. Cobb, Biotech. and Bioeng., 15, 1973, 441-445.
(14). S. P. O'Neill, J. R. Wykes, P. Dunnill, and M. D. Lilly, Biotech. and Bioeng., 13, 1971, 319-322.
(15). W. E. Hornby and H. Filippusson, Biochem. Acta, 220, 1970, 343-345.
(16). T. M. S. Chang, Science, 146, 1964, 524-525
(17). T. M. S. Chang, F. C. McIntosh, and S. G. Mason, Canad. J. Physiol. Pharmacol., 44, 1966, 115-128.
(18). T. M. S. Chang, Science J., July 1967, 62-68.
(19). P. R. Rony, Biotech. and Bioeng., 13, 1971, 431-447.
(20). P. J. Robinson, P. Dunnill, and M. D. Lilly, Biotech. and Bioeng., 15, 1973, 603-606.

(21). J. M. Reiner, Behavior of Enzyme Systems, Vorn Nostrand Reinhold Company, New York, New York, 1969.
(22). N. B. Havewala and W. H. Pitcher, Jr., "Immobilized Glucose Isomerase for the Production of High Fructose Syrups," presented at the Enzyme Engineering Conference, Henniker, New Hampshire, 1973.
(23). M. S. K. Chen, "Application of Enzyme Technology", University of Pennsylvania, Course notes, 1972.
(24). C. N. Satterfield, Nass Transfer in Heterogeneous Catalysis, MIT Press, Cambridge, Massachusetts, 1970.
(25). H. H. Weetall and N. B. Havewala, Biotech. and Bioeng. Symp., 3, 1972, 241-266.
(26). E. J. Wilson and C. J. Geankoplis, Ind. Eng. Chem. Foundamentals, 5, 1966, 9-14.
(27). P. L. T. Brian and H. B. Hales, AIChE Journal, 15, 1969, 419-425.
(28). S. P. O'Neill, Biotech. and Bioeng. 14, 1972, 675-678.
(29). W. H. Pitcher, Jr., Restricted Diffusion in Liquids within Fine Pores, Sc. D. Thesis, M.I.T., 1972.
(30). C. N. Satterfield, C. K. Colton, and W. H. Pitcher, Jr., AIChE Journal, 19, 1973, 628-635
(31). K. B. Bischoff, AIChE Journal, 11, 1965, 351-355.
(32). M. Moo-Young and T. Kobayashi, Can. J. Chem. Eng., 50, 1972, 162-167.
(33). A. Wheeler, Adv. in Cat., 3, 1951, 249-327.
(34). D. R. Marsh, Y. Y. Lee, and G. T. Tsao, Biotech. and Bioeng., 15, 1973, 483-492.
(35). P. S. Bunting and K. J. Laidler, Biochemistry, 11, 1972, 4477-4483.
(36). R. Goldman, O. Kedem, and E. Katchalski, Biochemistry, 7, 1968, 4518-4532.
(37). L. P. McMaster, Catalysis by Ion Exchange Resins and Liquid Phase Packed Bed Chemical Reactor Technology, Sc. D. Thesis, M.I.T. 1969.
(38). T. Kobayashi and M. Moo-Young, Biotech. and Bioeng., 13, 1971, 893-910.
(39). L. Goldstein, Y. Lenin, and E. Katchalski, Biochem., 3, 1964, 1913-1919.
(40). W. E. Hornby, M. D. Lilly, and E. M. Crook, Biochem. J., 107, 1968, 669-674.
(41). M. L. Shuler, R. Aris and H. M. Tsuchiya, J. Theor. Biol., 35, 67-76 (1972).
(42). B. K. Hamilton, L. J. Stockmeyer, and C. K. Colton, J. Theor. Biol., 41, 547 (1973).
(43). D. F. Ollis, "Diffusion Influences in Denaturable Insolubilized Enzyme Catalysts", presented at 64th Annual AIChE Meeting, San Francisco, California, 1972.
(44). E. R. G. Eckert and R. M. Drake, Jr., Heat and Mass Transfer, McGraw-Hill, New York, New York, 1959.
(45). J. H. Perry (Ed.), Chemical Engineers' Handbook, 4th Ed., McGraw-Hill, New York, New York, 1963.

Chapter 4

MODIFIED NYLONS IN ENZYME IMMOBILIZATION AND THEIR USE IN ANALYSIS

W. E. HORNBY & D. L. MORRIS

Department of Biochemistry
University Of St. Andrews
St. Andres, Fife
United Kingdom

I. INTRODUCTION

Over the past 10 years, many methods for preparing immobilized enzyme derivatives have been described (1).

A wide variety of materials have been used as support matrices for fabricating these artifacts. It's impossible to argue unequivocally from these reports that one support matrix is any better than another, since the ultimate application of the immobilized enzyme derivative often motivates the choice of support matrix. Nevertheless, it is possible to recognize some general characteristics, which are desirable in these support materials. For example, the support should be available commercially at moderate cost; it should be resistant to microbial attack and possess mechanical strength; if it is to be used for preparating covalently bound immobilized enzymes, it must have potential chemical reactivity; and, finally, it is advantageous if the support material is available in different configurations, so that different modes of the immobilized enzyme may be entertained. One support material used so far displays all of these characteristics.

Nylon is an unbranched polymer containing secondary amide bonds, typically synthesized by condensing a bis-aliphatic amine with a bis-aliphatic carboxylic acid. Thus the polycondensation of adipic acid with hexamethylene diamine yields nylon 66, so names because there are six carbon atoms in both the diamine and diacid components. Likewise, the polymerization of caprolactam produces nylon 6, so named because there are six carbon atoms in the single monomeric component. Polyamides of the nylon type are useful materials for preparing immobilized enzymes, since they display all the cited characteristics. Supports derived from these polymers are especially attractive, since they are available in a wide variety of configurations: e.g., nylon mesh, powder, membrane and tube are all readily available. This makes it possible to consider the preparation of multifarious immobilized enzyme structures, which have both a common support matrix and a common method of immobilization.

The use of nylon for preparating micro-encapsulated enzyme derivatives is well known (2). For example, Chang has described the preparation of micro-capsules of nylon 610, by the interfacial polymerization of sebacoyl chloride and hexamethylene diamine about droplets of an aqueous solution of the enzyme. The importance of these immobilized enzyme structures as tools in clinical therapy has been reviewed by Chang (3). Enzymes have also been immobilized in nylon membranes by crosslinking the enzyme molecules within the pores of the membranes by means of the bifunctional agent glutaraldehyde (4). Of more general applicability than the two methods already mentioned are those whereby the enzyme is covalently attached to the nylon structure. In this way, a greater choice of support configuration is realizable; consequently, the support can be tailor-made to suit any operational requirement. The nylon-supported enzymes discussed in this review are all confined to this latter category of derivative.

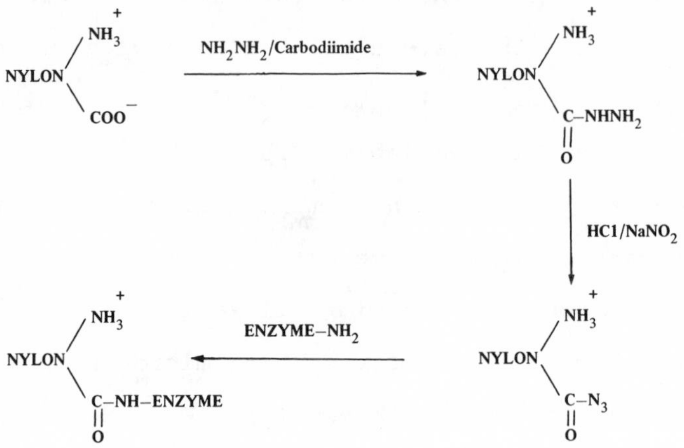

$$NYLON(NH_3^+)(COO^-) \xrightarrow{NH_2NH_2/\text{Carbodiimide}} NYLON(NH_3^+)(C(=O)-NHNH_2) \xrightarrow{HCl/NaNO_2} NYLON(NH_3^+)(C(=O)-N_3) \xrightarrow{ENZYME-NH_2} NYLON(NH_3^+)(C(=O)-NH-ENZYME)$$

Fig. 1a. Direct coupling of enzyme to the carboxyl groups of nylon.

$$NYLON(NH_3^+)(COO^-) \xrightarrow{\text{Benzidine/Carbodiimide}} NYLON(NH_3^+)(C(=O)-NH-C_6H_4-C_6H_4-NH_2) \xrightarrow{HCl/NaNO_2} NYLON(NH_3^+)(C-NH-C_6H_4-C_6H_4-N{\equiv}N^+\ Cl^-) \xrightarrow{ENZYME} NYLON(NH_3^+)(C(=O)-NH-C_6H_4-C_6H_4-N{=}N-ENZYME)$$

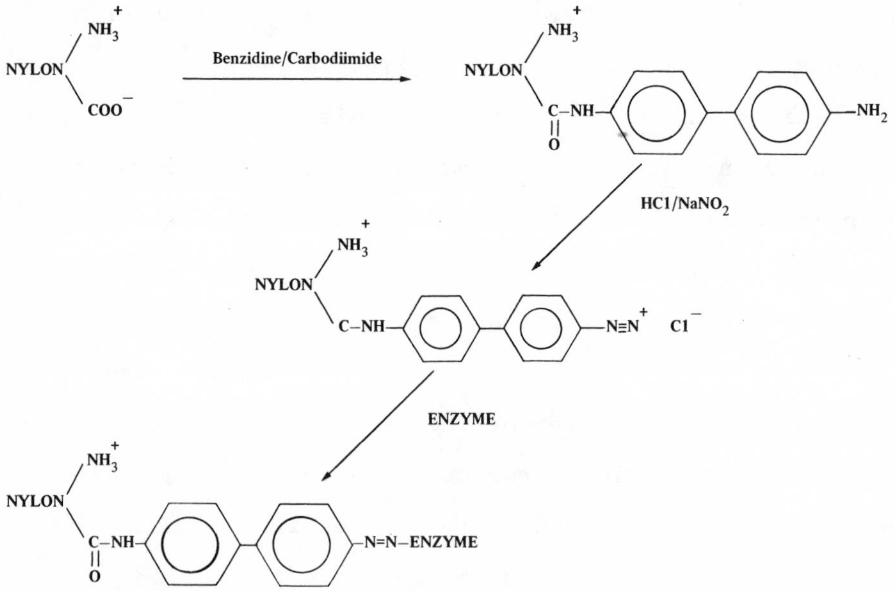

Fig. 1b. Coupling of enzyme to the carboxyl groups of nylon through benzidine.

Native nylon generally has a high molecular weight. Consequently, structures derived from this polymer have relatively few free end groups. However, it will be recalled that secondary amide linkages are characteristic features of nylon. If liberated, the components of these bonds, aliphatic amino and carboxyl groups, can be used as potentially reactive centres for the subsequent covalent binding of enzymes. For example, enzymes may be coupled to the carboxyl group components by a process involving the prior conversion of this moiety to its acid azide derivative (Fig. 1a). In this way, trypsin has been coupled to the inside surface of nylon tubing after release of some carboxyl groups through a partial acid hydrolysis (5). Alternatively, the carboxyl group can be used as a locus on which to attach some other reactive center. For instance, the aromatic diamine, benzidine, has been condensed with the carboxyl group component using a carbodiimide. This group was subsequently diazotized and used for the coupling of trypsin (Fig. 1b) (5). The chemical attachment of enzymes to nylon through the amino group component of the support is easily accomplished by using a bifunctional reagent. Thus, glutaraldehyde has been used in this capacity for immobilizing urease (6) and asparaginase (7) to the amino groups on the inside surface of partially hydrolyzed nylon tubing, as well as for the immobilization of uricase to low-molecular-weight nylon powder (80 (Fig. 2a).

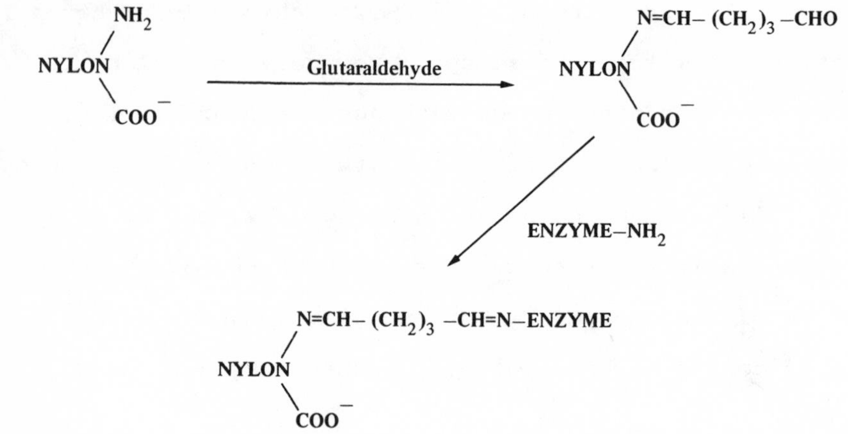

NYLON(NH_2)(COO^-) —Glutaraldehyde→ NYLON($N{=}CH{-}(CH_2)_3{-}CHO$)(COO^-)

—ENZYME–NH_2→ NYLON($N{=}CH{-}(CH_2)_3{-}CH{=}N{-}ENZYME$)($COO^-$)

Fig. 2a. Coupling of enzymes to the amino groups of nylon through glutaraldehyde

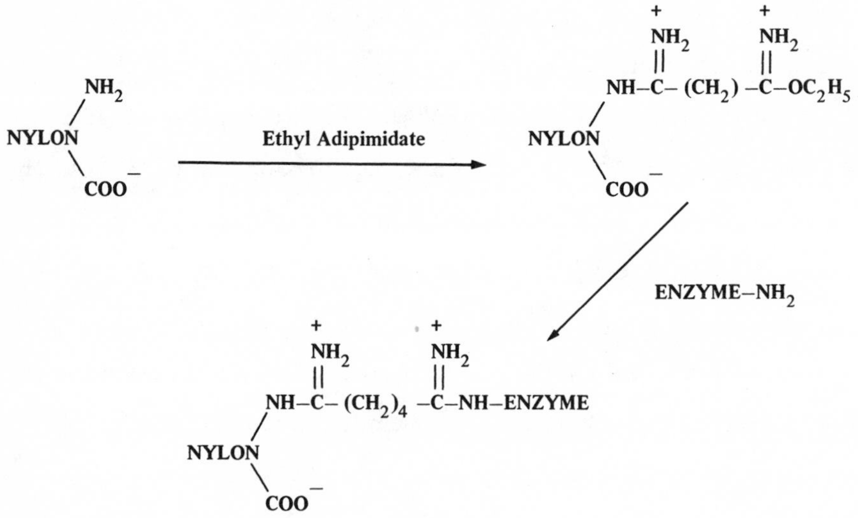

NYLON(NH_2)(COO^-) —Ethyl Adipimidate→ NYLON($NH{-}C(={}^{+}NH_2){-}(CH_2){-}C(={}^{+}NH_2){-}OC_2H_5$)($COO^-$)

—ENZYME–NH_2→ NYLON($NH{-}C(={}^{+}NH_2){-}(CH_2)_4{-}C(={}^{+}NH_2){-}NH{-}ENZYME$)($COO^-$)

Fig. 2b. Coupling of enzymes to the amino groups of nylon through ethyl adipimidate.

Clearly, a variety of bifunctional reagents can be used in a process such as this. For instance, bis-imido esters such as ethyl adipimidate can be employed to advantage (Fig. 2b) (9). The individual operations involved in some of these methods will now be discussed in more detail.

II. CLEAVAGE OF PEPTIDE BONDS

It has been suggested that the support material should be both mechanically strong and available in different configurations. Therefore, if the native support material displays these advantageous characteristics, it is imperative that the structure be conserved throughout the various operations involved in the activation of the support and the subsequent binding of enzyme. Clearly, the mechanical integrity of a polymer such as nylon will depend in part on its degree of polymerization. Consequently, any process which cleaves its monomeric linkages must affect its mechanical integrity. This means that the treatment adopted for affecting the release of the amino and carboxylic group components of the peptide bonds must represent a compromise. It must be sufficiently mild so that it impairs neither the morphology nor strength of the support.

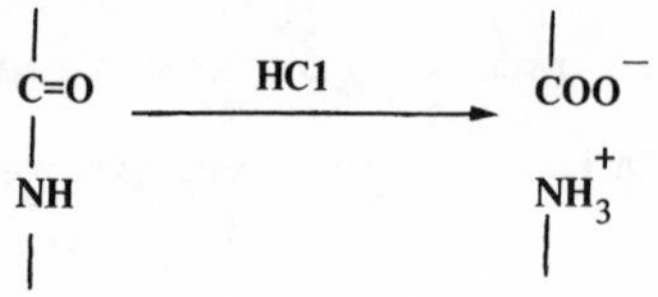

$$\begin{array}{c} | \\ C{=}O \\ | \\ NH \\ | \end{array} \xrightarrow{HCl} \begin{array}{c} | \\ COO^- \\ \\ NH_3^+ \\ | \end{array}$$

Fig. 3a. Hydrolytic cleavage by acid hydrolysis of peptide bonds.

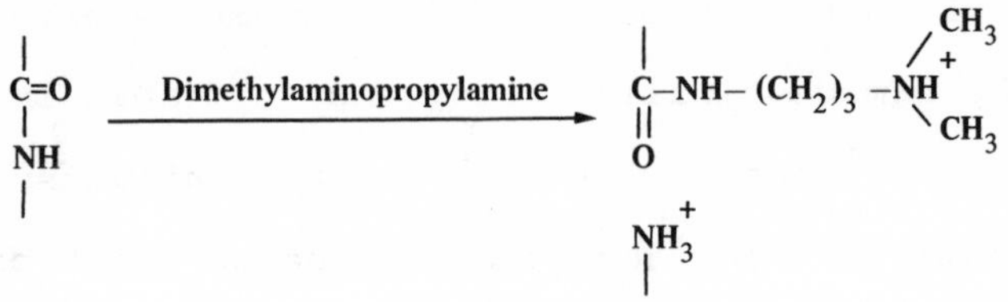

$$\begin{array}{c} | \\ C{=}O \\ | \\ NH \\ | \end{array} \xrightarrow{\text{Dimethylaminopropylamine}} \begin{array}{c} | \\ C{-}NH{-}(CH_2)_3{-}\overset{+}{N}H(CH_3)_2 \\ \| \\ O \\ \\ NH_3^+ \\ | \end{array}$$

Fig. 3b. Non-hydrolytic cleavage by amination in a non-aqueous solvent.

The peptide bonds of nylon can be cleaved either hydrolytically (4) or non-hydrolytically (10). The former process can be affected by hydrolysis of the nylon in acid conditions and yields free amino groups together with a complement of free carboxyl groups (Fig. 3a). The latter process involves treatment of the nylon under non-aqueous conditions with an amine. For example, if N, N-dimethylaminopropylamine is used, the treatment yields free primary amino groups together with a complement of tertiary amino groups, which are derived from the aminated carbonyl component of the peptide bond (Fig. 3b). Nylon structures thus treated can be used for binding enzymes, for example, by coupling through a bifunctional agent to the liberated primary amino groups. Since most coupling reactions take place at pH values where the carboxyl groups of hydrolytically cleaved nylon tube are ionized, then this derivative will carry a net negative charge after the amino group has been blocked with the bifunctional agent. On the other hand, after the amino groups of non-hydrolytically cleaved tubing have been blocked, this derivative will carry a net positive charge in the form of its protonated complement of tertiary amino groups. The following procedures have been successfully adopted for preparating hydrolytically and non-hydrolytically cleaved nylon, and satisfy the above criteria in that they liberate sufficient reactive centers without significantly impairing either the structure or strength of the nylon.

A. Hydrolytically Cleaved Nylon

Nylon tubing can be cleaved by perfusing 3-m lengths of the material with 3.65 M-HCl for 35 minutes at 45°C at a flow rate of 3-4 ml/min^{-1}. Immediately after this treatment, the tubes are washed free of acid by perfusing them with water for one hour at a flow rate of 10 ml/min^{-1}. Nylon powder can be similarly activated by stirring it as a suspension in 3.65 M-HCl for 45°C. The acid is then removed by filtration, and the powder is washed several times by suspending it in water, stirring and filtering.

B. Non-Hydrolytically Cleaved Nylon

Nylon tubing is activated by filling it with N, N-dimethylaminopropylamine, sealing the ends and immersing it in a water bath at 70°C for 15 hours. The amine is then pumped out of the tube and the latter is washed by perfusing it with water for one hour at a flow rate of 10 ml/min^{-1}. Likewise, non-hydrolytically cleaved nylon powder can be prepared by suspending the material with stirring in the amine for 15 hours at 70°C. The modified powder is then recovered by filtration and is washed as described above.

The cleavage of peptide bonds in the nylon by these two processes, and the consequent appearance of free amino groups can be confirmed by treating the products with 2, 4,6-trinitrobenzenesulphonic acid. The products are immersed in a solution of this reagent in water saturated with sodium tetraborate. A characteristic yellow colour indicates the presence of free amino groups.

III. COUPLING WITH BIFUNCTIONAL AGENTS

Bifunctional agents have been used extensively for covalently binding enzymes to insoluble supports. In particular, glutaraldehyde has found wide application in this respect (11). In principle, the support is treated with the bifunctional agent under conditions which favor reaction of only one of the functional groups of the agent with the support. This is most easily accomplished by treating the support with an excess of the agent. The activated support is then washed free of excess agent and is treated with the enzyme under conditions which favor reaction between the latter and the second functional group of the bound agent. Clearly, these conditions must be a compromise between those which favor the coupling reaction and those under which the enzyme is stable. In view of this, it is not possible to rigorously define the precise experimental conditions for affecting the optimal binding of enzymes in general to the support bifunctional agent adduct.

A. Coupling Through Glutaraldehyde

In a typical procedure for coupling enzymes to nylon tubing, 3-m length of modified nylon tubing containing free amino groups, is perfused at room temperature with 12.5% (w/v) glutaraldehyde in 0.1 M-borate buffer, pH 8.5, for 15 minutes at a flow rate of 3-4 ml/min^{-1}., followed by 25 ml of the buffer solution in which the enzyme is prepared. The nylon tube, so activated, is immediately used for the coupling of enzyme as follows. For the preparation of nylon tube-supported glucose oxidase, a 3-m length of type 6 nylon tube, hydrolytically cleaved and treated with glutaraldehyde, is filled with a 0.1% (w/v) solution of the enzyme in 0.1 M-phosphate buffer, pH 7.5, at 4°C and left for 3 to 4 hours. The tube is then washed free of any excess enzyme and non-covalently bound enzyme by perfusion with 1.0 M-NaCl in 0.1 M-phosphate buffer, pH 7.0, at 4°C for one hour at a flow rate of 10 ml/min^{-1}.

Linked enzyme systems can be immobilized on nylon tubing in an analogous manner. For example, for immobilizing a system comprising pyruvate kinase and lactate dehydrogenase, a 3-m length of non-hydrolytically cleaved nylon tubing is activated with glutaraldehyde as described earlier. The tubing is then filled with 0.1 M-phosphate buffer, 5.0 mM-EDTA, 5mM-mercaptoethanol, which is 0.1% (w/v) with respect to each of the enzymes. After 3 to 4 hours at 4°C, the tubing is washed by perfusion with a solution of 5 mM-EDTA, 0.5M-NaCl in 0.1 M-phosphate buffer, pH 7.5, for one hour at a flow rate of 10 ml/min^{-1}.

In many ways, glutaraldehyde is very acceptable as a bifunctional agent for immobilizing enzymes to nylon. Thus, the agent is cheap, easy to handle, and relatively stable under laboratory conditions. Although Fig. 2a depicts the reaction between glutaraldehyde and both support and enzyme as involving the formation of Schiff's bases, the exact mechanism of this reaction and the nature of the linkage is nevertheless uncertain (12). However, it is likely that this agent modifies lysine side chains without conserving the positive charge on these moieties. Clearly, if such lysine residues are required in their protonated form to maintain a stable and active conformation of the enzyme, then this agent may prove unsatisfactory.

B. Coupling Through Imido Esters

Some of the uncertainties associated with using glutaraldehyde can be obviated by using an alternative bifunctional agent. The chemistry of imidates is well documented (13); in particular, the way these compounds react with proteins has been shown to involve the formation of an amidine with the amino groups of lysine residues (14). Since these amidine groups are protonated at physiological pH values, the charge on the amino groups is conserved after their substitution in this way. The use of a bis-imidate, such as ethyl adipimidate, for coupling enzymes to derivatized nylon containing free amino groups is illustrated by the following examples.

A 3-m length of nylon tubing is filled with a 0.5% (w/v) solution of ethyl adipimidate in 20% (v/v) N-ethylmorpholine in ethanol and left at room temperature for 30 minutes. The tubing is then washed free of excess imidate by perfusion with 100 ml of ice-cold ethanol. Immediately thereafter, the tubing is filled with 0.1% (w/v) solution of the enzyme in 0.1 M-N-ethylmorpholine buffer, pH 8.0, and incubated in an ice bath for three hours. Non-covalently bound protein is then removed by perfusion with 250 ml of 0.5 M-NaCl in 0.1 M-phosphate buffer, pH 7.0.

Clearly, ethyl adipimidate can be used similarly to couple enzymes to other nylon structures which contain free amino groups. For example, immobilization to nylon powder can be affected in the following way. In a typical experiment, 200 mg of nylon powder (150 diam) are stirred in 5 ml of 0.5% (w/v) solution of the imidate in N-ethylmorpholine for 30 minutes at room temperature. The powder is recovered by centrifugation and washed five times with 10-ml aliquots of ethanol and immediately suspended with stirring in 4 ml of a 0.1% (w/v) solution of the enzyme in 0.1 M-N-ethylmorpholine buffer, pH 8.0, for three hours at 4°C. The product is finally collected by centrifugation and washed at least five times with 10-ml aliquots of 0.5 M-NaCl in 0.1 M-phosphate buffer, pH 7.0.

IV. ACTIVATION OF NYLON WITHOUT PEPTIDE BOND CLEAVAGE

The methods discussed thus far for activating nylon, all involve cleavage of the polymeric backbone of the support. As pointed out, inherent in these methods is the necessity to limit the degree of cleavage to a level at which the mechanical strength and structure of the nylon is not significantly impaired. The potential enzyme binding capacity of supports which can be modified in this way is therefore limited. In this way, a methodology which does not entail peptide bond cleavage is preferable when high support specific activities are required.

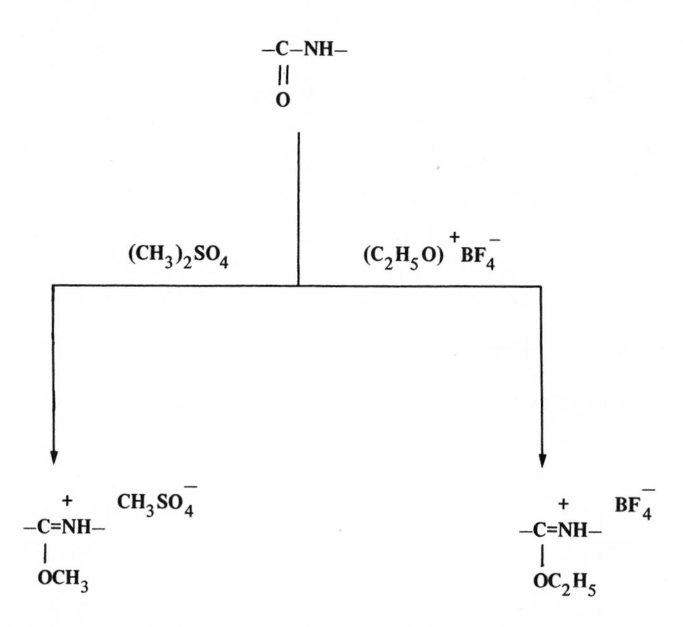

Fig. 4a. O-alkylation of peptide bonds with alkylating agents.

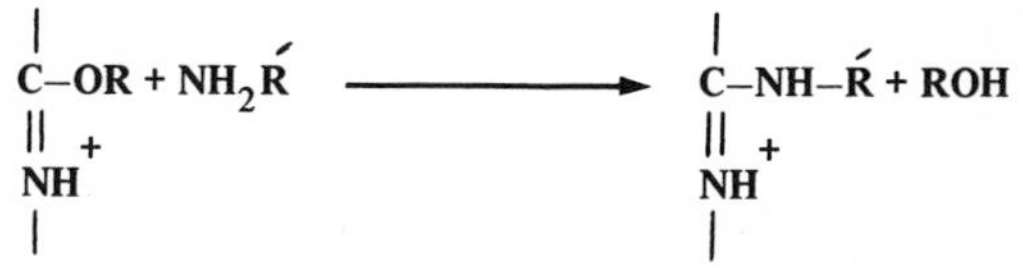

$$\begin{array}{c}|\\ \text{C–OR} + \text{NH}_2\text{R}' \\ \|\ {}^{+} \\ \text{NH} \\ |\end{array} \longrightarrow \begin{array}{c}|\\ \text{C–NH–R}' + \text{ROH} \\ \|\ {}^{+} \\ \text{NH} \\ |\end{array}$$

Fig. 4b. Amidine formation by reaction of amines with the imidate salt of nylon.

Secondary amides can be converted into imido esters by reacting them with a suitable alkylating reagent such as dimethyl sulphate or triethyloxonium tetrafluoroborate (15) (Fig. 4a). The resulting imidates react readily with amino groups under mildly alkaline conditions yielding amidines (Fig. 4b). Thus it can be readily appreciated that the preceding chemistry can be used to immobilize enzymes to nylon, thus overcoming the disadvantages of previous methods (16). For example, enzyme can be bound directly to the imidate salt of the nylon. Alternatively, a bifunctional spacer molecule, such as lysine or hexamethylene diamine, can be covalently bound to the support, and the enzyme is then coupled by using a cross-linking agent such as ethyl adipimidate or glutaraldehyde (Fig. 5).

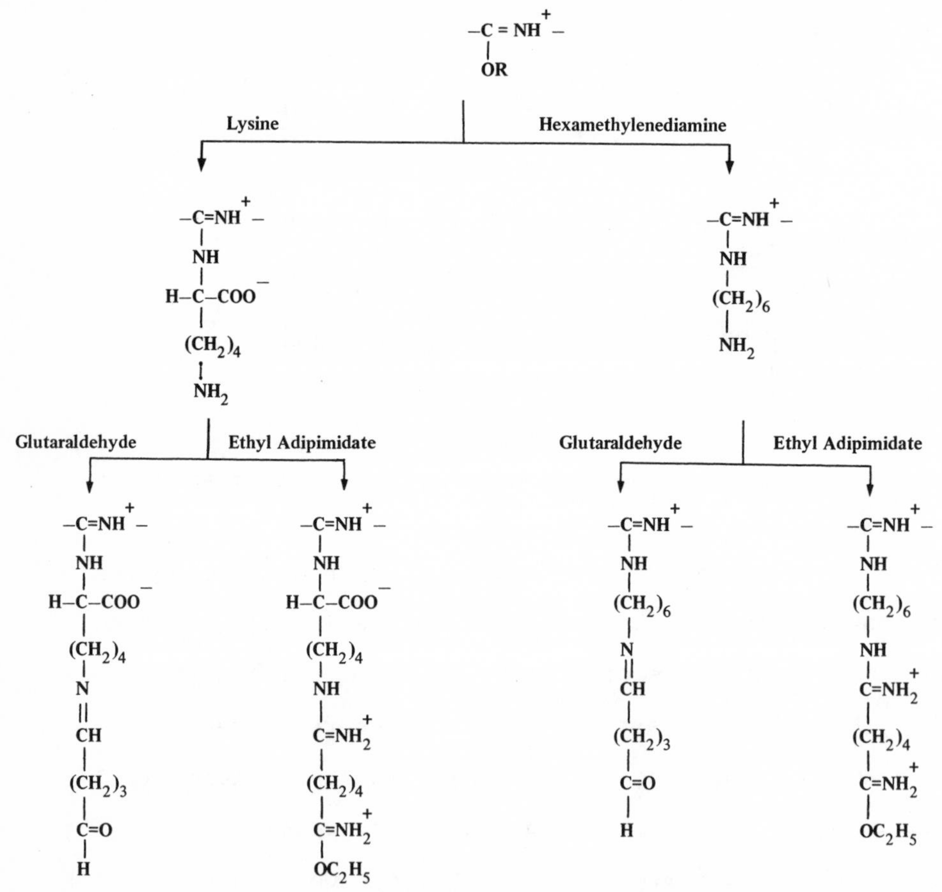

Fig. 5. The preparation of substituted nylons by the reaction of amines with the imidate salt of nylon.

At no point in these reaction sequences is the polymeric backbone broken by peptide bond cleavage. Hence, nylon structures can be activated to a greater extent than was possible by previous methods in which activation of the support was paralleled by a depolymerization of the support. Furthermore, the formation of these amidines results in the incorporation of positive charges into the nylon structure. This increases the polarity of the support and possibly renders it more favorable as an environment for bound enzymes.

Figure 6 shows the activity of two nylon tubing-supported glucose oxidase derivatives, which were prepared by coupling the enzyme directly to the imidate salt of the nylon tubing and by coupling it through ethyl adipimidate to nylon tubing containing a lysine spacer. The results show that the latter derivative is nearly twice as active as the former. These and other similar observations suggest that it is preferrable to couple enzymes to nylon structures after the insertion of a spacer in the support. In this way, the enzyme is attached to the support by relatively long flexible arms which reduce the possibility of the support sterically hindering the interaction between the enzyme and its substrate.

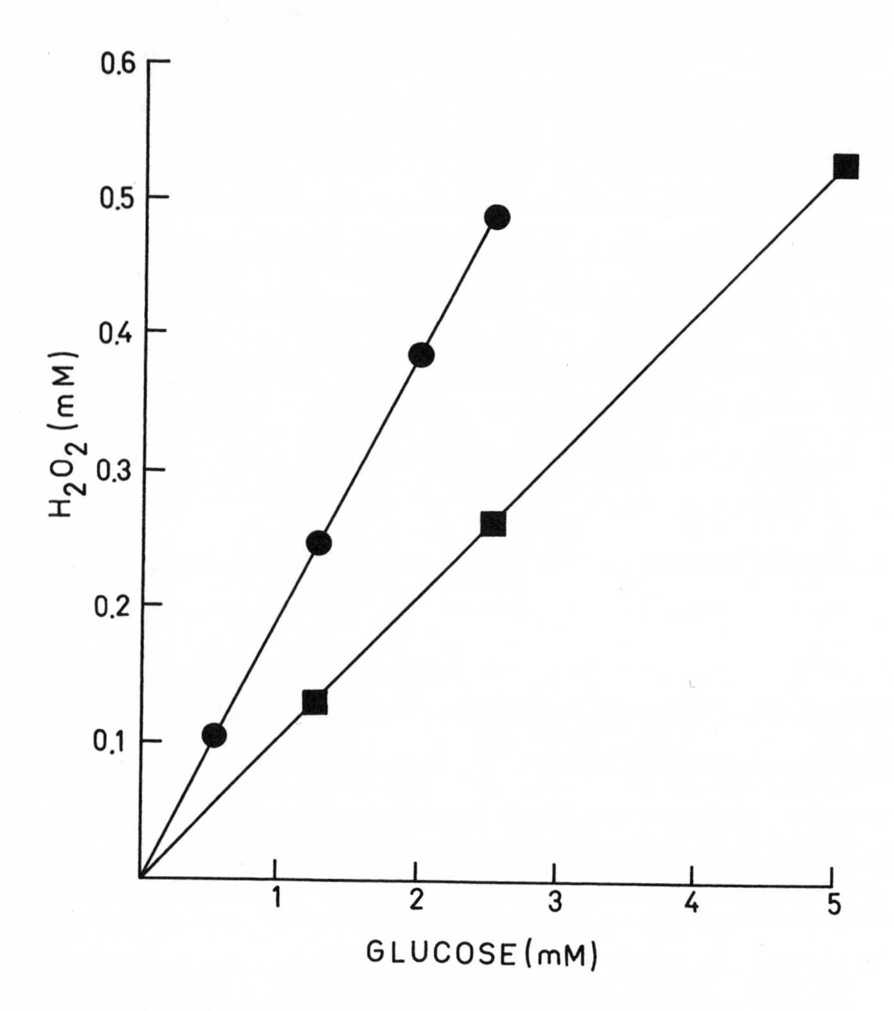

Fig. 6. The activity of nylon tubing-supported derivatives of glucose oxidase. (□) Glucose oxidase directly coupled to the imidate salt of nylon. (○) Glucose oxidase coupled through ethyl adipimidate to nylon tubing carrying a lysine spacer. Activities were assayed by the method of Inman and Hornby (4). Both derivatives were prepared under identical conditions with 1 m of the activated tubing and equivalent amounts of enzyme were coupled to each tube.

A. O-Alkylation of Nylon Tube

Nylon tubing is filled with dimethyl sulphate, the ends of the tubing are sealed, and it is immersed in a boiling water bath for exactly three minutes. The tubing is then transferred to an ice bath to stop the reaction. After five minutes in the ice bath, the dimethyl sulphate is sucked out of the tubing which is washed through with 100 ml of anhydrous methanol at a flow rate of 10 ml/min . The tubing is now filled with an aqueous solution of 0.5 M-lysine or 0.5 M-hexamethylene diamine at pH 9.0, and incubated at room temperature for two hours, after which it is washed through with 250 ml of 1.0 M-NaCl for 30 minutes. Finally, the tubing is washed through with the buffer in which the chosen crosslinking agent is to be used.

B. O-Alkylation of Nylon Powder

Activation of nylon powder is most conveniently accomplished with triethyloxonium tetrafluoroborate. Although this reagent is not available commercially, it is easily prepared by the method of Meerwein (17). In a typical experiment, 5 g of nylon powder are slowly added with stirring to 25 ml of 10% (w/v) triethyloxonium tetrafluoroborate in dichloromethane. After all the nylon has been added, the stirring is stopped and the mixture is allowed to stand at room temperature for one hour. The nylon powder may coalesce to some extent during this period. The dichloromethane and excess triethyloxonium tetrafluoroborate are decanted off, and the nylon is washed several times with 25-ml aliquots of methanol. The powder is then incubated in 25 ml of an aqueous solution of either 0.5 M-lysine or 0.5 M-hexamethylene diamine, pH 9.0, when any aggregates are dispersed. The suspension is stirred at room temperature for two hours, after which the product is thoroughly washed with 1.0 M-NaCl.

V. ANALYTICAL APPLICATIONS

The realization that immobilized enzyme derivatives can be prepared in a variety of configurations has catalyzed considerable interest in new applications of enzymes. For example, it is possible to use these materials in continuous flow processes in the form of packed beds, stirred tanks and open tubular reactors. Furthermore, since immobilized enzymes often display an enhanced stability relative to their espective free enzymes, it is possible to use them repetitively, thereby improving considerably upon the economics of enzyme utilization. In particular, the prospect of applying immobilized enzymes in analysis is very attractive, since many enzyme-based analyses are very expensive in terms of enzyme utilization and sometimes difficult to standardize because of the lability of the free enzyme under the operational conditions of the assay.

Enzymes are used extensively as analytical reagents for the estimation of their respective substrates. Their widespread application in analysis emanates from the specificity they exercise with respect to their substrates. Thus, the use of an enzyme in an analytical procedure is a very convenient means of introducing specificity into the operation. Generally, one of two strategies has been adopted for applying immobilized enzymes in analysis. The first approach entails the construction of the analytical system, both in terms of chemistry and instrumentation, around the immobilized enzyme preparation. On the other hand, the second approach involves the fabrication of the immobilized enzyme so that it is acceptable in existing analytical systems, without modification of either their chemistry or hardware. Some very interesting innovations in analysis have been made by the former approach. For example, several analytical systems have been described, which use immobilized enzymes in the form of packed beds. Thus, Updike and Hicks (17) described an automated method for the estimation of glucose. Their system was constructed around a small packed bed of glucose oxidase and employed a polarographic sensor for measuring the consumption of oxygen due to the oxidation of glucose in the sample. Membrane-immobilized enzyme derivatives have also been used to construct novel analytical tools. In general, these devices use an electrode which is sensitive to either the substrate or product of the immobilized enzyme membrane. Thus, electrodes which are sensitive to glucose have been prepared by shielding oxygen electrodes with glucose oxidase membranes (18), and electrodes sensitive to urea have been made by shielding ammonium ion-sensitive electrodes with urease membranes (19).

The second approach for applying immobilized enzymes in analysis considers these materials as modified enzyme reagents which must be tailor-made so that they are usable in the same analytical systems that utilize free enzymes. Consequently, prerequisites of this approach are the availability of support matrices in a variety of configurations and the ability to activate these structures for enzyme binding without impairing their morphology. For reasons previously discussed, the described methodology for covalently binding enzymes to nylon structures satisfies these conditions. Automated analytical techniques using enzymes as specific reagents for the estimation of their respective substrates have been in use for many years, and several instrument types are commercially available for executing such operations. Clearly, because of the diverse nature of these analytical devices, no general format can be described for the application of immobilized enzymes in such systems. However, the strategy that has been adopted for the design and application of immobilized enzyme derivatives in one particular commercial analyzer illustrates the type of approach that is necessary.

Continuous flow analyzers, such as that pioneered by Technicon, are used extensively in many routine analytical laboratories. Analysis with this type of instrument involves the proportional pumping of enzymes, reagents and substrates in an air-segmented stream. Therefore, in principle any immobilized enzyme structure, which can be operated in the continuous flow mode, can be employed in such an analytical process. For example, a powder-supported immobilized enzyme derivative could be used in the form of a packed bed; a membrane-supported derivative could be used in a continuous flow mode by its incorporation into a conventional dialyzer module, and a tube-supported derivative could be used by its insertion in series with the routine flow system.

The three types of immobilized enzyme structures, prepared with nylon support matrices, have been evaluated as potential enzyme reagents for the automated determination of their substrates using routine Technicon hardware (4). On practical considerations, the application of the powder-supported derivatives in the form of packed beds was unsatisfactory, since it was not possible to perfuse these structures with an air-segmented stream. Consequently, their use involved the prior de-bubbling of the sample stream, which enhanced the likelihood of carry-over, and limited the rate at which samples could be analyzed. Likewise, because of lateral diffusion in the membrane-supported derivatives, these materials did not permit a sufficiently high rate of sample analysis. However, derivatives prepared by covalently binding enzyme to the inside surface of open nylon tubing, had adequate catalytic activity for producing the required sensitivity, and they could also be perfused with an air-segmented stream, which meant that an acceptable rate of sample analysis could be achieved.

The application of nylon tubing-supported immobilized enzymes in a Technicon Auto-Analyzer, for the estimation of their substrates, can be executed with the same flow system as that used for analysis with the corresponding free enzymes. The tube-supported enzyme is fabricated from nylon tubing, the bore of which is comparable to that of the transmission tubing used in the system. The derivative is then inserted at the place in the flow system where the reaction between the sample and the free enzyme takes place. The reaction products, which issue from the tube, are then analyzed in the conventional way. Figure 7 shows a typical flow system for determining either urea or glucose using nylon tubing-supported derivatives of either urease or glucose oxidase, respectively. In these systems, the urea is determined by measuring the ammonia, which is produced in the urease reaction, by the method of Chaney and Marbach (20); the glucose is determined by using acid potassium iodine to measure the hydrogen peroxide produced in the glucose oxidase reaction.

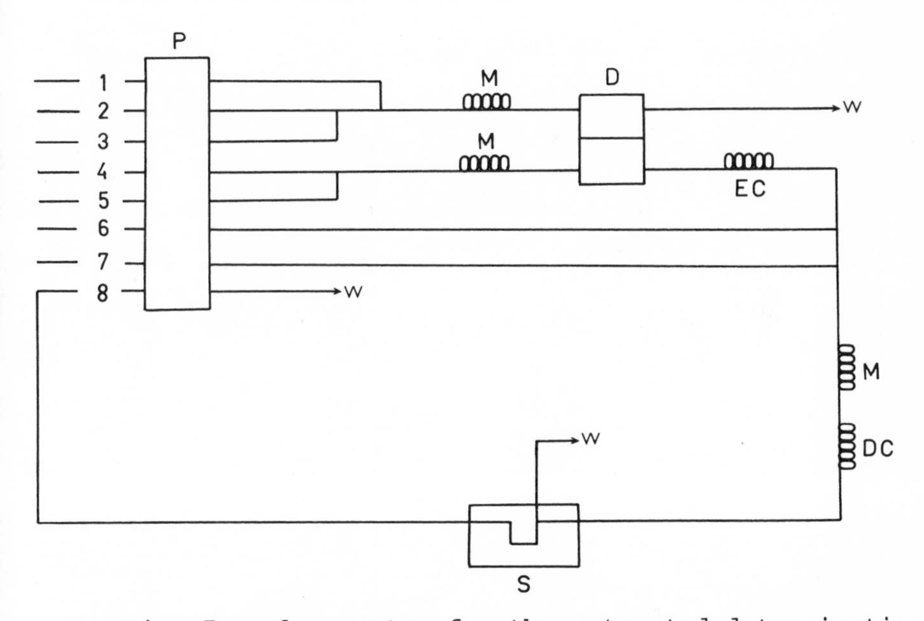

Fig. 7. Flow system for the automated determination of glucose and urea. For the determination of glucose the pump tubing lines 1, 2, 3, 4, 5, 6, 7 and 8 gave flow rates of 0.23, 1.00, 0.80, 1.20, 0.80, 1.20, 1.20 and ml/min^{-1}, respectively. Sample, 0.1 M-acetate, pH 5.0, air, 0.1M-acetate, pH 5.0, air, 1.20 M-HCl and 0.25 M-KI were pumped through the lines 1, 2, 3, 4, 5, 6 and 7 respectively. Three m of nylon tubing-supported glucose oxidase was inserted at position EC in the flow system and maintained at 25°C. For the determination of urea, the pump tubing lines 1, 2, 3, 4, 5, 6, 7 and 8 gave flow rates of 0.03, 2.00, 0.80, 2.00, 0.80, 0.60, 0.60 and 2.50 ml/min^{-1} respectively. Sample, 10 mM-EDTA, pH 7.0, air, 10 mM-EDTA, pH 7.0, air, 0.006% (w/v) sodium nitroprusside in aqueous 4.7% (w/v) phenol, NaOCl in 0.5 M-NaOH were pumped through the lines 1, 2, 3, 4, 5, 6 and 7, respectively. Three meters of nylon tubing-supported

urease was inserted at position EC in the flow system and maintained at 25°C. The pump (P), mixing coils (M), dialyzer (D) and delay coil (DC) were all Standard Technicon Equipment. The spectrophotometer (S) was a Beckman DBGT Spectrophotometer fitted with 1-cm light path flow-through cuvettes. The mixing coils, dialyzer module and delay coil were all maintained at 37°C. Other symbol in the test (W), waste.

Typical standard curves for estimating these substances in this way are shown in Fig. 8.

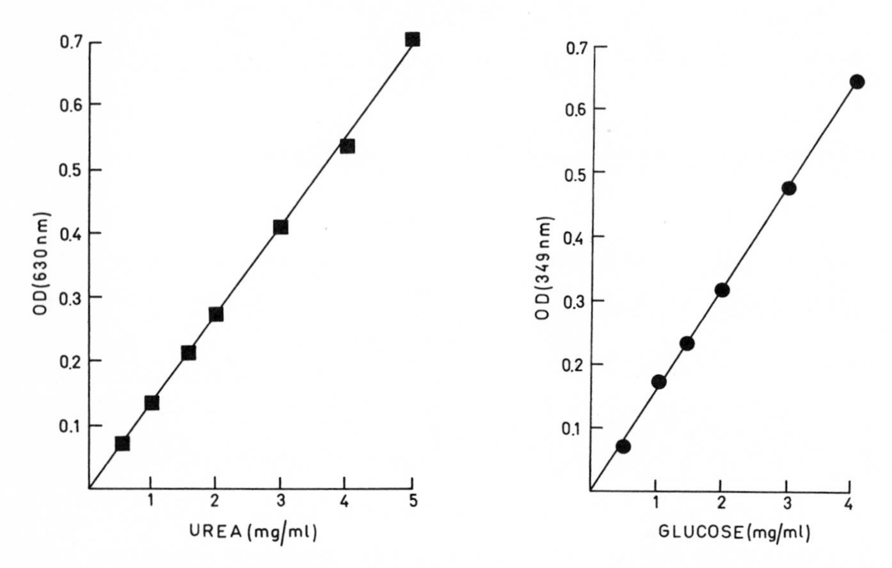

Fig. 8. Standard curves for the automated determination of glucose with three m of nylon tubing-supported glucose oxidase (O) and urea with three m of nylon tubing-supported urea (□). The derivatives were prepared by coupling the respective enzymes through glutaraldehyde to nylon tubing carrying a lysine spacer. Assays were performed using the flow system described in Fig. 7.

In principle, any analytical method, which uses enzymes as analytical reagents, can be performed on a continuous flow analyzer using tubing-supported derivatives of the enzymes. For example, analyses which require a linked enzyme system for determining a substrate, can be immobilized and used in the same way. In cases such as this, either the component enzymes of the linked system can be co-immobilized on the same tube, or the individual enzymes can be immobilized on separated tubing and operated in series with each other. Thus, the linked enzyme systems comprising nylon tubing-supported derivatives of invertase with glucose oxidase, β-galactosidase with glucose oxidase and amyloglucosidase with glucose oxidase have been prepared and used in the automated determination of the disaccharides, sucrose, lactose and maltose respectively (21). It is also possible to use immobilized enzymes in analysis for purposes other than the determination of their substrates. For example, some enzyme-based analyses require the provision of expensive coenzymes, which makes them even more expensive. In some such cases, it is possible to consider using an immobilized enzyme for the continuous generation of the expensive cofactor from a cheaper precursor. Thus NADH can be produced continuously by perfusion through nylon tubing-supported alcohol dehydrogenase with a solution containing ethanol and the less expensive coenzyme NAD^+. This system has been used *in situ* in continuous flow analysis to supply NADH for the assay of pyruvate and oxalacetate with lactate and malate dehydrogenases, respectively (10).

There are considerable advantages to be derived from using an immobilized enzyme in analysis instead of the same enzyme in solution. It is possible to realize a far greater economy in terms of enzyme utilization. For example, many of these preparations can be used for the execution of more than 10^4 separate analyses. Furthermore, in many instances, the stability of the immobilized enzyme surpasses that of the enzyme from which it was derived. Consequently, uncertainties in the standardization of the assay due to the lability of the free enzyme are minimized. Finally, the application of an analytical enzyme in its immobilized form obviates the need to periodically prepare solutions of this reagent. Thus, this contributes toward the ultimate goal of the reagentless assay as proposed by Updike and Hicks (17).

REFERENCES

(1). G. J. H. Melrose, Rev. Pure Appl. Chem., 21, 83, (1971).
(2). T. M. S. Chang, Science Journal, 3, 62, (1967).
(3). T. M. S. Chang, Science Tools, 16, 35, (1969)
(4). D. J. Inman and W. E. Hornby, Biochem. J., 129, 255, (1972).
(5). W. E. Hornby and H. Filippusson, Bioch. Biophys. Acta, 220, 343 (1970).
(6). P. V. Sundaram and W. E. Hornby, FEBS Lett., 10, 325, (1970).
(7). J. P. Allison, L. Davidson, A. Gutierrez-Hartman and G. B. Kitto, Biochem. Biophys. Res. Commun., 47, 66, (1972).
(8). H. Filippusson, W. E. Hornby and A. McDonald, FEBS Lett., 20, 291, (1972).
(9). Subject of U. K. patent application 16148/73
(10). W. E. Hornby, D. J. Inman and A. McDonald, FEBS Lett., 23, 114 (1973).

(11). A. F. S. A. Habeeb, Arch. Biochem. Biophys., 119, 264, (1967).
(12). F. M. Richards and J. R. Knowles, J. Mol. Biol, 37, 231, (1968).
(13). R. Roger and D. G. Neilson, Chem. Rev., 61, 179 (1961).
(14). M. J. Hunter and M. L. Ludwig, in Methods Enzymol., (C. H. W. Hirs and Serge N. Timasheff, eds.), Part B, Vol. 27, Academic Press Inc., 1972, pp. 585 - 596.
(15). H. Meerwein, Org. Syn., 46, 113, (1966).
(16). Subject of U. K. Patent application 37230/73.
(17). S. J. Updike and G. P. Hicks, Science, 158, 270 (1967).
(18). G. G. Guilbault and G. J. Lubrano, Anal. Chim. Acta., 60, 254 (1972).
(19). G. G. Guilbault and J. G. Montalvo, Jr., J. Amer. Chem. Sol., 92, 2533 (1970).
(20). A. L. Chaney and E. P. Marbach, Clin. Chem., 8, 130 (1962).
(21). D. J. Inman and W. E. Hornby, Biochem. J., 137, 25 (1974).

Chapter 5

FIBER-ENTRAPPED ENZYMES

D. Dinelli, W. Marconi and F. Morisi

Laboratory for Microbiological Processes,
SNAM Progetti, Monterotondo
Rome, Italy

I. INTRODUCTION

The new system of enzymes immobilization developed in our laboratory consists of the physical entrapment of enzymes, as aqueous solutions, within microcavities contained inside polymeric filamentous structures. This is carried out by emulsifying an enzyme aqueous solution with the solution of a polymer dissolved into a solvent immiscible with water. This emulsion is then extruded through a spinneret into a coagulation bath. The polymer coagulates as a filament which includes the microscopic droplets of aqueous enzymic solution initially present in the emulsion. The filament structure is similar to that of a porous gel, allowing low-molecular-weight molecules to enter the enzymic microcavities while preventing higher-molecular-weight molecules from escaping the structure. These filaments can be considered enzymic catalysts of heterogeneous type.

This immobilization technique has many advantages: low cost; several different enzymes may be entrapped contemporarily; enzymes are protected from attack by microorganisms and exocellular proteases; very large amounts of enzyme (up to 30%, w/w, with respect to the polymer) may be entrapped; the enzymes to be entrapped do not generally require careful purification; and the resulting enzymic preparations generally possess elevated stability and activity. In addition, it is also possible to entrap microsomes, mitochondria, spores, or organ homogenates. There are also some limitations with this technique owing to the physical nature of the entrapment. These include: diffusion-controlled reaction kinetics; high-molecular-weight substrates cannot reach the enzymic microcavities to be transformed; during the spinning, solvents or coagulants can denature particulate or lipoproteic enzyme systems.

Several industrial applications of these enzymic catalysts were studied. In the field of carbohydrates, hydrolysis reactions were particularly investigated; e.g., sucrose inversion by invertase; saccharification of liquified starch by amyloglucosidase; hydrolysis of lactose in milk and whey by β-galactosidase; debittering of fruit juices by hydrolysis of the bitter principle (naringin) by the enzyme naringinase. In the field of carbohydrate transformations, the isomerization of glucose to fructose by entrapped glucose isomerase was also studied. In the field of antibiotics, transformations the hydrolysis of penicillins and cephalosporins, and the synthesis of new semisynthetic penicillins and cephalosporins through entrapped penicillin acylase were investigated. The use of highly active catalysts, such as those obtainable by the present technique, allowed us to drastically reduce the reaction times, and thus minimize losses due to decomposition of these moderately stable substances.

The enzymic preparation of optically active (levo) aminoacids by direct synthesis or by stereoselective hydrolysis of derivatives of the racemic aminoacids was also carried out. Examples of this are the synthesis of L-tryptophan from L-serine and indole by tryptophan synthetase, and the resolution of DL-methionine by enzymic hydrolysis of the corresponding N-acyl-derivatives: Acylase (from organs) hydrolyzes the N-acyl-derivative of the series L, while the unreacted D-N-acyl-derivative is chemically racemized and recycled.

In two cases, the study of enzymic reactions was carried to the point of development of an industrial process: a) hydrolysis, through entrapped penicillin acylase, of penicillin G to 6-amino penicillanic acid; b) preparation of a dietetic milk with low lactose content using entrapped β-galactosidase.

Fiber-entrapped enzymes have proved particularly suitable in analytical uses, where enzymic specificity is an essential requirement. Furthermore, the stability and repeatability characteristics of these enzymic preparations permit the use of the same analytical kit an enormous number of times. This noticeably lowers the cost of each analysis and enables the use of enzymes that might otherwise be considered too expensive.

Other interesting applications of fiber-entrapped enzymes were found in the biomedical field: Fiber-entrapped urease was used in an extracorporeal shunt to reduce urea level in blood; fiber-entrapped asparaginase was found effective in suppressing the growth of some asparagine-dependent tumors. An interesting biomedical application of this technique was in a series of researches on enzymic systems typical of the liver metabolic activity (such as transaminases, phosphatases, dehydrogenases, etc), which were fiber-entrapped in an attempt to try to build an artificial liver.

II. PREPARATION OF FIBER-ENTRAPPED ENZYMES

Fiber-entrapped enzymes are obtained by following a slightly modified wet-spinning procedure (1,2). The literature dealing with methods for preparing fibers is very extensive and relevant information can be found in the book "Man made fibres" edited by R. W. Moncrieff (3).

Figure 1 illustrates the system for preparing enzyme-containing fibers.

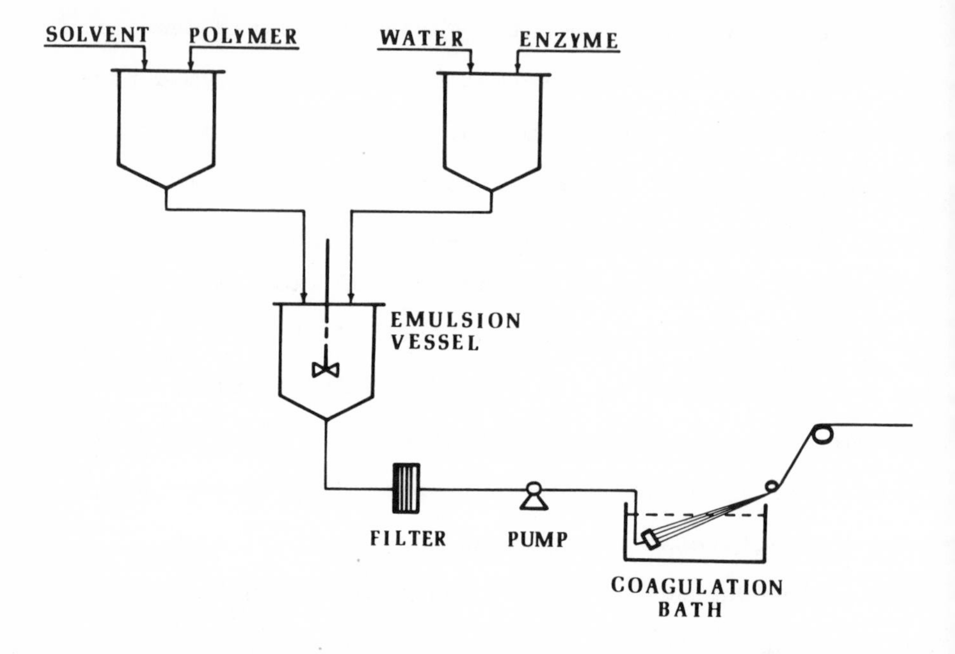

Fig. 1. Wet-spinning apparatus for preparating fiber-entrapped enzymes.

It consists of four steps:

1) In the first step, the polymer and enzyme solutions are prepared. The polymer is dissolved into an organic solvent not miscible with water. The enzyme is dissolved into water or appropriate buffer containing stabilizing agents such as substrate, inhibitors, chelating agents, sulphydryl groups protecting compounds, etc.

2) The second step consists of preparing a finely dispersed emulsion by vigorously stirring the polymer and enzyme solutions. This is an important step, because

the better the emulsion, the more active and stable the corresponding entrapped enzyme.

3) In the third step, the emulsion is extruded through a spinneret into a non-solvent for the polymer and the freshly precipitated polymer is pulled away from the spinneret holes at a controlled rate.

4) The fibers, collected on rollers, are dried by blowing air over them to remove organic solvents and are stored under wet conditions.

The principal advantages of this technique are:

1) Simplicity and low cost of fixation. Emulsifying and wet spinning are common, economical operations used mainly in industry. In a pilot plant (Fig. 2) fibers made of 10,000 filaments are usually spun (about 5 g of polymer in one hour) and there appears to be no difficulty in scaling-up this procedure to industrial production levels.

Fig. 2. Pilot plant for preparating enzyme fibers.

2) High yields of protein entrapment. For maximum economy, all of the enzyme employed should be entrapped, thus avoiding losses and complicated recovery operations. Usually, all of the protein introduced in the preparation of the emulsion is found in the fibers. If the enzyme is stable under the spinning conditions, all of the enzymic activity is retained.

3) Use of the same technique to entrap different enzymes. Table 1 shows the many enzymes that have been entrapped using the same facilities and under very similar conditions. On the other hand, some enzymes, especially particulated or lipid containing enzymes, are damaged by the contact of the solvents to be employed.

4) Possibility of entrapping enzymes at any level of purity. This point is important in choosing the most economic level of purity. Insoluble particles large enough to occlude the orifices of the spinneret or cause fiber breakage must be eliminated by filtration.

5) Possibility of entrapping high percentages of enzymes in the support. This enables high activity levels even with enzymes which have low specific activity. It is possible to entrap up to 300 mg of protein per gram of polymer. The only limitation is the solubility of the enzyme in the aqueous solution employed for the emulsion.

Table 1

Enzymes Entrapped in Cellulose Triacetate Fibers

Enzyme	Stability[a]
Oxidoreductases	
Alanine dehydrogenase	+++
Alcohol dehydrogenase	++
Catalase	+++
D-amino-acid oxidase	+
Glucose oxidase	++
Glucose-6-phosphate dehydrogenase	++
Glutamate dehydrogenase	+++
Glyceraldehyde-3-phosphate dehydrogenase	+
3α, 20β hydroxysteroid dehydrogenase	++
Lactate dehydrogenase	+++
Peroxidase	++
Tyrosinase	+
Uricase	++
Transferases	
Adenosylmethionine transferase	++
D-Fructose-6-phosphate kinase	++
Glutamic-pyruvic transaminase	+++
Hexokinase	++
Pyruvate kinase	++
Ribonuclease	++
Tyrosine aminotransferase	+++

Table 1 (continued)

Enzyme	Stability[a]
Hydrolases	
Acylase	+++
β-amylase	+
Arginase	++
Asparaginase	+++
Chymotrypsin A	+++
β-Galactosidase	+++
Glucoamylase	+++
Invertase	+++
Leucine aminopeptidase	++
Naringinase	+++
Papain	++
Penicillin acylase	+++
Trypsin	+++
Urease	+++
Lyases	
Aldolase	+
Fumarase	++
Hydroxynitrile lyase	++
Tryptophan synthetase	+++
Isomerases	
Glucose isomerase	+++
Phosphoglucose isomerase	++

a) +++ very stable; ++ stable; + unstable

6) Possibility of simultaneously entrapping more than one enzyme in the same fiber. Multienzyme systems enable reactions to be accomplished simultaneously. This is important when one of the reactions has an unfavorable equilibrium or when treating solutions containing more than one substance to be transformed. Multienzyme systems utilizing fiber entrapment can easily be made using the same procedure as for a single enzyme.

7) Easy handling and sufficient mechanical stability. Because the mechanical resistance of enzyme-containing fibers is only slightly lower than that of normal wet-spun fibers obtained with the same polymer, the enzyme containing fibers can be weaved. The fibers are easy to handle and can be used in several forms; continuous filaments, short staple fibers, woven and non-woven fabrics, felts, and sheets.

A. Entrapment of Enzymes in Cellulose Triacetate Fibers

Seven grams of cellulose triacetate is dissolved in 93.0 g of methylene chloride under gentle stirring at room temperature. When all the polymer has been dissolved, the solution is cooled to 1°C, and 14.0 g of the enzyme solution is added slowly under gentle stirring. Stirring is continued for 30 minutes at 400 rpm to obtain finely dispersed emulsion. The emulsion is allowed to stand for 30 minutes to eliminate air bubbles. Metered by a gear wheel pump at a constant flow rate of 15 ml/min, the emulsion passes through a filter and then through the holes of the spinneret. The spinneret is submerged in the coagulation bath containing toluene, which is maintained at 0°C. As the emulsion passes through the orifices of the spinneret into the coagulation bath, it solidifies into filaments because toluene causes the coagulation of cellulose triacetate. As the filaments emerge from the bath, they are guided around a pair of receiving rollers and collected on bobbins. The fibers are dried by flowing air over them. Depending on the type of enzyme to be entrapped, the fibers can be washed with water or buffers and stored under wet conditions.

B. Entrapment of Enzymes in Cellulose Fibers

Eight grams of cellulose nitrate is dissolved in 92.0 g of a mixture of n-butyl acetate and toluene (60 : 40, v/v) saturated with water. The solution is cooled to 1°C, and 12.0 g of aqueous enzyme solution is added. Spinning is done according to the procedure used for cellulose triacetate. The coagulation bath contains petroleum ether. The fibers of cellulose nitrate are treated with a 2% aqueous solution of ammonium hydrosulphide at pH 8.5 for six hours. The denitrated fibers are then washed with water.

Table 2 lists conditions under which several enzyme-entrapping polymers were wet-spun. The technique was the same for cellulose triacetate, with some modifications of the polymer solution and the coagulating solvents. It must be pointed out that though better conditions for preparating normal textile fibers than those reported here can be found in the literature, these were used because they offered conditions compatible with enzyme stability.

Table 2

Wet-Spinning Conditions for Some Polymers

Polymer	Solvent	Polymer Conc. (%, w/w)	Coagulant
Cellulose diacetate	Acetone	20	Petroleum ether
Ethyl cellulose	Toulene	8	Petroleum ether
Polyacrylonitrile	γ-butyrolactone	10	Glycerol
Polybutadiene	Toulene	16	Metanol
Polycarbonate	Methylene chloride	10	Petroleum ether
Poly-γ-methyl-glutamate	sym-Dichloro ethane	5.5	Petroleum ether
Polystyrene	Chloroform	25	Petroleum ether
Polyvinyl chloride	Methylene chloride	15	Petroleum ether

C. Fiber-forming polymers

The ability of several fiber-forming polymers to entrap enzyme was studied. A broad choice of polymers allows a correspondingly wide choice of solvents which can be used during the spinning process. As a result, it is possible to choose the organic solvent couple (polymer solvent and coagulating agent) most suitable for the enzyme to be entrapped; i.e., the solvents will not inactivate the enzyme. In theory, any wet-spinnable polymer can be used to prepare insoluble enzyme fibers. Some of the polymers currently used in the textile industry (e.g. polyacrylonitrile) were not suitable for spinning under conditions compatible with enzyme stability; others (e.g. polyvinyl chloride, polyolefins) could be spun, but the resulting fibers showed very poor mechanical resistance and/or too high hydrophobic properties.

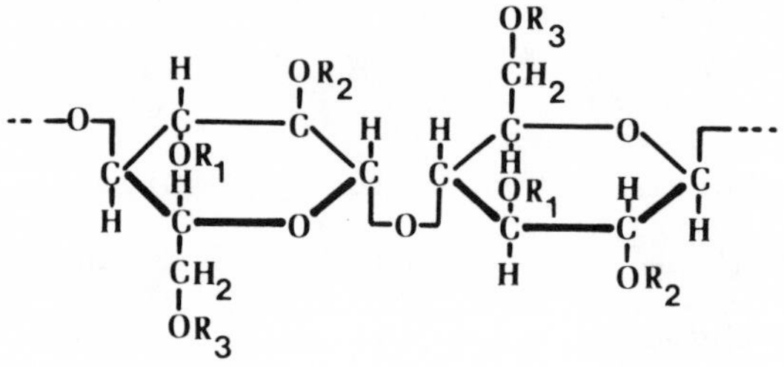

$R_1 = R_2 = R_3 = H$	CELLULOSE
$R_1 = R_2 - COCH_3$, $R_3 = H$	CELLULOSE DIACETATE
$R_1 = R_2 = R_3 - COCH_3$	CELLULOSE TRIACETATE
$R_1 = R_2 = R_3 - C_2H_5$	ETHYL CELLULOSE
$R_1 = R_2 = R_3 - NO_2$	CELLULOSE NITRATE

--HN-CH(R)-CO-NH-CH(R)-CO-- POLY-α-AMINOACIDS

Fig. 3. Some fiber-forming polymers.

Among the polymers tested, some gave good results (Fig. 3). Two classes of polymers are indicated: cellulosic and polyamino acids. Cellulose fibers are easily obtained by converting cellulose ester fibers back to cellulose without destroying the fibrous structure. Cellulose triacetate is the product of nearly complete acetylation of cellulose. Cellulose nitrate is the only inorganic ester of cellulose produced on a commercial basis. The second class of polymers consists of synthetic polyaminoacids such as poly γ-methyl-glutamate, a polymer used for the production of artificial silk.

Most of the studies at SNAM Progetti have concentrated on the use of cellulose triacetate for the following reasons:

1) low cost
2) good biological resistance microbial attack.
3) good resistance to dilute solutions of weak acids and many common solvents.
4) good mechanical properties of enzyme cellulose triacetate fibers. (In fact, the fibers maintain about 65% of the mechanical strength of textile triacetate fibers. This is important because the fibers can be used in different types of reactors, sometimes under severe flow or mechanical stress conditions.

A disadvantage of cellulose triacetate fibers is their hydrophobic character and hence their low water regain.

III. ENZYME FIBERS STRUCTURE

Optical microscopy was used in studying the emulsions and the gross structure of the fibers; electron microscopy was very useful in investigating the details of the fine structure. The usual technique of cross and longitudinal sectioning was followed. The method consists of cutting sections about 100 Å thick, mounting them on slides, and embedding them in polyester.

To investigate the external surface of the enzyme fibers, the scanning electron microscope was used. Figures 4 and 5 are electron micrographs of sections of a single filament. In Figure 4, the dark areas are polymer gel and the light areas are the microcavities. The microcavities are surrounded by a porous material that allows low-molecular-weight compounds to enter and does not allow the enzyme to escape. The cavities show elongated shapes in the longitudinal direction because of the stretching during spinning. In Figure 6, a scanning electron micrograph of the external surface of a fiber elongated voids interconnected with each other can be seen on the fiber surface.

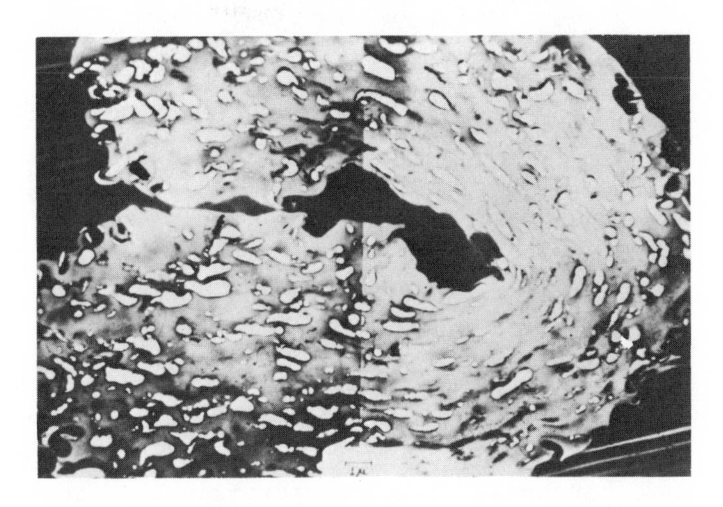

Fig. 4. Electron micrograph of a monofilament cross-section.

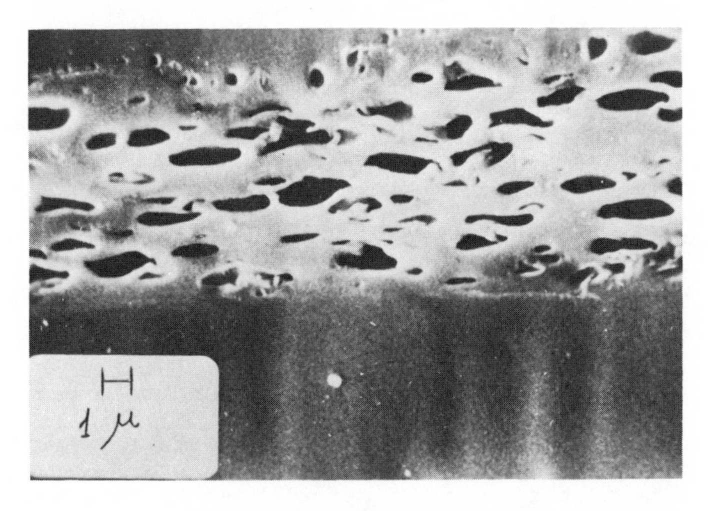

Fig. 5. Electron micrograph of a monofilament longitudinal section.

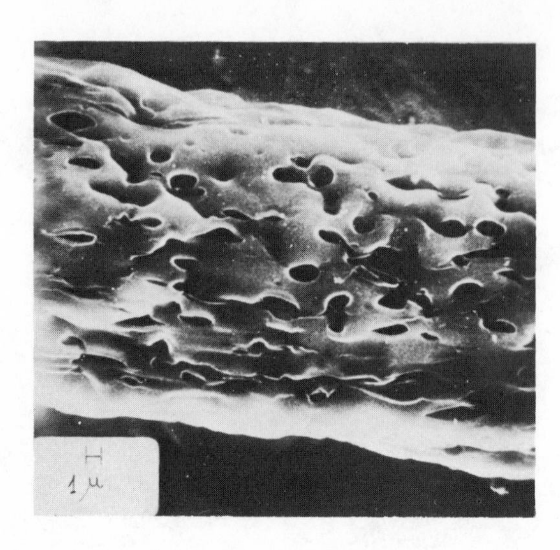

Fig. 6. Scanning electron micrography of the monofilament surface.

By varying the spinning conditions (e.g. polymer concentration, ratio of aqueous to organic phase, composition and temperature of the coagulation bath), it is possible to control the pore size, the number and the diameter of the microcavities, and the overall porosity of the obtained yarns. The porosity of the fibers is a controlling parameter for the activity and stability of the entrapment. Therefore, the porosity of the fibers has been particularly investigated in order to better understand mass transport phenomena and to improve the performance of the enzyme fibers. Some traditional test methods, such as mercury intrusion porosimetry, were used, but the results were erroneous because the physical properties of the fibers and their cavities became altered under compression during the measurement. Direct microscopic observation of the pore structure yields much more useful information, but it is usually difficult to measure pore volume. By analyzing several photographs of fiber cross sections, the diameter of the microcavities was measured and the size distribution curve was statistically calculated. Figure 7 shows a typical size distribution curve of the microcavities, 75% of which have diameters below 0.4 microns. The void volume of the fibers was measured by examing the elution of different molecular weight compounds through a column packed with fibers.

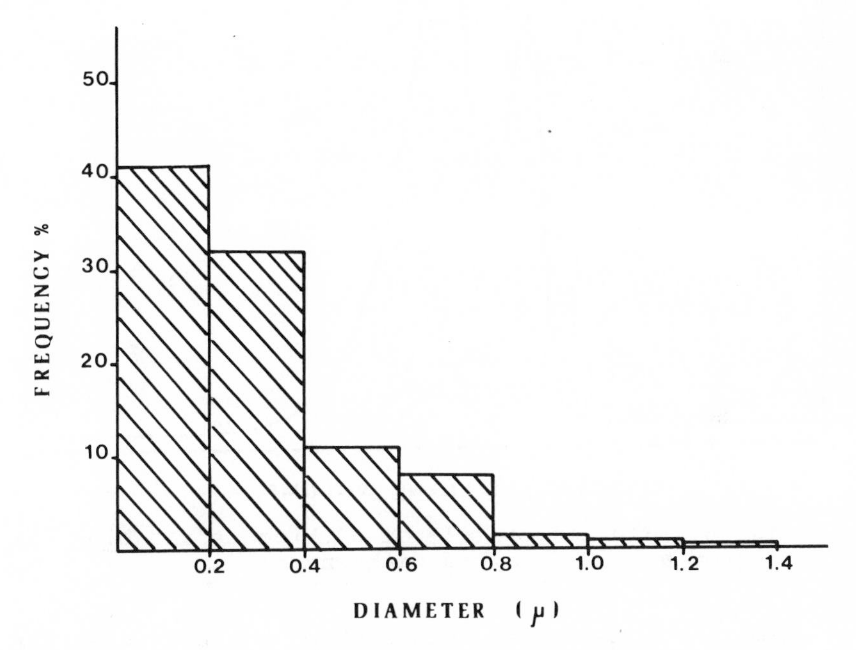

Fig. 7. Size distribution curve of fiber microcavities.

Figure 8 shows the different elution of dichromate and blue dextran. The small molecules of dichromate penetrate the pores and microcavities and flow more slowly down the column than blue dextran molecules which are too large to penetrate the pores.

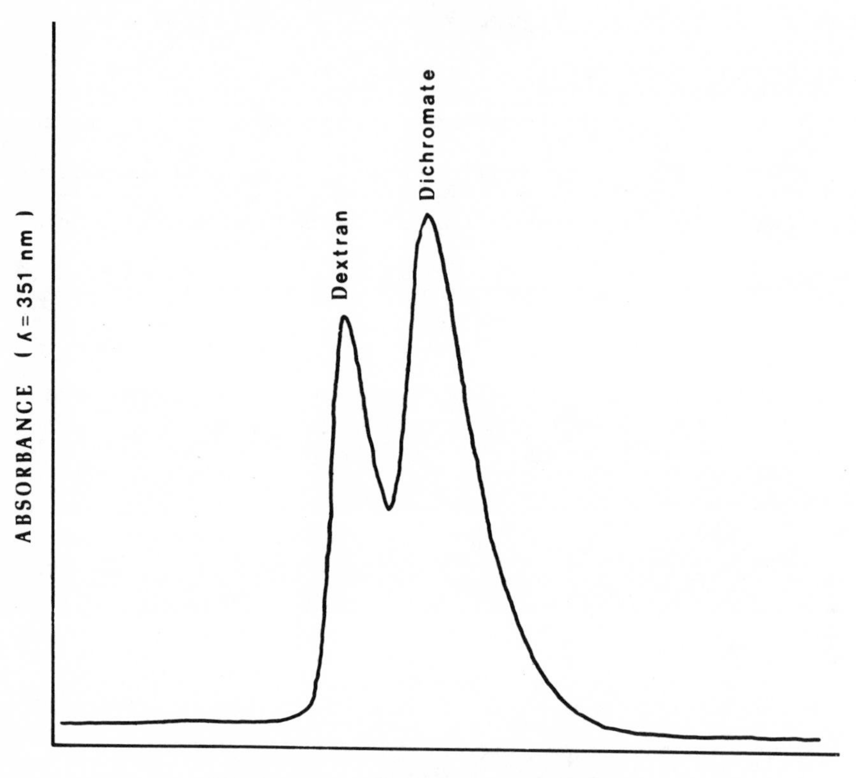

Fig. 8. Elution profile of dichromate and blue dextran from a column packed with enzyme fibers.

A more sophisticated method which has been used for determining void volume combines flow through the porous fibers with the simultaneous diffusion of different molecular weight compounds. By treating mathematically the elution profiles of the solutes, it was observed that they could be described with good approimation by a function of the type:

$$y - y_o \exp (wx) \qquad (1)$$

where w is the flow rate of the solvent and x is a function of the molecular weight of the solute (4). The use of this technique makes it possible to follow the changes in the void volume of the fibers during their use. Using this technique, it was possible to establish that the enzyme fibers undergo structural changes: Under operating conditions, some very small cavities disappear, while the volume of others is reduced. The findings were in good agreement with microscope examination.

IV. PROPERTIES OF FIBER-ENTRAPPED ENZYMES

The catalytic properties of fiber-entrapped enzymes are as dependent on diffusion as those of immobilized enzymes prepared by other procedures (5). For heterogeneous catalysts, Nernst (6) was the first to describe an external diffusion effect due to the formation of an unstirred layer at the interface between the enzyme fibers and the outer solution. The influence of the external diffusion is a function of the stirring speed in the reaction vessel (7, 8). By increasing the stirring speed, the activity of enzyme fibers increases approximating an asymptote. When the support is porous, as in the case of the fibers, internal diffusion of the reactants and products through the pores of the support must be considered. The effect of internal diffusion on the kinetics of the reactions catalyzed by porous catalysts were studied by Wheeler and Thiele (9, 10). The reaction rate in a solid support is influenced by the availability of the reactants. During transformation, the diffusion process replaces depleted reactants to maintain the original reactant concentration.

Other factors, such as the chemical structure of the matrix and the high enzyme concentration existing inside the microcavities, may influence the properties of the immobilized enzymes.

A. Effect of the Nature of the Matrix

The behavior of the same enzyme entrapped in different polymers seems to indicate that the effects of the nature of the support are related more to the morphological structure of the resulting fibers than to the chemical composition of the polymer. Some properties of invertase entrapped in different polymers are illustrated in Table 3. The differences appear to be due mainly to the different porosity of the fibers. For cellulose fibers, the improved activity may be explained by the more hydrophilic character of the support. In this same case, it is worth emphasizing that some degradation of the polymer doubtless occurs after denitrations, resulting in simultaneous modification of the fiber structure. The scarce influence of the chemical nature of the fibrous matrix on the properties of the entrapped enzyme might be explained in view of the fact that the cavities of the fibers are quite large compared to the enzyme molecules and the probability that an active site of the enzyme interacts with the polymer surface is low.

Table 3

Properties of Invertase Entrapped in Different Polymers

Polymer	Efficiency (η)	Enzyme washed-out[a] (%)
Ethyl cellulose	10	14
Polyvinyl chloride	8	14
Poly-methyl glutamate	15	31
Nitrocellulose	41	9
Cellulose	73	20
Cellulose triacetate	41	1

a) Percent of enzyme, washed out in 24 hr. washing, calculated on the amount originally entrapped.

B. Effect of the Enzyme Density in the Support

Figures 9 and 10 show the effect that enzyme density in the support has on activity. Only a fraction of the activity put into the emulsion is displayed by the fibers.

This is expressed by the efficiency, which is obtained by:

$$\eta = 100 \frac{\text{Units of enzyme activity put into the emulsion/g of polymer}}{\text{Units of enzyme activity measured in the fibers/g of polymer}} \qquad (2)$$

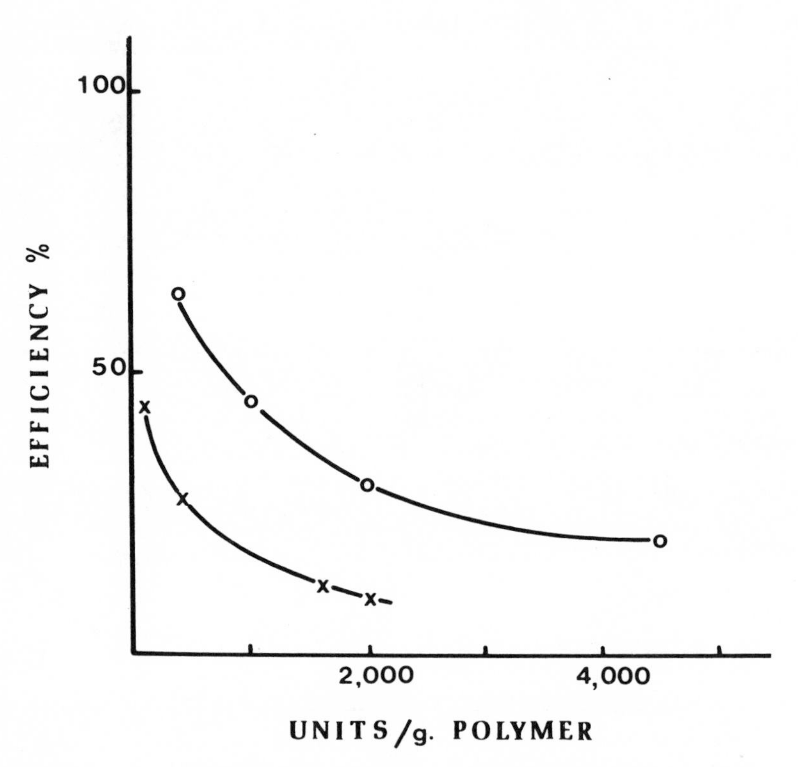

Fig. 9. Effect of enzyme density in the support on the expressed activity:

O —— O, Invertase in cellulose triacetate.

x —— x, Invertase in ethyl cellulose.

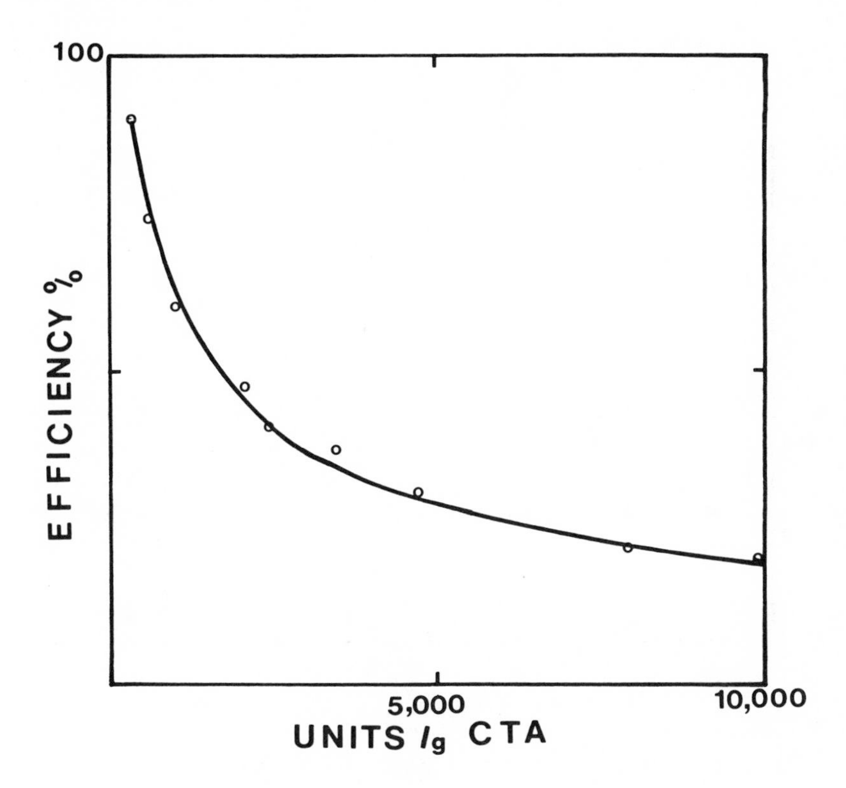

Fig. 10. Effect of penicillin acylase density in cellulose triacetate fibers (CTA) on the expressed activity.

Increasing the concentration of the enzyme in the fibers, decreases efficiency. At low enzyme content the activity approximates 100%. Of course, the efficiency takes into account: (1) the extent of the denaturation of the enzyme during spinning; (2) the yield of entrapment which normally is higher than 90%; (3) the effect of diffusion on the measured activity of the fibers; and (4) other factors related to the possible enzyme-polymer interactions (e.g. adsorption of enzyme molecules).

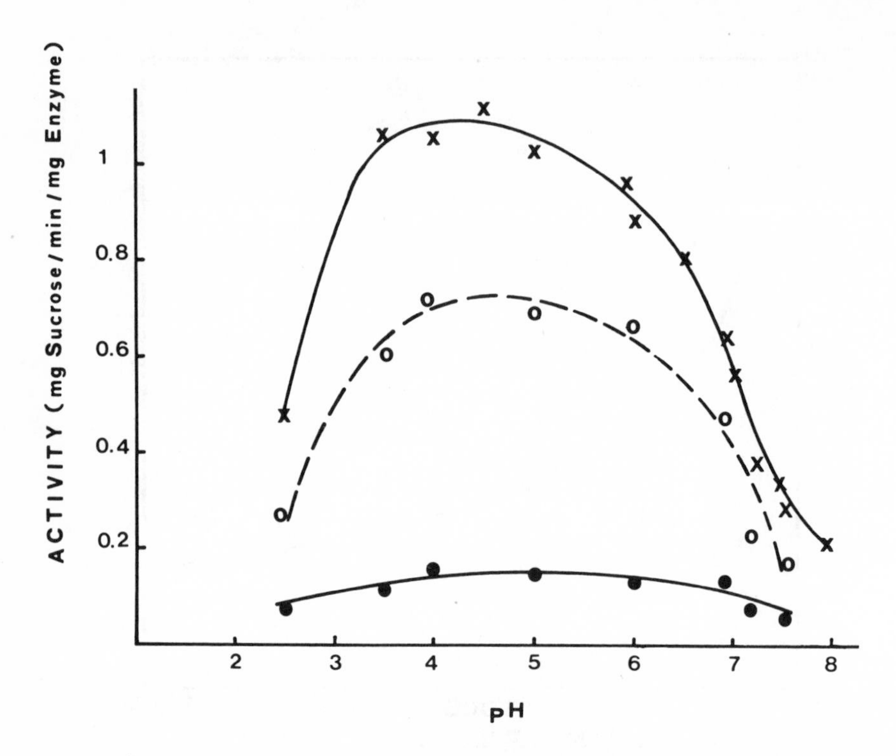

Fig. 11. pH-activity profiles of free and entrapped invertase:

x —— x, free invertase

o —— o, entrapped invertase, 100 units per g. of cellulose triacetate.

● —— ●, entrapped invertase, 14,500 units per g. of cellulose triacetate.

C. pH-Activity Profile

The pH-activity profile of the entrapped enzyme is quite similar to that of the free enzyme when substrates and products are uncharged. When cellulose triacetate is used as the support, no shifts of optimum pH values are expected because of its electrostatically neutral character. Examining its pH activity curve (Fig. 11), it is interesting to note that, with fibers entrapping large amounts of enzyme, a flattening of the curve around the optimum pH is observed. From a practical standpoint the fiber-entrapped enzyme maintains a constant activity in a wide range of pH values.

With enzymes that act on charged substrates or lead to charged products, shifts of optimum pH are observed. Figure 12 shows the pH-activity profiles of free entrapped urease. An apparent shift of optimum pH is the result of an unequal distribution of hydrogen ions; in fact, the pH inside the microcavities, where the enzyme is located, will be higher than that of the external bulk solution, due to accumulation of ammonia. This finding is in agreement with the results obtained with bound urease (11). Figure 13 shows the effect of entrapping polyacrylic acid and invertase in the same fiber. Polyacrylic acid changes the microenvironment in which the enzyme acts and hence its pH-activity profile. The results are compatible with the interpretation given by Katchalski *et al.* (5) in their review on the effect of the microenvironment on the mode of action of immobilized enzymes. Another important result is the stabilization of the enzyme at alkaline ph-values due to the presence of polyacrylic acid. In fact, the fiber containing only invertase and held at pH 9.0 for 100 hours lost about 75% of its original activity, whereas the fiber containing invertase and polyacrylic acid lost only 15% of its activity under identical conditions.

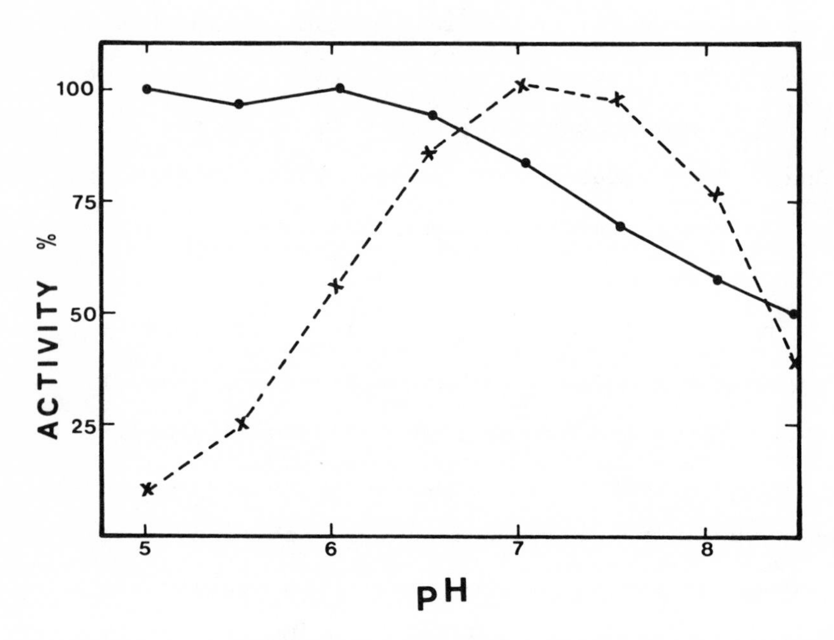

Fig. 12. pH-activity profiles of free and entrapped urease:

x ----- x, free.

● ——— ●, entrapped.

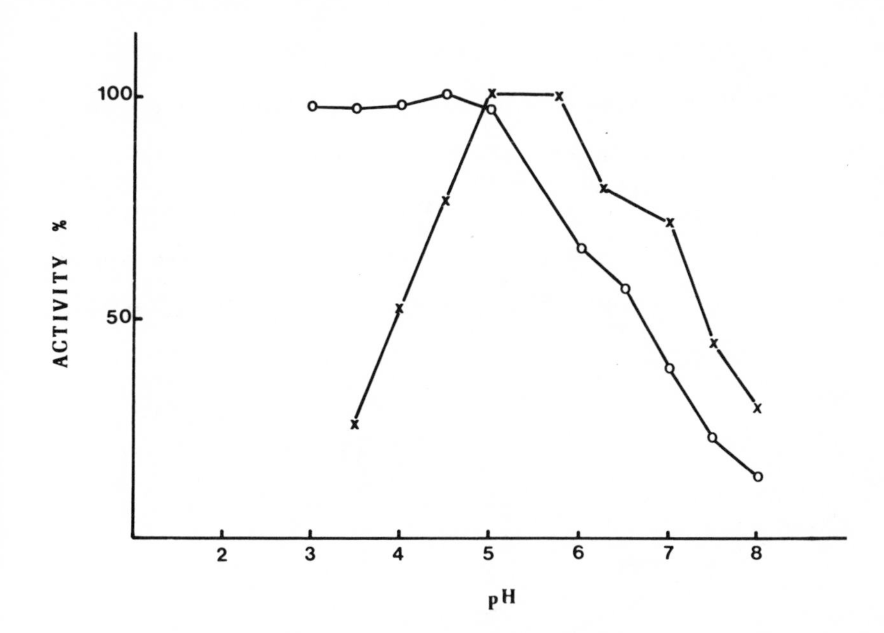

Fig. 13. Effect of entrapping polyacrylic acid on pH-activity profile of entrapped invertase:

O —— O, free invertase.

x —— x, fibers containing polyacrylic acid and invertase.

D. Effect of Temperature on the Reaction Rate

Figure 14 shows the effect of temperature on the rate of the reaction catalyzed by free and entrapped glucose isomerase. The reaction temperature coefficient for the entrapped enzyme is lower than that of the free enzyme. In other words, the efficiency of the entrapped enzyme increases as temperature decreases. This makes it more favorable to use fiber-entrapped enzymes whenever operating at low temperature is desirable either because the substrates or products are unstable or because it is necessary to minimize the effects of bacterial contamination. The lower apparent activation energy found for enzyme fibers was interpreted according to the Wheeler-Thiele (9, 10) analysis for simultaneous diffusion and chemical reaction in a porous catalyst. The same approach was used also to correlate the sucrose inversion data in resin particles (12).

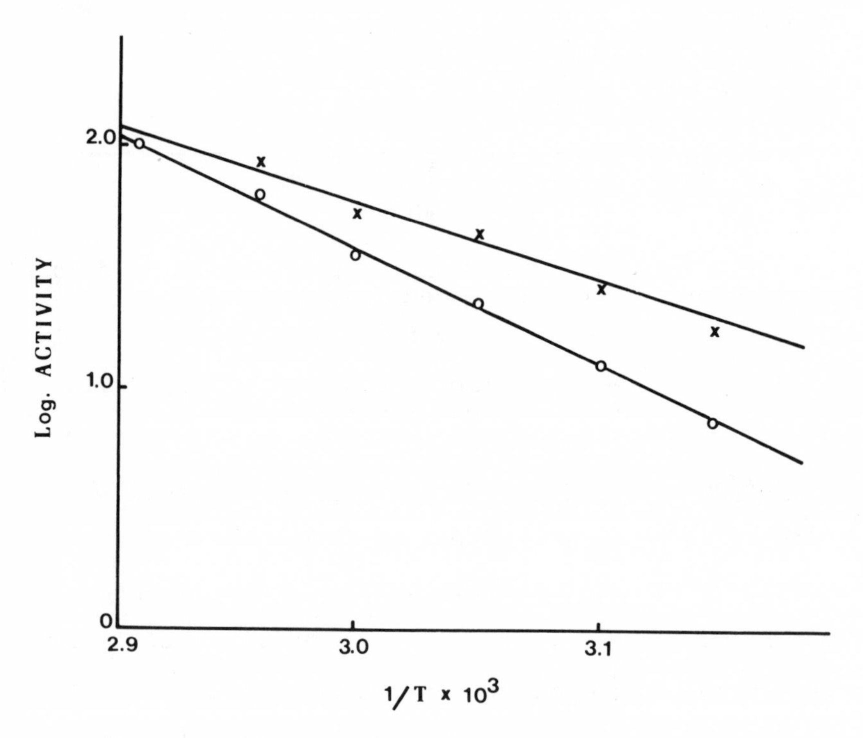

Fig. 14. Effect of temperature on the rate of the reaction catalyzed by glucose isomerase:

o —— o, free enzyme.

x —— x, entrapped enzyme.

E. Effect of Substrate Concentration

Figure 15 shows the effect of substrate concentration on the activity of free and entrapped invertase. Two entrapped preparations containing a low and a large amount of enzyme per g. of polymer, repestively, were compared with the free enzyme. The bulk substrate concentration required to produce saturation is different in the three cases and is higher for the high-enzyme-content fiber. The increased apparent Michaelis-Menten constant found for the entrapped enzyme is the result of the effect of diffusion. Similar behavior was observed for other types of immobilized enzymes (13), and equations to account for the effect of diffusion have been derived (14,15). It must be pointed out that the fibers with heavier concentrations of enzyme show almost no substrate inhibition because, due to diffusional restrictions, the substrate concentration inside the microcavities never reaches inhibitory values. This is a real advantage offered by entrapped invertase in the sugar industry, where operating at high sucrose concentrations is desirable to avoid microbial contamination and to reduce reactor volume.

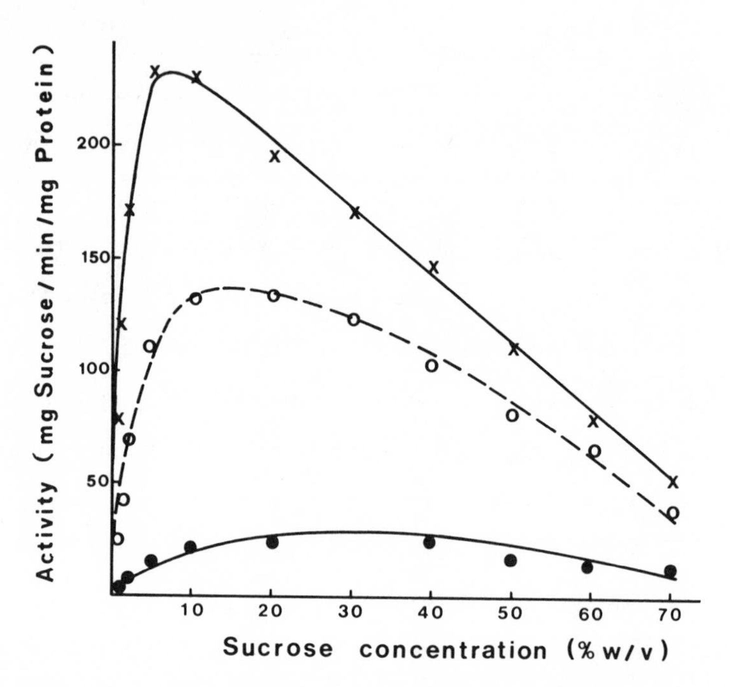

Fig. 15. Effect of substrate concentration on invertase activity:

x ——— x, free enzyme.

o ——— o, entrapped enzyme, 100 units per g. of cellulose triacetate.

● ——— ●, entrapped enzyme 14,500 units per g. of cellulose triacetate.

F. Effect of Diffusion

All the mentioned properties of enzyme fibers demonstrate the great influence of diffusion. If the rate of the enzymic reaction is high, there will be a significant depletion of the substrate inside the microcavities, and the overall reaction will be diffusion-controlled. The reaction rate is increased by: (1) increasing the enzyme density in the support -- even though fibers containing large amounts of enzyme show low efficiency; (2) increasing temperature; (3) approximating optimal conditions of pH substrate concentration. The overall reaction will be more diffusion-controlled at those pH, temperature or substrate concentration values at which the enzymic activity is higher.

G. Behavior of Enzymes at High Concentration

Though many of the characteristics of the enzyme fibers have been attributed to diffusion phenomena, the high concentration of enzyme inside the microcavities of the fibers must also be considered. The situation is very similar to the "in vivo" situation described by Srere (16) who emphasized that the intracellular concentrations of enzymes are generally high. Other authors have discussed the importance of this fact in determining the different behavior of the enzymes between the "in vitro" and "in vivo" conditions (17, 18). Also, the diffusion of the substrates through the concentrated protein solution of the microcavities might be effected, and some enzyme molecules might be not available to the substrate, thus reducing the enzyme efficiency. It is worth emphasizing that a better understanding of the behavior of enzymes at high concentration might be obtained by the study of the enzyme fibers. Recently the use of permeated cells in the study of enzyme activity and its regulation has been proposed (19). Fiber-entrapped enzymes are likely to be very useful for these studies.

V. STABILITY OF FIBER-ENTRAPPED ENZYMES

The "life" of an enzyme is the result of many factors related to its structure and to its use conditions. Fiber-entrapment does not change the structure or the thermodynamic stability of the enzyme. The thermal stability of many enzymes entrapped in cellulose triacetate fibers was similar to that of the enzymic solution used for the entrapment. This finding is in contrast with the expected denaturation effect of the support because of its hydrophobic character, and indicates a negligible interaction between protein and support. Cellulose triacetate is largely employed for the electrophoresis of proteins and enzymes and no evidence of strong interactions has been accumulated (20).

The fiber-entrapped enzymes are protected from external agents such as microorganisms and proteases. Finally, entrapping improves the stability of those enzymes which undergo denaturation by dilution because they can be entrapped as an aqueous solution at high protein concentration.

A. Storage Stability

Many fiber-entrapped enzymes stored at 4°C, as they came out of the spinning apparatus, i.e. wet and containing some toluene, remained perfectly stable for many months. In the absence of toluene, the growth of molds was often observed without any change in enzyme activity. Other fiber-entrapped enzymes, unstable under such conditions, showed good stability even at room temperature when treated with water-glycerol and air-dried. Dramatic changes in fiber permeability were observed by drying the fibers exhaustively or by storing them under dry conditions. Lyophilization mechanically damages fibers and results in rapid leakage of enzyme.

B. Operational Stability

Some fiber-entrapped enzymes have maintained an unchanged activity for very long time under operating conditions (Table 4). The most important point is the control of those factors which influence the "life of fixed enzymes under working conditions". Often a low thermodynamic stability of the enzyme structure is supposed, as well as that of the enzyme-substrate complex. Measurements of the inactivation rate as a function of the temperature permit not only the identification of such cases but also the establishment of optimal conditions under which the enzyme can be used.

In the continuous use of fixed enzymes, continuing dialysis may also take place. The enzymic activity can decrease rapidly when the activity is dependent upon the presence of a low-molecular-weight compound (ions, coenzymes, etc.) not irreversibly bound to the enzyme. Such a phenomenon may be demonstrated by a decrease in activity, which may be related to the quantity of solution treated rather then operation time or to the quantity of product produced. β-galactosidase from yeast entrapped in fibers rapidly looses its activity if the fibers are washed with water; but washing with water or buffer containing Mg^{++} ions preserves the activity.

Table 4

Stability of Some Fiber-Entrapped Enzymes Under "Working Conditions"

Enzyme	Operational Stability (Days)
β-Galactosidase	150
Glucoamylase	120
Glucose Isomerase	150
Glucose oxidase	60
Invertase	1,900
Lactate dehydrogenase	360
Penicillin acylase	180

In some cases, it was noticed that the enzyme-substrate complex dissociates following partially a secondary reaction inactivating the enzyme (21). If the rate of enzyme inactivation in the presence of substrate is far greater then in the absence of substrate, this may be a sign of such a phenomenon.

In many cases, fiber-entrapped enzymes work for long periods with no apparent inactivation, then activity drops suddenly. Disregarding routine causes (like overheating, pH variations, etc.) the most frequent cause of dropping enzymatic activity is the introduction of poisons and growth of microorganisms in the system. As an example of poisoning, consider the behavior of glucose isomerase. The fiber-entrapped enzyme is not damaged by contact with air. The substrate and the product are slowly altered by air, but the resulting impurities do not alter the activity of the enzyme. However, when oxygen contacts the enzyme-substrate complex, rapid inactivation is observed. By operating under carefully controlled conditions, the activity remains practically constant. Occasional contaminations with oxygen-bearing air, which may take place because of imperfect control, cause sudden activity decrease.

Growth of microorganisms is probably the most serious problem, because the solutions being treated are commonly good nutrient media for microorganisms. Although the enzyme solutions to be entrapped are generally not sterile, this is not a major cause of contamination. Apart from the sterilizing action of the solvents employed, only 1% of the droplets of the enzyme solution present in the emulsion (having a mean volume of about 10 μ^3, 10^{-11} cm^3) would contain one microorganism, and the microorganisms would not have room enough to duplicate. The growth of microorganisms in the substrate solution may damage the enzyme by microbial attack on the enzyme; destruction of the enzyme by proteases produced by the same microbes; or by poisoning the enzyme by low-molecular-weight metabolites. Fiber-entrapped enzymes are protected against the first two mechanisms of inactivation, and in some cases this is sufficient. For instance, microorganisms sometimes grew over fibers containing invertase which had worked more than five years treating 20% sucrose solutions. Washing with an aqueous solution of glycerol was sufficient to almost completely eliminate the microorganisms and the impurities retained by the fibers. In other cases, such as β-galactosidase, bacterial growth caused a loss of activity, but washing with glycerol and water re-stabilized activity at a lower constant level until another bacterial contamination occured.

C. Leakage of Enzyme from the Fibers

Another reason for entrapped enzyme instability may be continuous leakage from the support. The study of enzyme leakage from a fibrous matrix has been particularly investigated to help determine spinning parameters. By examining the results of a typical washing experiment, it is apparent that the leakage of the enzyme takes place in two steps. The first step is characterized by a high release rate for loosely adsorbed enzymes from the external surface and from open surface microcavities (see electron micrographs). The second step, during which enzyme is lost at a much lower rate than observed in the first step, is related solely to the porosity of the fibers. The amount of loosely adsorbed enzyme lost by washing accounts for from 2 to 5% of the entrapped activity. Table 5 shows the initial efficiency, the loss of activity, and the half-life of invertase fibers of differing porosities. Figure 16 illustrates the leakage of invertase as a function of time. The leakage from the fibers of dextran of varying molecular weight was investigated to establish whether the shape of the entrapped molecules is an important factor in leakage rate. Figure 17 illustrates the results.

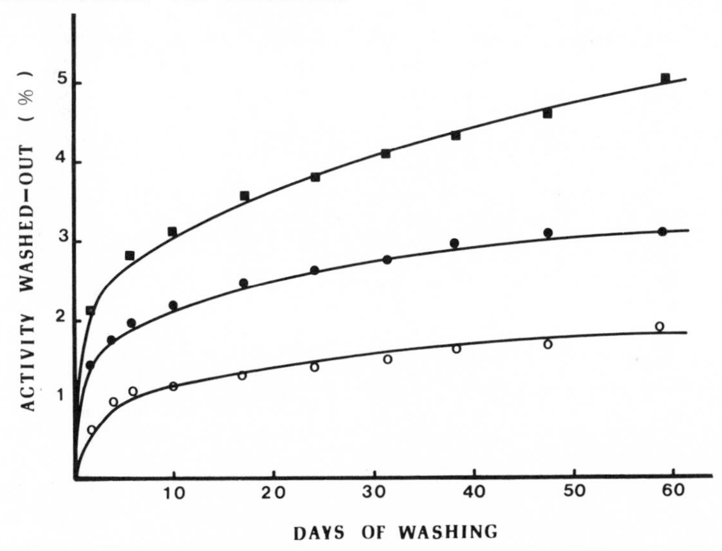

Fig. 16. Leakage of invertase from fibers with different porosity; % of the originally entrapped activity:

■-----■, porosity, 2.0, ● ----- ●, porosity, 1.0, O ——— O, porosity, 0.5.

Table 5

Properties of Fiber-Entrapped Invertase Preparations At Different Porosity Degree

Prep. No.	Porosity Degree[a]	Initial Efficiency	Loss of Activity[b] After 100 Days of Washing %	Half-life Days
1	4	67.20	48.80	77
2	3	29.70	9.14	330
3	2	28.30	5.43	2,547
4	1	27.20	3.0	5,330
5	0.5	21.13	0.53	--

[a]Arbitrary

[b]Percentage calculated on the entrapped activity. The data are mean average values obtained examining five fiber-entrapped invertase preparations.

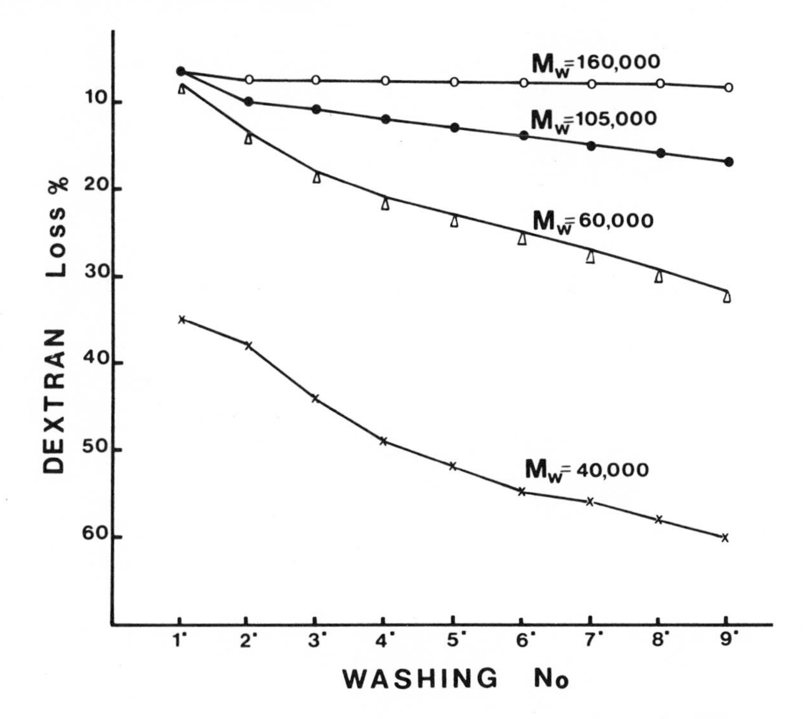

Fig. 17. Leakage of dextrans from fibers.

VI. APPLICATIONS OF FIBER-ENTRAPPED ENZYMES

A. Synthesis and Resolution of Aminoacids

Aminoacylase was entrapped in cellulose triacetate fibers to form an insoluble enzyme preparation for the resolution of the racemic mixtures of aminoacids. The source of enzyme was either animal (hog kidney) or microbial (Aspergillus sp.). Both entrapped enzymes are very stable under operating conditions as shown in Table 6. The apparent relative activities of the entrapped aminoacylase toward different substrates differ from those of the soluble counter-parts because diffusion is more limiting for the better substrates. Using the enzyme from hog kidney, aminoacylase fibers showing an activity of 17,700 units per gram of polymer were obtained (1 unit = 1 μ mole of methionine per hour at 37°C.)

Table 6

Operating Stability of Fiber-Entrapped Aminoacylase

Operation Day	Retained Activity %	
	Kidney Acylase	Aspergillus Acylase
1	100	100
2	98	102
3	103	100
7	109	100
9	109	98
11	100	105
14	98	100
16	98	102
18	100	96
22	104	100
23	92	97
25	80	95
37	82	78
42	77	80
50	71	75

D-aminoacid oxidase was entrapped in fibers to transform the D-form of the aminoacids while leaving the natural form unaltered. The entrapped enzyme was not stable, and progressively lost its activity, as shown in Figure 18. Most probably, the enzyme's stability is related to its cofactor, FAD.

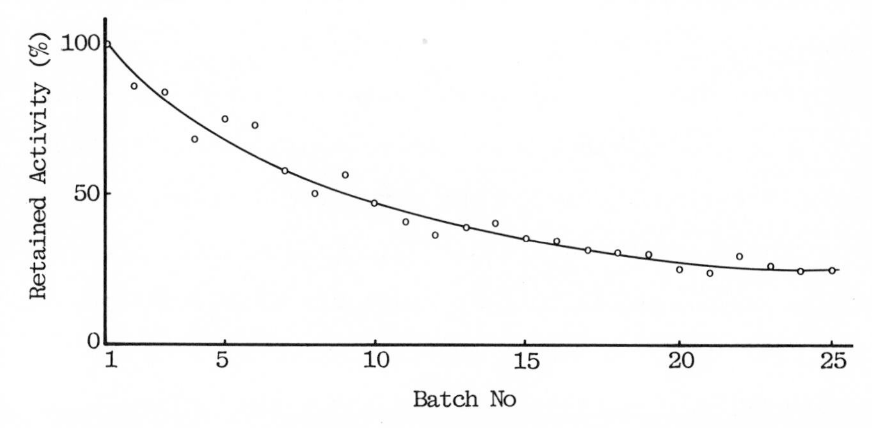

Fig. 18. Operational stability of aminoacids oxidase fibers.

The synthesis of L-tryptophan from indole and DL-serine has been extensively investigated using tryptophan synthetase from E. coli (22). Though this enzyme is composed of two types of readily separable subunits, no preferential leakage of either subunit from cellulose triacetate fibers was observed, and the entrapped enzyme showed very good operational and storage stabilities. In contrast with the behavior of many other enzymes, the efficience of the tryptophan synthetase fibers remains practically constant at a very high value (about 75%) by varying the enzyme density in the support as shown in Table 7. Also, for the pH-activity profile, the entrapped tryptophan synthetase gave an anomalous result. Similar pH-activity profiles were expected, because no hydrogen ions are produced or consumed in the course of the reaction. However, the pH optimum for the free and entrapped enzyme was 9.0 and 8.0, respectively. Although other explanations might be found, it seems likely that at pH values higher than 8.0, because of the relatively low purity of the enzyme preparation, a precipitate occurs inside the microcavities and occludes the pore of the fiber, thus preventing the substrate's diffusion. Different type reactors were studied for the synthesis of L-tryptophan using the entrapped enzyme (23). Figure 19 shows the time course of the reaction using the batch method. Indole, which is strongly adsorbed by cellulose triacetate, is added stepwise to prevent its accumulation on the fibers. The enzyme is specific for L-serine; the D-serine, present in the final reaction mixture, after isolation of tryptophan, is racemized and recycled. The continuous synthesis of tryptophan using a continuous feed recycle reactor is also practical. A concentration of tryptophan of 500 mg/l is obtained in the outlet solution, and indole is almost completely reacted (Figure 20).

Table 7

Efficiency of Entrapped Tryptophan Synthetase (TS) as a Function of the Amount of Enzyme Entrapped per Gram of Cellulose Triacetate (CTA).

Entrapped TS Preparation	Entrapped Activity (Units/g CTA)	Displayed Activity (Units/g CTA)	Efficiency = $\frac{\text{Displayed}}{\text{Entrapped}}$ Activity %
1	1,125	850	75
2	2,070	1,570	76
3	3,710	2,940	79
4	4,100	3,030	74

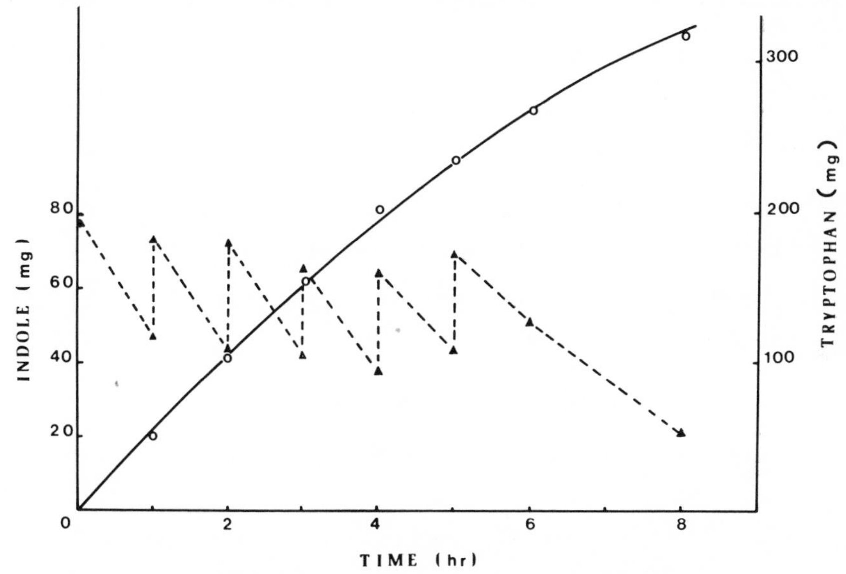

Fig. 19. Batch synthesis of L-tryptophan using tryptophan synthetase fibers:
▲ —— ▲, indole
O —— O, L-tryptophan.

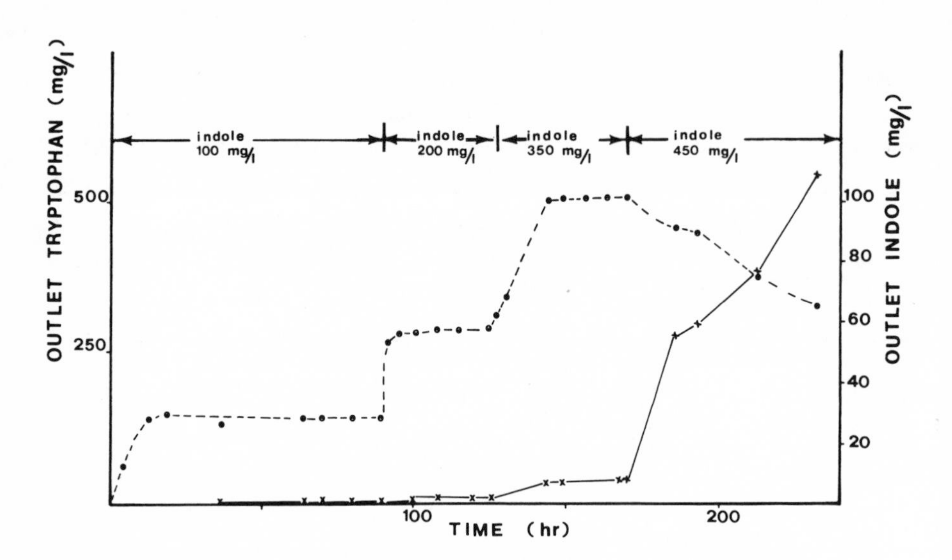

Fig. 20. Synthesis of L-tryptophan catalyzed by entrapped tryptophan synthetase using a continuous feed recycle reactor:

● ----- ●, tryptophan concentration.
x ——— x, indole concentration.

The synthesis of L-alanine from lactic acid and ammonia was studied using two dehydrogenases, lactic acid and alanine dehydrogenases, entrapped together in the same fibers (24). The two enzymes catalyze the following reactions:

$$\text{Lactate} + NAD^+ \rightleftharpoons \text{Pyruvate} + NADH + H^+, \tag{2}$$

$$\text{Pyruvate} + NADH + H^+ + NH_3 \rightleftharpoons \text{Alanine} + NAD^+ + H_2O. \tag{3}$$

Figure 21 shows a typical synthesis of L-alanine. The accumulation of the aminoacid is linear with time and the concentration of NADH reaches a steady-state value, indicating that the amount consumed in the second reaction is is replaced by the amount formed in the first one.

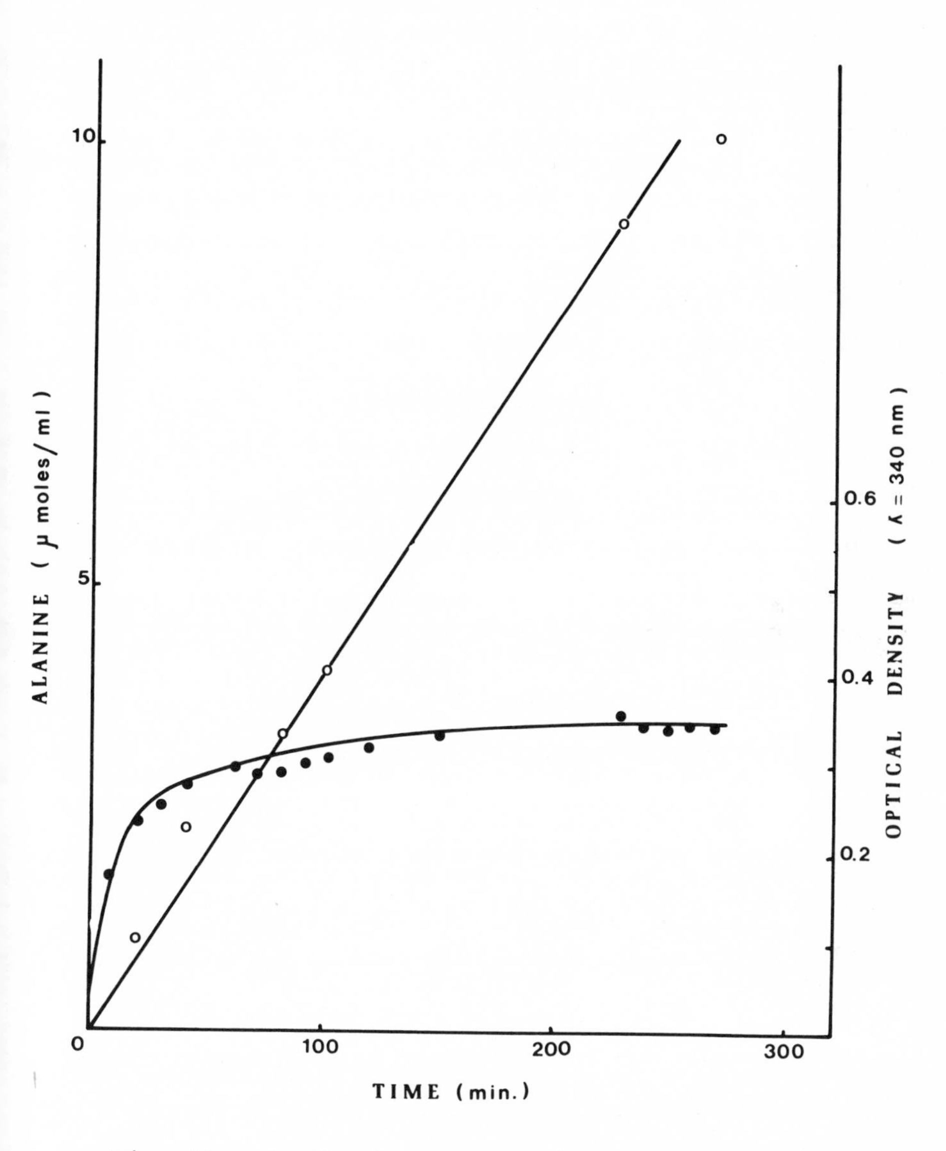

Fig. 21. Synthesis of L-alanine from ammonia and lactic acid:

O ——— O, L-alanine concentration.

● ——— ●, NADH absorption at 340 nm.

Two two entrapped enzymes have been operating for one year at 18°C showing a 10% loss of activity. The efficiency of this bienzymatic system is related to the concentration of the cofactor employed, which being of low molecular weight, is diffusible and thus distributed in the reaction mixture. The enzymes are localized inside the microcavities of the fibers and are not diffusable.

B. Carbohydrates

The properties of fiber-entrapped β-galactosidase have been intensively investigated with an eye toward using the entrapped enzyme for hydrolysing lactose in milk and dairy products. A comparative study of β-galactosidase from E. coli and yeast was done (25). The bacterial enzyme is more stable than the yeast enzyme, as shown in Table 8. However, the latter was used to develop an industrial process because it hydrolyzes milk and pure lactose at approximately the same rate, while the E. coli enzyme seems to be inhibited by some inhibitors present in milk, reducing the hydrolysis rate of lactose in milk to about one third of that for pure lactose. Moreover, the yeast enzyme can be considered GRAS (Generally Recognized As Safe). This is important even if no detectable leakage of enzyme from the fibers was determined (Table 9). The operational stability of the entrapped yeast enzyme depends mainly on bacterial contaminations. The preparation and use of very active enzyme fibers can shorten hydrolysis time, even at temperatures near 0°C, thus reducing the risks of microbial contamination and leaving the organoleptic properties of milk unchanged (26). A pilot plant was operated (27). The activity displayed by the entrapped β-galactosidase in this reactor was only slightly less than that obtained in laboratory, probably due to channelling and diffusional gradients (Figure 22). Figure 23 shows the stability of the entrapped enzyme in the pilot plant, and the bacterial count of the treated milk. The experimental data indicate good stability for β-galactosidase fibers. After 50 days of operation, 80% of the original activity is retained. Insoluble material remained on the fibers and probably affected their permeability. The total bacterial count at 32°C was within the limits fixed for pasteurized milk and always less than 10,000 per ml. Furthermore, the contaminant bacteria were generally psychrophiles rapidly destroyed at the usual temperatures of pasteurization.

Table 8

Residual Activity of Entrapped β-galactosidase from E. coli and from Yeast after Five Days' Operation on Milk at Various Temperatures.

Temperature (°C)	Residual Activity %	
	E. coli Enzyme	Yeast Enzyme
25	100.0	94.0
35	100.0	98.5
45	98.2	25.0

Table 9

Leakage of β-galactosidase from Fibers: Nitrogen Content of Fibers Before and After Use.

Fibers Sample	Initial Nitrogen[a] (mg/g)	Nitrogen After Washing (mg/g)	Nitrogen After 15 Days (mg/g)
1	5.0	4.70	4.67
2	5.15	4.65	4.60
3	5.25	4.75	4.65

[a]Total Nitrogen (according Kjeldahl) per gram of cellulose triacetate.

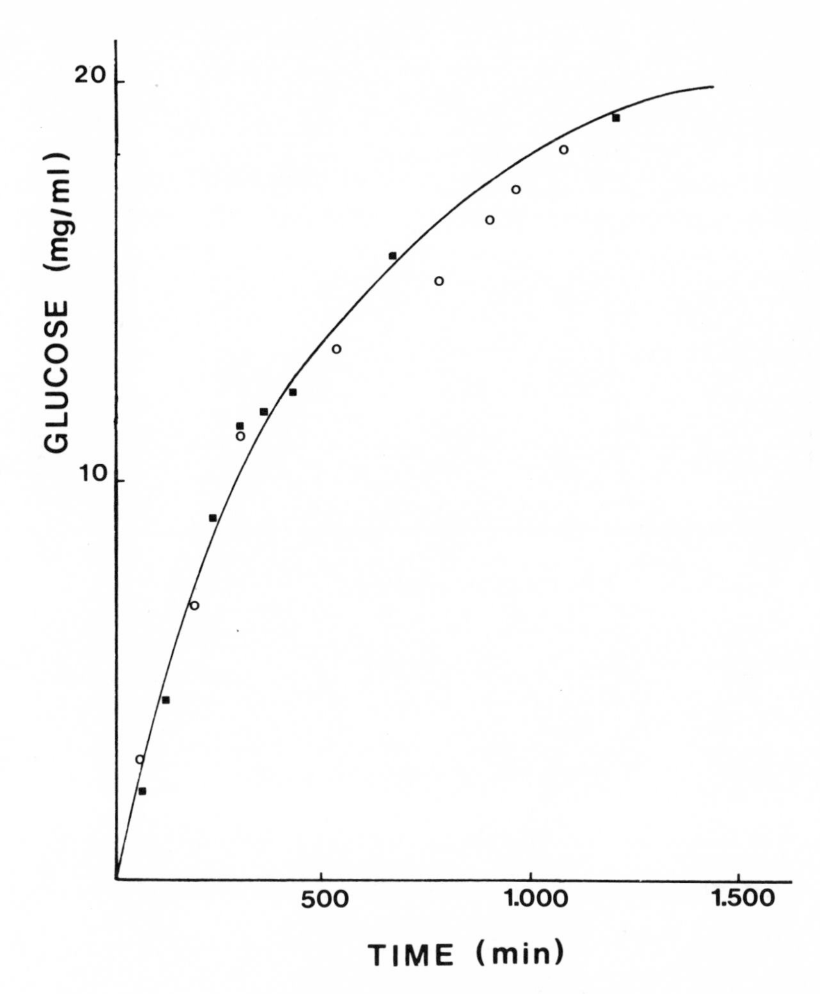

Fig. 22. The kinetics of milk lactose hydrolysis:
■ —— ■, laboratory experiments.
O —— O, pilot-plant experiments.

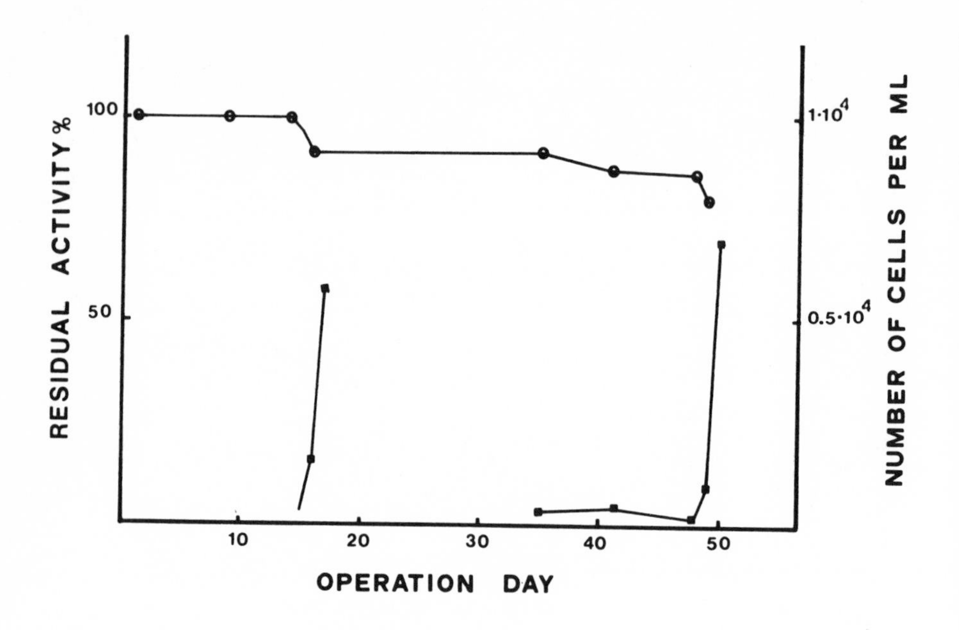

Fig. 23. Operational stability of β-galactosidase fibers and bacterial count in pilot plant experiments:
O ——— O, residual activity.
■ ——— ■, bacterial count.

Besides its use in the industrial production of glucose from starch, the entrapped glucoamylase is of particular interest because it displays its activity toward substrates of different molecular weight (28). The study of the properties of glucoamylase entrapped in fibers gave significant information about the diffusion of compounds having the same structure but different molecular weights. Table 10 gives an idea of the dramatic drop in activity passing from maltose to liquified starch (dextrins, 8-20 glucose units) as substrate. Nevertheless, dextrose equivalents, very similar to those observed for the free enzyme can be obtained in the saccharification of dextrins by the entrapped enzyme, indicating a complete accessibility of the enzyme to substrate molecules. Of course, this depends strictly on the porosity of the fibers. This can be varied choosing the optimal conditions for good activity and low enzyme leakage rates. The activity of the glucoamylase fibers remained constant during six months of operation on maltose at 25°C. Working on a 30% solution of liquified starch at 45°C, the fibers maintained unchanged activity after 300 hours. An advantage the fibers offer is the possibility that liquified starch slurries can be used without filtration. Using enzyme reactors in which the fibers are placed in an orderly way, no blockage is observed even if the reaction mixture contains particles in suspension.

Table 10

Activity of Immobilized Glucoamylase Toward Substrates of Different Molecular Weight

Substrate	Relative Velocity
Maltose	1.0
Liquified Starch	0.105
Potato Starch	0.05
Soluble Starch	0.025

Glucose isomerase catalyzes the isomerization of glucose to fructose to give a mixture of approximately the same composition as the invert sugar. Glucose isomerase is an economically attractive enzyme because it allows the preparation of invert sugar from glucose syrups obtained by saccharification of starch. Glucose isomerase from streptomyces sp. was entrapped in cellulose triacetate fibers (29). The efficiency of the enzyme fibers was about 60%. Very active entrapped preparations (30 m moles of fructose/hr/g of polymer, at 70°C) were used to carry out the isomerization of 50% (w/w) glucose solutions. Working at near-neutral pH values and at relatively low temperatures (45°+55°C) under strictly anaerobic conditions, the stability of the glucose isomerase fibers was very good (Fig. 24). The product was colorless and contained negligible amount of by-products. As mentioned earlier, the stability of the entrapped enzyme depends mainly on the simultaneous presence of oxygen, substrates and enzyme in the reaction mixture. These findings suggest an irreversibly inactivating effect of oxygen on the enzyme-substrate complex. In contrast with the normal behavior of enzymes, glucose isomerase seems to be less stable when substrate is present than when it is absent at least as far as the effect of oxygen is concerned. As with the free enzyme, entrapped glucose isomerase requires Co^{++} and Mg^{++} ions for its activity.

This is shown in Fig. 25 where the reversible inactivation of the enzyme by the absence of metal ions and the irreversible inactivation by oxygen are illustrated.

Glucose oxidase and catalase co-entrapped in the same fibers were used for preparating gluconic acid from glucose. The stability of glucose oxidase was not completely satisfactory (Fig. 26), and it was difficult to develop a process competitive with the very efficient fermentation method.

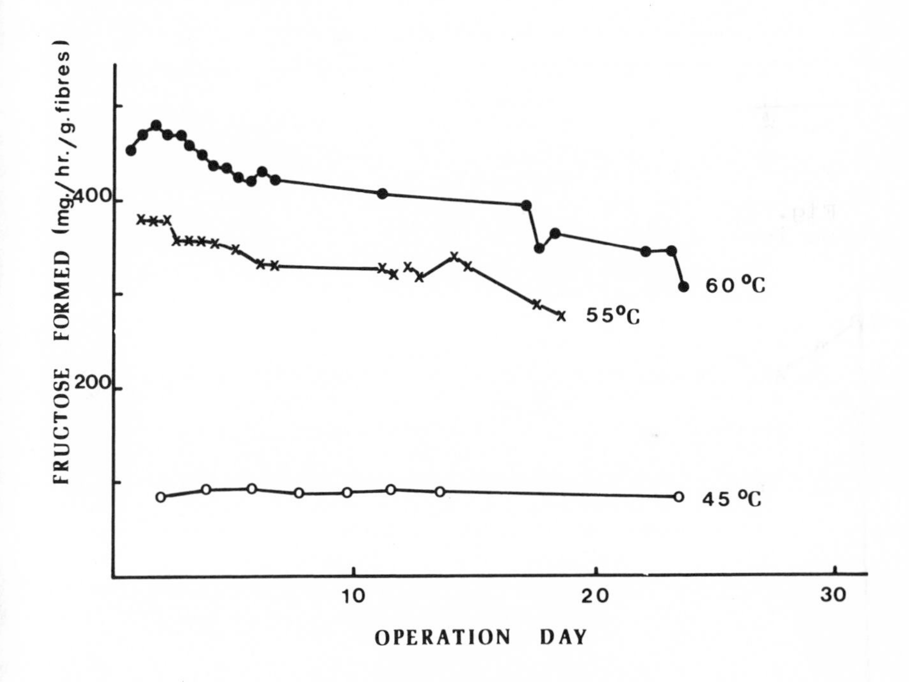

Fig. 24. Operational stability of entrapped glucose isomerase at different temperatures:

● —— ●, 60°C.
x —— x, 55°C.
o —— o, 45°C.

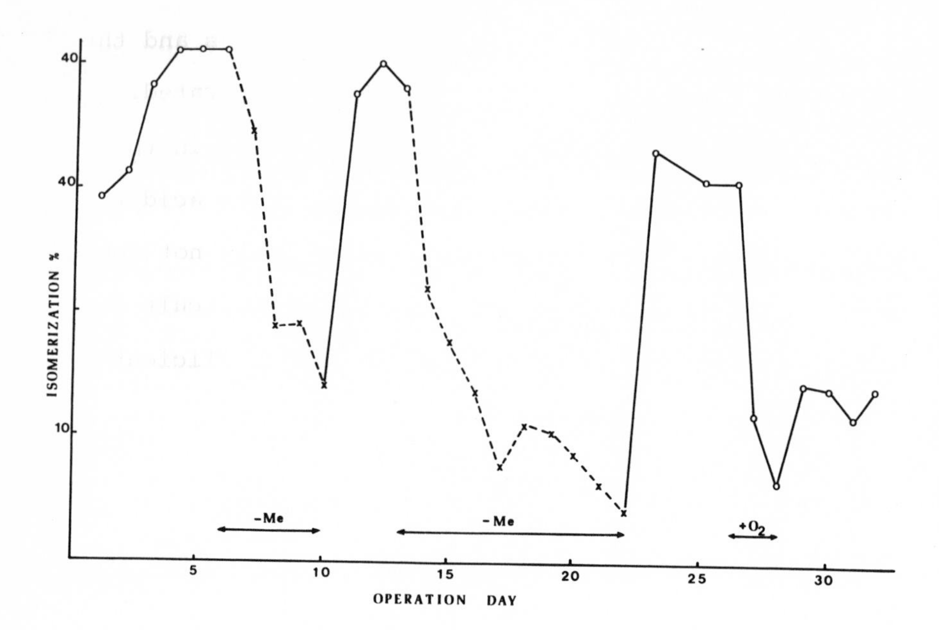

Fig. 25. Effect of metals and oxygen on entrapped glucose isomerase activity.

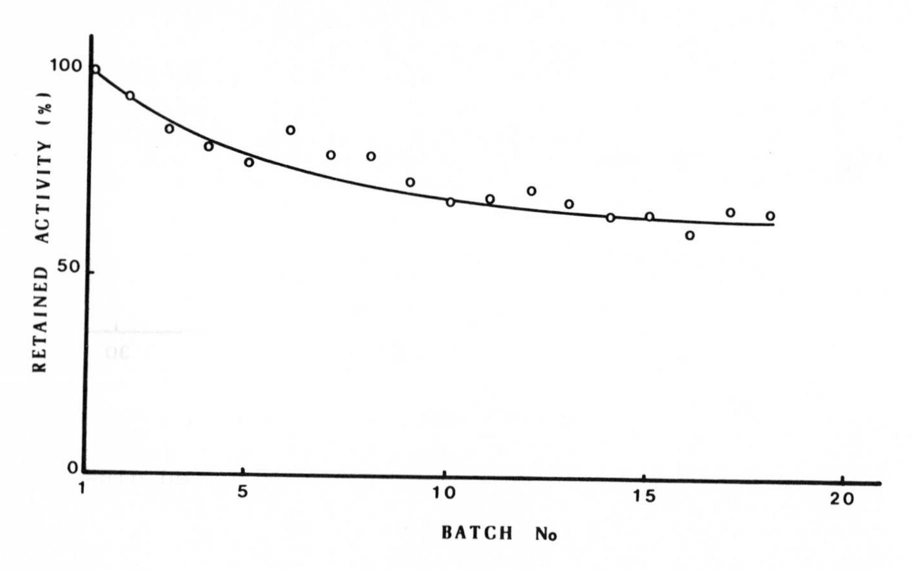

Fig. 26. Operational stability of glucose oxidase fibers.

Fibers containing three enzymes (invertase, glucose oxidase and catalase) were studied with respect to producing gluconate and fructose from sucrose. Figure 27 shows the kinetics of the reaction catalyzed by the entrapped multienzyme system.

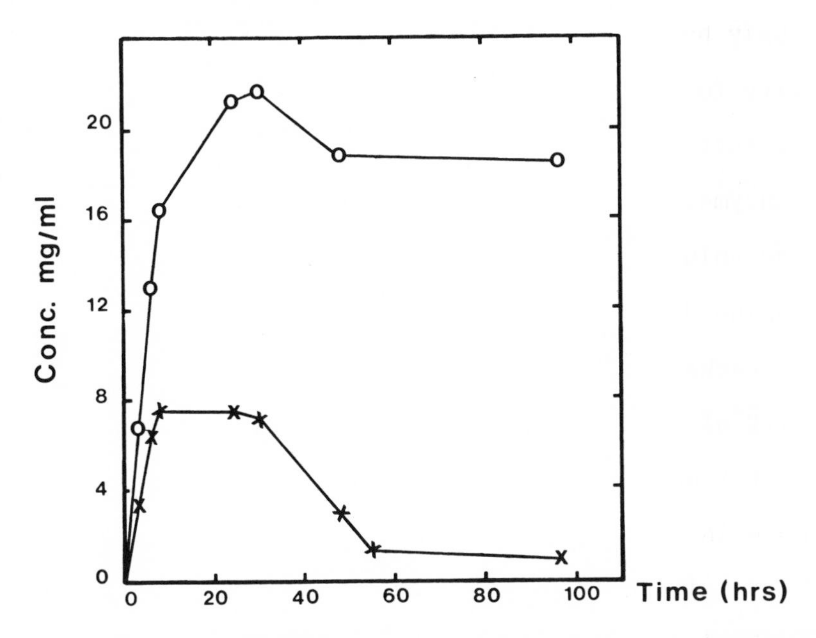

Fig. 27. Production of fructose and gluconate by using invertase, glucose oxidase and catalase entrapped in the same fibers:

o —— o, total reducing sugars.
x —— x, glucose.

working conditions: 0.1 M sucrose, pH 5.6, 25°C.

A great deal of information about the properties of enzyme fibers has been obtained by entrapping invertase (30). The excellent stability and the good efficiency of invertase fibers, especially at high substrate concentrations, increase their industrial potential for the production of invert sugar, despite the low cost of the traditional process. In one case, invertase fibers continuously hydrolyzing sucrose at 25°C maintained their activity for five and a half years. The thermal stability of the entrapped enzyme does not differ from that of the free enzyme. The continuous hydrolysis of 50% (w/w) sucrose solutions is possible. Figure 28 shows the effects of packing degree and linear velocity on the activity of fibers packed in a column. By increasing the linear velocity of the substrate and the amount of invertase fibers per unit volume of the reactor, a considerable increase in enzyme efficiency was observed. Under the best conditions, the efficience of invertase fibers was about 55%.

A crude preparation of naringinase from aspergillus niger was entrapped in cellulose triacetate fibers (31). Naringinase is a complex enzyme containing two glycosidases which hydrolyze the bitter principle of grapefruit juice, naringin, to naringenin, rhamnose and glucose. Fiber-entrapped naringinase was continuously used for hydrolyzing naringin solutions at 25°C. After six months of operation, the activity was still about 90% of the original value. The reaction kinetics is complicated by the adsorption of naringin on the cellulose triacetate fibers. Because of the continuous action of the enzyme on the substrate present in the aqueous phase, naringin is gradually desorbed from the fibers, passes in the aqueous phase and is hydrolyzed. This was confirmed by simultaneously determining naringin according to the method of Davis (32) and the liberated reducing sugars. A typical kinetics of naringin hydrolysis by naringinase fibers is shown in Figure 29. Entrapped naringinase was used to commercially treat grapefruit juices. The enzyme fibers showed good stability after working at 45°C for about two months. However, the activity was markedly lowered by the presence of high concentrations of glucose, which inhibits the enzyme.

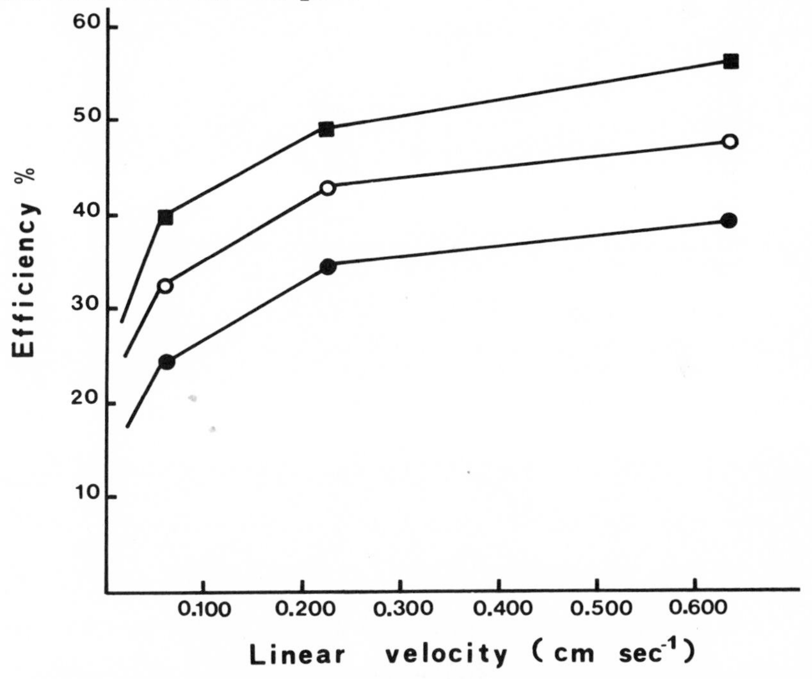

Fig. 28. Effect of linear velocity and packing degree on invertase fibers activity: ■——■ , 190 mg of invertase fibers per ml of column volume, O —— O, 90 mg of invertase fibers per ml of column volume, ● —— ●, 35 mg of invertase fibers per ml of column volume.

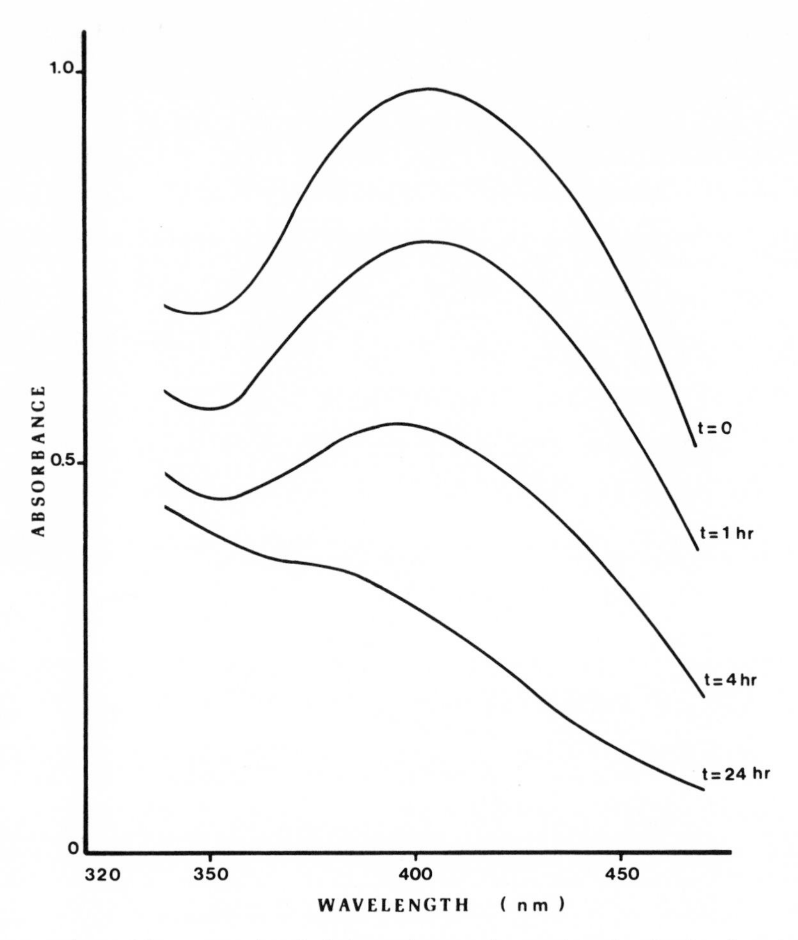

Fig. 29. Typical kinetics of naringin hydrolysis by naringinase fibers: absorption of color produced by alkaline diethyleneglycol on naringin solution.

C. Penicillin Acylase

The hydrolysis of penicillin G to 6-amino penicillanic acid (6-APA) by fiber-entrapped penicillin acylase has been intensively investigated (33). The enzyme was extracted from E. coli and partially purified. Very active enzyme fibers containing up to 670 units per gram of cellulose triacetate could be prepared (1 unit = 1 μ mole 6-APA/min, 37°C). The high activity of the entrapped enzyme enables the hydrolysis of penicillin G within very short times, and has the advantages of: negligible substrate and product decomposition; high yield of product recovery; and absence of by-products. The operational stability of penicillin acylase fibers is quite good (Fig. 30). While the free enzyme shows a severe inhibition by excess of substrate, the entrapped enzyme maintains a constant activity over a wide range of penicillin G concentration. Figure 31 shows the effect of substrate concentration on the activity of free and entrapped enzyme. The behavior of the enzyme fibers enables operation at high penicillin G concentration and direct recovery of the product by acidification.

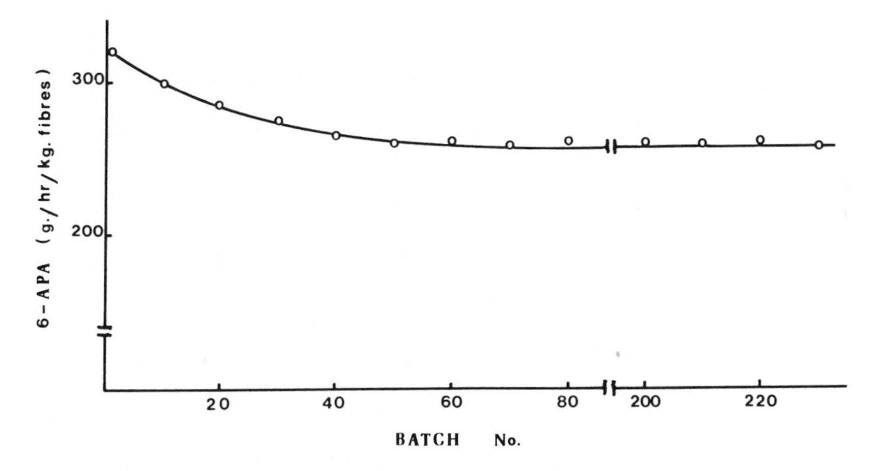

Fig. 30. Operational stability of entrapped penicillin acylase.

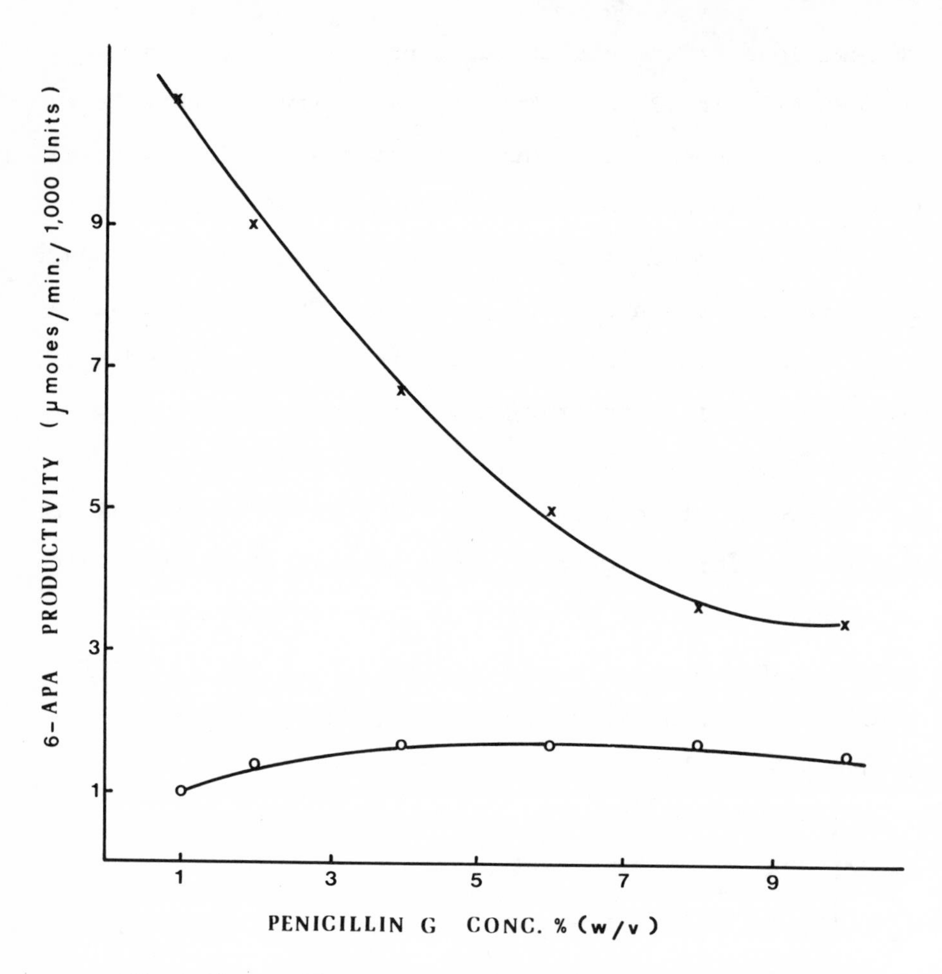

Fig. 31. Effect of substrate concentration on penicillin acylase activity:

x —— x, free enzyme.
O —— O, entrapped enzyme.

In a pilot-scale plant process, the hydrolysis of penicillin G is carried out batchwise in a recirculation reactor. With the pH held at 8.0, circulation of substrate solution through the catalyst bed is carried out at 37°C for the time necessary to achieve 98% hydrolysis.

The product is free from protein impurities which could lead to allergenic semisynthetic penicillins, and no subsequent purification is needed. Other reactions catalyzed by penicillin acylase were studied. The hydrolysis of cephalosporins derived from the chemical enlargement of the penicillin G ring can be carried out using the entrapped enzyme at approximately the same rate at which the enzymatic hydrolysis of Penicillin G occurred. Hydrolysis of penicillin V is also possible at lower rates. It is interesting to note that free penicillin acylase hydrolyzes penicillin G 17 times faster than penicillin V. For the entrapped enzyme, this ratio is about 7 to 1 because diffusion is more limiting for the better substrate, penicillin G. As a result, E. coli may be used as the entrapped enzyme even when more specific enzymes for penicillin V are available. It is well known that penicillin acylase catalyzes the reverse reaction at acidic pH values (34). Also with the fiber-entrapped enzyme, it is possible to carry out the synthesis of ampicillin from 6-APA and esters of phenylglycine (35). Yields of about 70% have been obtained. Figure 32 shows a typical synthesis of ampicillin carried out at 37°C with pH held at 6.0 by adding sodium hydroxide. By using some derivatives of phenylglycine substituted in the aromatic ring, e.g. p-hydroxy-phenylglycine, the synthesis takes place with a lower yield of about 25%. Similar results were noted in field syntheses of cephalosporins. This suggests that the enzyme activity is affected maily by the acylating agent.

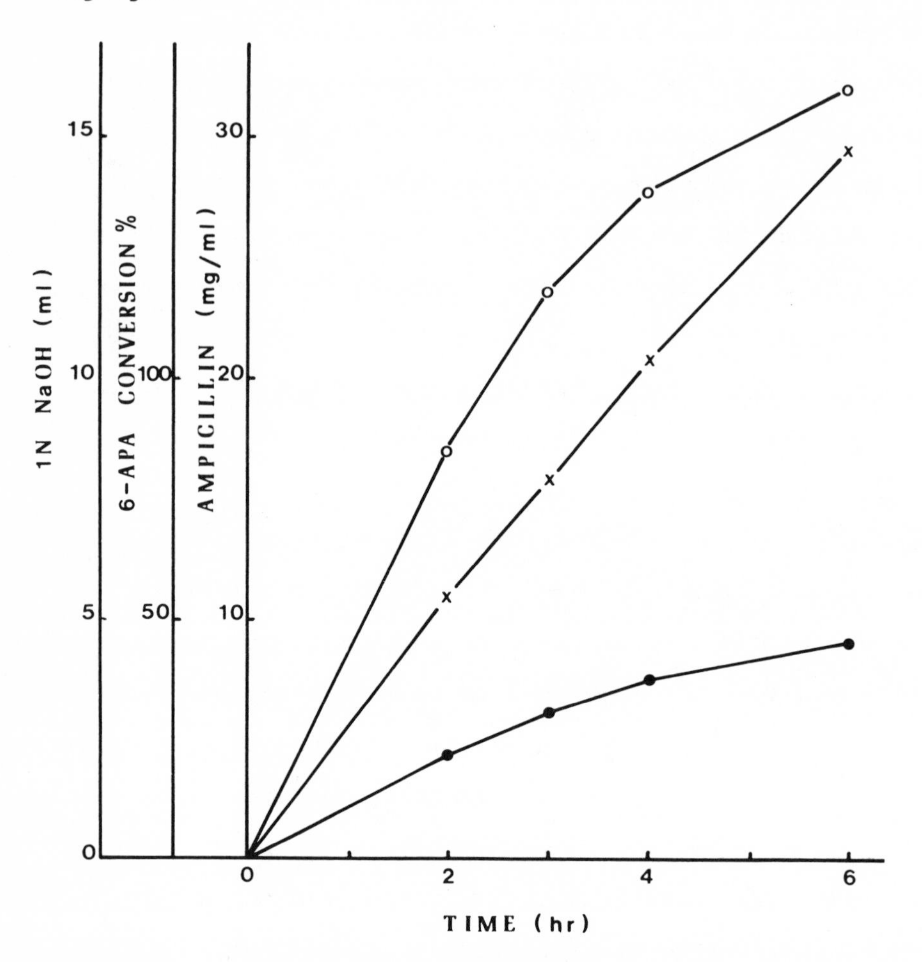

Fig. 32. Time course synthesis of ampicillin catalyzed by entrapped penicillin acylase:

● —— ●, 6-APA conversion.

x —— x, NaOH comsumption.

o —— o, synthetized ampicillin.

Penicillin acylase and urease were entrapped in the same fibers to obtain a self-controlling enzymic system. Penicillin acylase hydrolyzes penicillin G to 6-APA and phenylacetic acids. The latter, because of diffusional restrictions, accumulates inside the microcavities and results in a local pH lower than the bulk pH. To achieve maximum activity, the pH must be controlled at 8.5, which may be done by operating at high buffer concentration and adding sodium hydroxide. Better results are obtained by entrapping urease and penicillin acylase in the same fibers. Urease directly controls the pH inside the microcavities by liberating ammonia (Fig. 33).

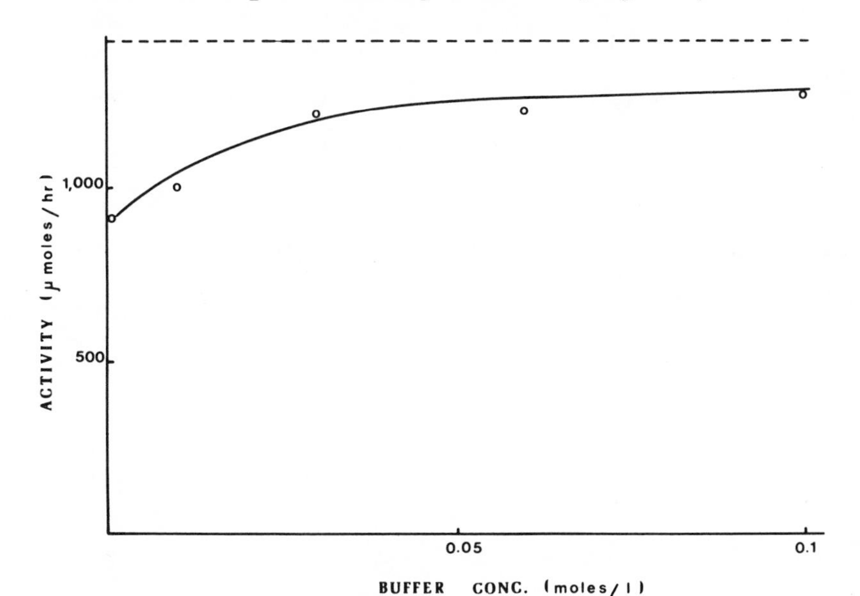

Fig. 33. Activity of entrapped penicillin acylase:

- - - - -, fibers containing penicillin acylase and urease.

O ——— O, fibers containing only penicillin acylase

D. Fumarase

Fumarase from pseudomonas sp. was entrapped in cellulose triacetate fibers (36). The expressed activity of the enzyme fibers was about 35% with respect to the free enzyme. The stability of the free enzyme depends on protein concentration; the enzyme is totally active at protein concentrations higher than 15 mg/ml. The extrapped fumarase has good stability both in storage or under operating conditions. This is probably because the highly concentrated solution of the enzyme used for the entrapment left highly concentrated aqueous enzyme solution within the fiber microcavities. Entrapped fumarase was used to continuously hydrate fumaric acid to L-malic acid. Figure 34 shows that the continuous use of the fiber-entrapped enzyme demonstrated no activity reduction. At pH 8.0, a decrease of the activity was observed. It should be noted that no activity loss was observed due to microbial contamination.

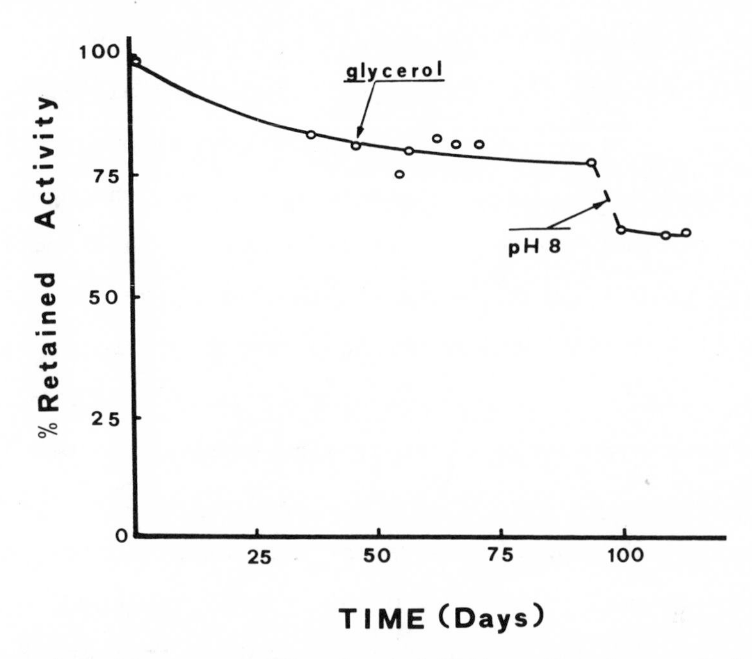

Fig. 34. Operational stability of entrapped fumarase during continuous use. A solution of sodium fumarate (0.5 M in 0.001 M phosphate buffer, pH 7.0) was pumped at a flow rate of 50 ml/hr through a column (3.4 x 45 cm) packed with 30 g. of enzyme fibers. The temperature was kept at 25°C.

E. Analytical Applications.

Some fiber-entrapped enzymes may be useful in analytical chemistry to reduce the cost of routine analyses being repeatedly used (37). Particular emphasis was given to potential applications for enzyme fibers in automated apparatus. Glucose oxidase + peroxidase, urease, asparaginase, penicillin acylase entrapped in cellulose triacetate gave satisfactory results from the point of view of reproducibility, accuracy and stability. The use of insoluble enzyme preparations in particle, sheet and pad forms proved generally troublesome in automatic analyzers because they destroyed the uniform distribution of air segments and resulted in sample overlapping. Enzyme fibers, on the other hand, can be run through silicon plastic tubing that is large enough that the fiber occupies only a small portion of the tubing volume. As a result, the air bubbles can pass properly through the tubing. Figure 35 shows typical recordings obtained in the automated analysis of asparagine using entrapped asparaginase. Figure 36 is a standard curve that shows the linear relation of asparagine concentration to optical density. A small sample of urease fibers was used for 20 days in carrying out urea determination on standards and biological specimens without detectable loss of activity. Other fiber-entrapped enzymes, such as alcohol dehydrogenase, uricase and lactic acid dehydrogenase were used in manual analyses with good results.

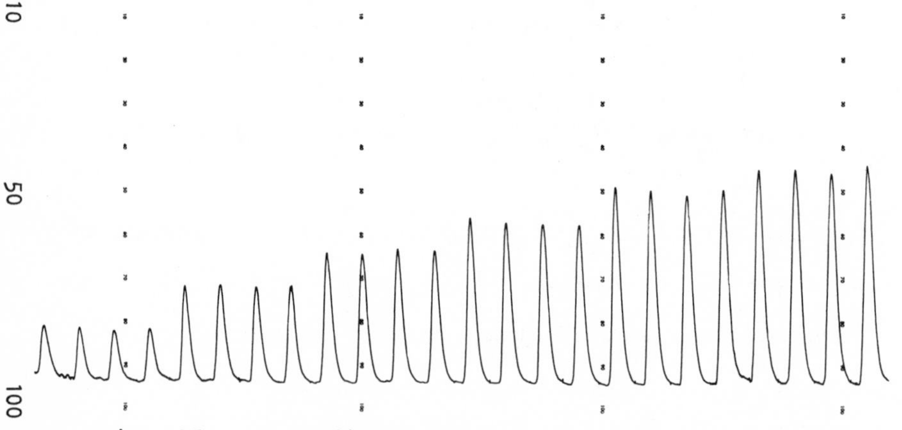

Fig. 35. Recordings for the asparagine determinations.

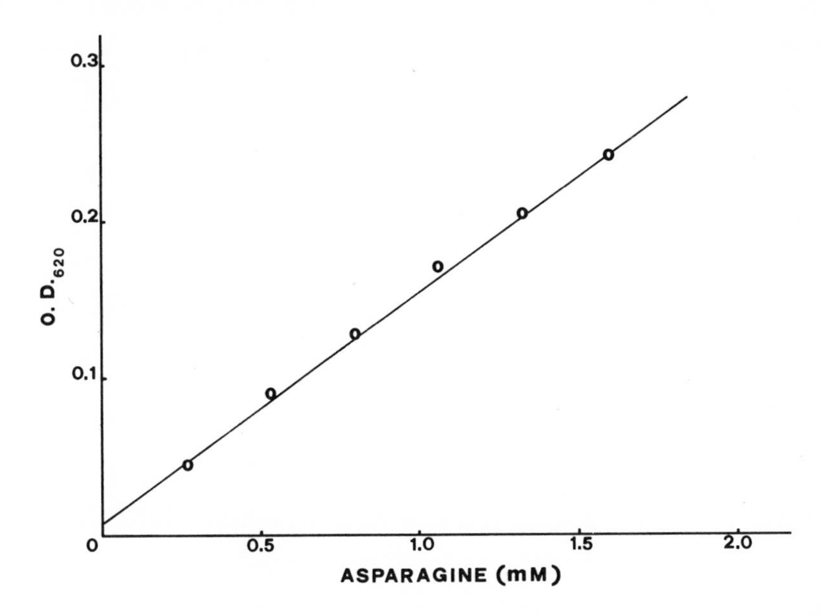

Fig. 36. Standard curve for asparagine determination.

F. Biomedical applications

Fiber-entrapped asparaginase was used with encouraging results in reducing asparagine levels in blood for the treatment of some forms of leukemia (38). Presently, the use of soluble enzymes is limited by the appearance of antibodies. Entrapped enzyme, however, might be used continuously. A small sample of asparaginase fibers was put in the peritoneal cavity of rats. Preliminary results indicate that the entrapped enzyme maintains its immunosuppressive activity and the level of asparagine is drastically lowered, as shown in Fig. 37.

Urease fibers were used to reduce the urea content of the blood of perfused rat liver (38). Figure 38 shows the urea and ammonia concentrations of untreated blood and blood treated with urease fibers.

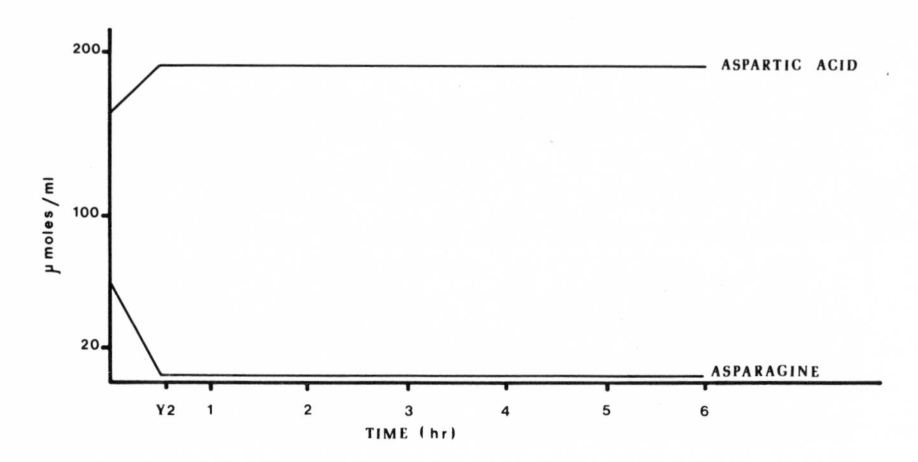

Fig. 37. Hydrolysis of asparagine in blood using Asparaginase fibers.

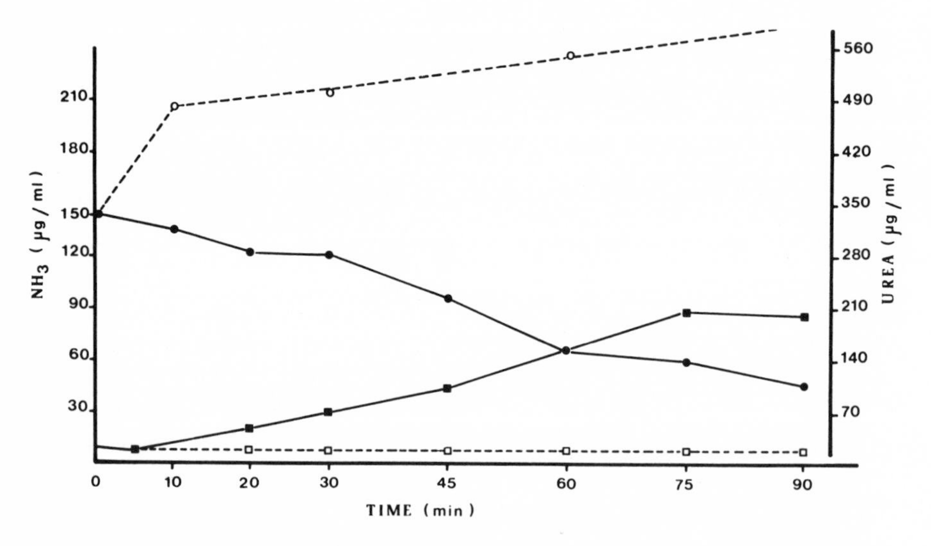

Fig. 38. Hydrolysis of urea in blood using Urease fibers.

O - - - - O, urea, without enzyme fibers.
● ——— ●, urea, with enzyme fibers.
□ ——— □, ammonia, without enzyme fibers.
■ ——— ■, ammonia, with enzyme fibers.

In studying an artificial liver model, some enzymic activities of the liver were observed (39). For instance, lactic and glutamic dehydrogenases were entrapped in cellulose triacetate fibers. The two entrapped enzymes were found quite stable, with activity losses ranging from 5 to 10% after 6 months. The effects of NAD concentration and flow rate were investigated. A whole extract of rat liver was entrapped in fibers in order to study the enzymes which remained active after entrapment. Several enzymes found active in the fibers, are listed in Table 11. It must be stressed that entrapping such a complex mixture of enzymes provides a unique opportunity to study enzyme action under near-*in vivo* conditions; i.e. in the presence of many other proteins that can interact with each other.

Table 11

Enzymatic Activities of a Whole Liver Extract Entrapped in Fibers

Malate dehydrogenase	Glucose phosphate isomerase
Glutamic pyruvic transaminase	Phosphofructokinase
Glutamic oxaloacetic transaminase	Aldolase
Cholinesterase	Pyruvate kinase
Acid phosphatase	Lactate dehydrogenase
Alkaline phosphatase	ATPase
Glutamate dehydrogenase	Alcohol dehydrogenase
β-Glucuronidase	

REFERENCES

(1). D. Dinelli, Process Biochem., 7, 9 (1972).
(2). D. Dinelli and F. Cognigni (to SNAM Progetti S. p.A.), US Pat. 3,715,277 (1973).
(3). R. W. Moncrieff, in Man-made fibres (R. W. Moncrieff, ed.), Heywood Books, London 1970.
(4). A. Marani, F. Bartoli, W. Marconi and F. Morisi, submitted for publication.
(5). E. Katchalski, I. Silman and R. Goldman, In Advance in Enzymology (F. F. Nord, ed.) vol. 34, Interscience Publishers, New York, 1971, pp. 445-536.
(6). W. Nernst, Z. Physik. Chem., 47, 52 (1904).
(7). W. E. Hornby, M. D. Lilly and E. M. Crook, Biochem. J., 98, 420 (1966).
(8). I. H. Silman, M. Albu-Weissenberg and E. Katchalski, Biopolymers, 4, 441 (1966).
(9). A. Wheeler, in Advances in Catalysis (W. G. Frank-enburg, V. I. Komarewsky and E. K. Rideal, eds.), Vol. III, Academic Press, Inc., New York, 1951, pp. 249-327.
(10). E. W. Thiele, Ind. Eng. Chem., 31, 916 (1939).
(11). H. H. Weetall and L. S. Hersh, Biochim. Biophys. Acta, 185, 464 (1969).
(12). E. R. Gilliland, H. J. Bixler and J. E. O'Connell Ind. Eng. Chem. Fundam., 10 (2), 185 (1971).
(13). G. J. H. Melrose, Rev. Pure and Appl. Chem., 21, 83 (1971).
(14). W. E. Hornby, M. D. Lilly and E. M. Crook, Biochem J., 107, 669 (1968).
(15). R. Goldman, O. Kedem and E. Katchalski, Biochem., 7, 4518 (1968).
(16). P. A. Srere, Science, 158, 936 (1967).
(17). C. Frieden and R. F. Colman, J. Biol. Chem., 242, 1705 (1967).
(18). A. Sols and R. Marco, in Current topics in cellular regulation (B. L. Horecker and E. R. Stadtman, eds), Vol. 2, Academic Press, New York, 1970, p. 227.
(19). P. D. J. Weitzman, Febs. Letters, 32 (2), 247 (1973).
(20). I. Smith, in Chromatographic and electrophoretic techniques, (I. Smith, ed.), Vol. II, Heinemann, London, (1960).
(21). P. Fasella and F. Rossi Fanelli, Acta vitaminol. et enzymol., 5-6, 159 (1971).
(22). P. Zaffaroni, V. Vitobello, F. Cecere, E. Giacomozzi and F. Morisi, Agr. Biol. Chem., in press.
(23). W. Marconi, F. Bartoli, F. Cecere and F. Morisi, Agr. Biol. Chem., in press.
(24). R. Gianna, F. Morisi, G. Prosperi and G. Spotorno, unpublished data.
(25). F. Morisi, M. Pastore and A. Viglia, J. of Dairy Science, 56, 1123 (1973).

(26). M. Pastore, F. Morisi and A. Viglia, J. of Dairy Science, in press.
(27). M. Pastore, F. Morisi, A. Viglia and D. Zaccardelli, presentation at 1° Int. Symp. on Insolubilized enzymes, Milan, Italy, 15-16 June 1973.
(28). C. Corno, G. Galli, F. Morisi, M. Bettonte and A. Stopponi, Die Stärke, 24, 420 (1972).
(29). S. Giovenco, F. Morisi and P. Pansolli, Febs Letters, 36 (1), 57 (1973).
(30). W. Marconi, S. Gulinelli and F. Morisi, Biotech. Bioeng., in press.
(31). W. Marconi, C. Corno, G. Galli and F. Morisi, unpublished work, 1969.
(32). W. B. Davis, Anal. Chem., 19, 476 (1947).
(33). W. Marconi, F. Cecere, F. Morisi, G. Della Penna and B. Rappuoli, J. of Antibiotics, 26 (4), 228 (1973).
(34). W. Kaufmann and K. Bauer, Naturwiss., 20, 474 (1960).
(35). W. Marconi, F. Bartoli, F. Cecere, G. Galli and F. Morisi, submitted for publication.
(36). W. Marconi, F. Morisi and R. Mosti, presentation at FEBS special meeting, Dublin, Ireland, 15-19 April 1973.
(37). W. Marconi, F. Bartoli, S. Gulinelli and F. Morisi, in press.
(38). M. Salmona, C. Saronio, I. Bartosek, F. Spreafico and E. Mussini, presentation at 1° Int. Symp. on Insolubilized enzymes, Milan, Italy, 15-16 June 1973.
(39). N. Dioguardi, private communication.

Chapter 6

IMMOBILIZED ENZYMES AND THEIR BIOMEDICAL APPLICATIONS

Thomas Ming Swi Chang, M. D., Ph. D., F.R.C.P. (C)

McGill University
Montreal, Quebec, Canada

I. INTRODUCTION

Immobilized enzymes have recently become an important part of enzyme engineering. This is due to the recognition that immobilized enzymes have much potential for basic research and for practical applications in many areas. Workers in this area (1) have classified the numerous methods available for immobilizing enzymes into four basic approaches (Figure 1): covalently bound (CVB); adsorbed (ADS); matrix-entrapped (ENT); and microencapsulated (MEC). Detailed reviews and monographs (2-22) have been written on the different approaches and modifications for preparing immobilized enzymes. This chapter starts with a brief description of each of the four basic approaches, pointing out features important in biomedical applications. This is followed by examples of biomedical applications of immobilized enzymes with emphasis on enzyme administration and artificial organs.

II. COVALENTLY BOUND (CVB) ENZYMES

One of the most widely studied basic approaches is the immobilization of enzymes by covalent bonds (CVB). This involves covalent linkage between enzymes and functional groups of polymers or other enzyme molecules. The first examples were described by Grubhofer and Schleith (23, 24) in 1953. Since then, a great deal of work has been carried out in this area. However, since this approach has been described in great detail elsewhere in this book and is well reviewed (2-22), only a brief summary will be made here in order to lead to the discussion on the biomedical aspects.

Enzymes can be covalently bound to water-insoluble synthetic polymers like polyacrylamide, polystyrene, maleic anhydride polymers, methacrylic acid polymers, and polypeptides. Enzymes can be bound to a water-insoluble natural carrier like agarose, cellulose, sephadex, glass, and starch. For example, enzymes have been covalently linked with the azide of carboxymethylcellulose (25).

Polyaminostyrene can be activated with nitrous acid or phosgene to be converted to the active derivative for binding enzymes (23, 24). Water-insoluble polypeptides in the form of co-polymers of p-aminophenylalanine can be used to form the carrier for enzymes (26). Aminoaryl-containing celluloses and s-triazinyl derivatives of cellulose have been used to immobilize enzymes (27). Aminoalkyl and aminoaryl derivatives of porous silica glass have been used to covalently attach enzymes (28,29).

Enzymes can be immobilized with nitrated co-polymers of methacrylic acid and methacrylic acid-m-fluoroanilide (30). Maleic anhydride co-polymers have been used (30). Cyanogen bromide-activated agarose and dextran is another commonly used approach in covalently binding enzymes. Other types of support include derivatives of acrylamide (28, 31), modified polyacrylamide gel beads (32), poliodal and copoliodal supports (8), dialdehyde starch-methyl-enedianiline resin (33), transition metal-activated cellulose, nylon or glass (34). Another approach is the activation of supports like sephadex and agarose by the Ugi reaction (35). Enzymes may also be covalently bound by co-polymerization with low molecular weight monomers (26, 36).

Another approach to covalent binding of enzymes is the intermolecular cross-linking of enzymes using multi-functional reagents (37, 38, 39). In the intermolecular cross-linking, there is also a certain degree of intra-molecular cross-linking. One method of intermolecular cross-linkage involves using a bi-functional reagent like glutaraldehyde (39, 40). Other bi-functional reagents include bisdiazobenzidine 2, 2'-disulfonic acid, trichloro-s-triazine, and others. There may be some loss of enzyme activity.

Enzyme crystals could also be treated with solutions of multi-functional reagents (41). The cross-linking occurs throughout the entire crystal. Microdroplets of enzyme and protein solution dispersed in an organic phase, could be cross-linked using a bi-functional agent, sebacoyl choride, in the organic phase (37). Depending on the reaction condition, complete or partial cross-linkage of the whole microdroplet is possible, especially for smaller microdroplets. The final product would then be resuspended into an aqueous phase.

Enzymes could also be treated with a bi-functional reagent like glutaraldehyde after microencapsulation (42). Enzymes could also be adsorbed first to the surface of a carrier and then intermolecularly cross-linked with multi-functional reagents like glutaraldehyde (43). Enzymes have also been cross-linked after they have been impregnated in porous membranes (44, 45).

Another interesting approach is to covalently bind enzymes to water-soluble macromolecules. For example, enzymes have been covalently bound to water-soluble dextrans, CM-cellulose (46), or to a dextran conjugate (47) to form soluble immobilized enzymes.

A. Features of Significance in Biomedical Applications

The following features of covalently bound enzymes have important bearings on their feasibility for biomedical applications. First, the enzyme-to-carrier ratio in this system is usually variable. For instance, in the case of enzymes covalently bound to insoluble carriers, it could range from as low as microgram of enzyme per gram of carrier, to as high as 3.5 gram of enzyme per gram of carrier (14). Thus, one could generally say that, in the case of enzymes covalently bound to insoluble carriers, the carrier content is quite appreciable. On the other hand, in the case of enzymes immobilized by intermolecular cross-linging, most of the system could be made to consist of the enzyme.

The soluble type of covalently bound enzymes and co-enzymes have potential advantages for certain biomedical applications. In the case of enzymes bound on insoluble carriers, the pH optimal can be the same, higher, or lower. Interesting basic research on the microenvironment of enzyme action has come out these studies (33, 48, 49). The variations in pH optimal may depend on the charge of the carrier (49). The change in pH optimal may also be due to chemical modifications (33), since the change in pH optimal could also take place in carrier matrice which are electrically neutral. In the case of intermolecular linkage, an interesting observation has been made on papain (50). Here, the cross-linked papain has a pH activity curve which increases with increasing pH to pH 9.5. It also has a relatively higher activity than the native enzyme down to pH 3. This will have important significance, since enzymes in the gastrointestinal tract pass through a region of acidic pH in the stomach to regions of very alkaline pH in the intestine.

One of the most important considerations of covalently bound enzymes in biomedical applications is their stability. Covalently bound enzymes have increased thermal and storage stability. However, this is not a general rule, since some have lower stability. Some of the covalently bound enzymes also have increased stability toward high concentration of urea, quanidine, and organic solvents. The covalently bound enzyme systems can come in direct contact with both small molecules and large molecules like protein, with steric factors favoring the smaller molecules to different extents. While this may be advantageous if the substrate is a macromolecule; it is a disadvantage if the immobilized enzymes are exposed to proteolytic enzymes, antibodies, or leucocytes. To circumvent these problems and to make use of their other advantages, covalently bound enzymes are used in combination with microencapsulation (3, 4, 42). This will be discussed later.

III. IMMOBILIZATION OF ENZYMES BY ADSORPTION

This type of immobilized enzyme is formed by the adsorption of enzymes in aqueous solution onto the surface of carriers. The various types of carriers used includes carbon, alumina, collodion, clay, glass, cellulose, collogens, and others (51, 52, 53). After adsorption, the activity retained is variable from negligible to extremely high. Another way is the adsorption of enzyme to ion-exchange resins by electrostatic interactions (54).

A. Features of Significance in Biomedical Applications

The pH optimal could be the same, increased, or decreased. The biggest advantages of this method are easy preparation and the different types of carriers that could be used and the fact that adsorption does not involve chemical reaction. The stability of adsorbed enzymes is increased significantly in many cases. However, the carrier-to-enzyme ratio is usually quite high, and the adsorbed enzyme is exposed to proteolytic enzymes, antibodies, and leucocytes.

One of the features of adsorbing enzyme on carriers is the reversibility factor. In some cases, the enzyme could be desorbed from the carrier if there is sufficient change in an environmental factor like pH, electrolyte concentration, or temperature. This feature could be a problem in therapeutic applications, but advantageous in other biomedical applications. For instance, this property has been used in the regeneration and reuse of support in enzyme columns (55), or for the purification of crude enzyme preparations (56). On the other hand, the relatively large amount of carrier, the reversible adsorption, and the exposure of the adsorbed enzymes to proteolytic enzymes, antibodies, and leucocytes, could be disadvantageous for therapeutic uses when they have to be introduced into the body. Here agin, microencapsulation of the adsorbed enzymes may prevent these problems while utilizing their advantages (3, 4).

IV. IMMOBILIZATION OF ENZYMES BY MATRIX-ENTRAPMENT

Immobilization of enzymes by matrix-entrapment (57) is the entrapment of enzyme molecules within the interstitial space of a network of cross-linked water insoluble polymer (Figure 1). This is molecular entrapment, in that each enzyme molecule is entrapped within a separate space in the polymer network.

The most common type of polymeric material used is the polyacrylamide gel system in which the enzymes are mixed with acrylamide and N, N'-methylenebisacrylamide. On polymerization, a gel is formed and the enzyme molecules are entrapped in the cross-linked network of the polymer. The formed gel is then broken down mechanically into small, solid particles.

The nature of the formation of this type of system is such that a large amount of polymer network is required to entrap a rather small amount of enzyme molecules. The enzyme-to-polymer ratio is as much as 1:100. The nature of the entrapment is also such that there is a solid network of polymer material through which the substrate must diffuse to reach the enzyme system. As a result, it was found that only the entrapped enzymes near the particle surface can act on substrates (58). Though the most popular polymer used is polyacrylamide (57), others like starch gels (59), silica gels and silicone rubber (60) can also be used.

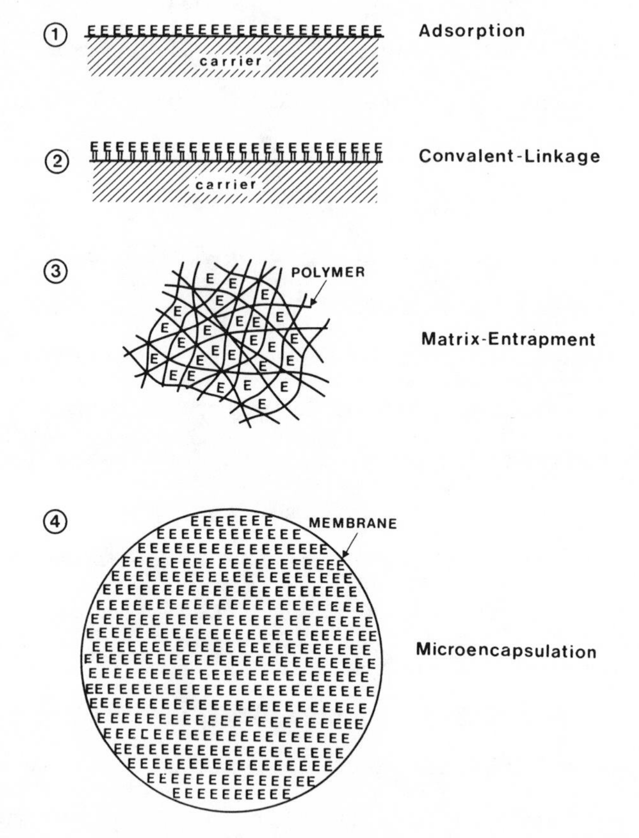

Fig. 1. Four basic approaches in the immobilization of enzymes (1) E-enzyme molecule.

A. Features of Significance in Biomedical Applications

Because of the large amount of polymer and the diffusion problem, the relative activity of the immobilized enzyme is usually low; i.e., about 10% of the enzyme activity in solution in most cases. The pH optimal could be displaced toward the alkaline or acidic pH. Report on the thermal stability and the storage stability is variable: cholinesterase (61) and urease (62) show improved stability, and asparaginase immobilized in this form (63) also shows some improvement in thermal and storage stability.

The gel-entrapped approach has important biomedical applications in a number of *in vitro* uses such as in columns, enzyme electrodes, or elsewhere. However, where the immobilized enzyme has to be introduced into the body, or where it must act quickly on body substrates, there are a number of problems. For instance, the extremely large polymer component in comparison to the enzyme component (100 parts polymer to 1 part enzyme) would result in introducing a large amount of exogenous polymers. In addition, the large amount of solid polymer greatly reduces the rate of diffusion of substrate into the immobilized enzyme.

Another problem is the presence of some leakage of enzyme from the cross-linked polymeric network. With more cross-linking of the polymers, the leakage could be reduced (63). However, this would greatly restrict the already slow entry of substrate into the solid particles. Enzymes entrapped this way are affected by proteolytic enzymes only to a negligible extent, when compared to enzyme in free solution (63). Enzyme entrapped in matrices are also protected from antibodies and leucocytes (57).

V. IMMOBILIZATION OF ENZYMES BY MICROENCAPSULATION (ARTIFICIAL CELLS)

In nature, most enzymes are located in an intracellular environment, either in solution or associated with membranes. As a result, they are stabilized and separated from the extracellular environment, and remain able to act on external and internal metabolites. Research in enzymes has passed through different stages (Table 1). Stage I involved the extraction, purification, crystallization, and kinetic studies of enzymes in dilute environments. Stage II involved binding enzymes to polymers either covalently, by adsorption, or by matrix-entrapment in model systems for studying enzymes attached to membranes. Stage III was the study of enzymes in an artificial intracellular environment. The preparation of these artificial cells was initiated by Chang in 1957 (64-67). Since then, a great deal of work has been done, and detailed review of this work is available in a monograph (4).

TABLE 1

Stages of Enzyme Research Development

Stage I.	Enzyme in dilute solution extracted from biological cells.
Stage II.	Enzyme in microenvironment - e.g. covalent linkage, adsorption, gel-entrapment - model for enzyme in biological membranes.
Stage III.	Enzyme in "intracellular environment" - e.g. microencapsulation (artificial cells) - model for enzymes in biological cells.

The first artificially prepared cells retained the same red blood cell content, but the red blood cell membrane was replaced with an ultrathin synthetic polymer membrane. In this initial approach, the preparation consisted of three main steps (Fig. 1) (64-66): 1) The aqueous phase containing the enzyme solution was emulsified in a water-immiscible organic solution to form aqueous microdroplets; 2) The addition of a polymer to the stirred emulsion resulted in the formation of a membrane around the surface of each microdroplet; 3) The artificial cells were then transferred from the organic solution into an aqueous solution. This general procedure has been modified and extended for preparing artificial cells with great variations in membrane material, contents, size, and configuration (Fig. 3) (65-67). The details have been updated and described (4).

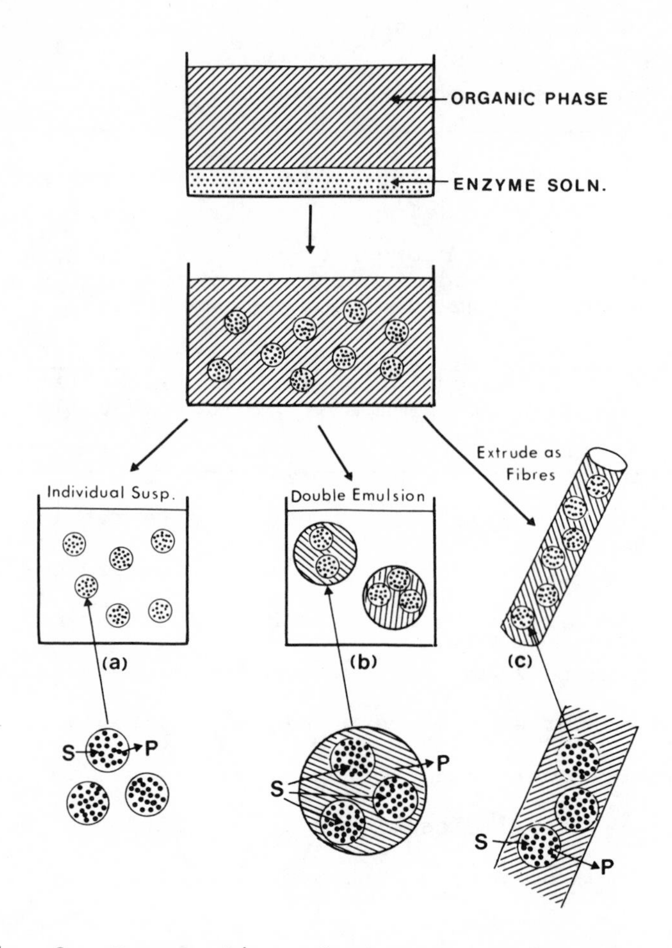

Fig. 2. Preparation of microencapsulated enzyme. (a) Individual microcapsules with spherical ultrathin membranes enclosing enzyme. (b) Double emulsion - solid spheres with microdroplets. (c) Fibre extrusion.

With the organic phase separation approach, cellulose nitrate is the most convenient way of preparing microencapsulated enzymes (4, 64-67). This is a physical process which does not inactivate enzymes. With this approach, many other types of polymers, like polystyrene and others, which can take part in organic phase separation, could also be used.

A second approach is to use interfacial polymerization in which a diamine or a polyamine dissolved in the aqueous microdroplets reacts at the interface with a dibasic acid in the organic phase to form a membrane (Fig. 2) (4, 65-67). Thought nylon is most commonly used, other membranes such as cross-linked protein (65-67), polyamines (68), polyurea (69), and others (4) could also be formed. However, interfacial polymerization is a chemical process which, unlike the organic phase separation procedure, may inactivate some enzyme systems.

A third approach is by secondary emulsion, in which a fine emulsion of enzyme solution is first dispersed in an organic phase of polymer solution; the organic phase containing the aqueous microdroplets is then emulsified into larger microdroplets. This way, each larger microdroplet of organic polymer solution contains a number of aqueous microdroplets. As the organic polymer solution is solidified to form solid spheres, the microdroplets of enzyme solution are retained within the solid spheres (Fig. 2). One such example is the use of silicone rubber to encapsulate enzyme solution (70). Other polymers such as cellulose (4) and other membranes (4, 71).

Another extension is the emulsification of the enzyme solution in the polymer organic solution as above, but instead of forming micro-spheres, the polymer solution is extruded in the form of fibers (Fig. 2). The fibers then contain microdroplets of enzyme solution (72). Another example of double emulsion is to emulsify enzyme microdroplets in a water-immiscible phase consisting of surfactant, additives, and a high-molecular-weight paraffin, and then emulsifying the organic phase. This way, a liquid-surfactant membrane is formed (73).

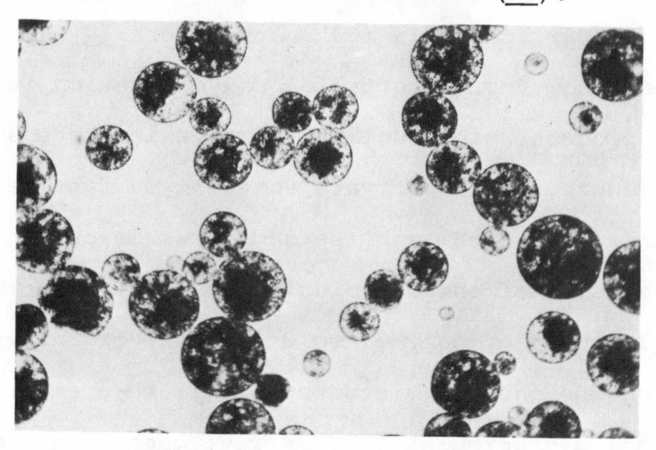

Fig. 3. Semipermeable microcapsules with ultrathin membrane containing enzyme. Mean diameter - about 100 micron.

Another method of preparing artificial cells is to use biological materials instead of synthetic polymers for forming the artificial cell membrane. The initial attempt is to use protein cross-linked by the interfacial polymerization approach (65-67).

An artificial cell membrane of lipids has also been studied. Mueller and Rudin (74) used a modification of the standard procedure (66) to prepare artificial cells with lipid membranes enveloping red blood cell content. A complex of lipids and proteins has also been formed to microencapsulate red blood cell hemolysate (75).

Liposomes are onionskin-like multiple laminar lipid crystals, used by Bangham (76) to study membrane transport mechanisms. Recently, this type of liposome was used to entrap enzymes, (77, 78). Unlike the microcapsule system, the enzymes in liposomes can only act after they are released from the liposomes. Even more recently, red blood cells have been used to entrap external enzymes by reverse hemolysis (79).

The contents of the artificial cells can be varied at will over a wide extent (Fig. 4) (4). Thus, single enzyme systems, multiple enzyme systems, or complex enzyme systems can be easily microencapsulated using the same procedure (4). All that is required is to dissolve or suspend the enzyme systems in the aqueous emulsion around which a membrane is then formed. Any of the immobilized enzyme systems covalently bound, adsorbed, matrix-entrapped, or microencapsulated, could be endorsed with artificial cells (3,4). Microencapsulated enzyme solution could also be subsequently treated by cross-linking agents like glutaraldehyde to form a high concentration of cross-linked enzymes within the microcapsules (42). Enzymes can also be microencapsulated with suitable adsorbent so that the product of the enzymatic reaction can be removed by the adsorbent (4, 70). A typical example is the combination of urease with ammonium adsorbent (4, 70). Whole cells could also be microencapsulated with all the cell content (4, 66, 67).

In summary, the artificial cell is a concept (4). The artificial cells prepared by the procedure described are examples that demonstrate this concept. Extension of the procedures described, or completely new procedures, could be used to further demonstrate the principle of artificial cells.

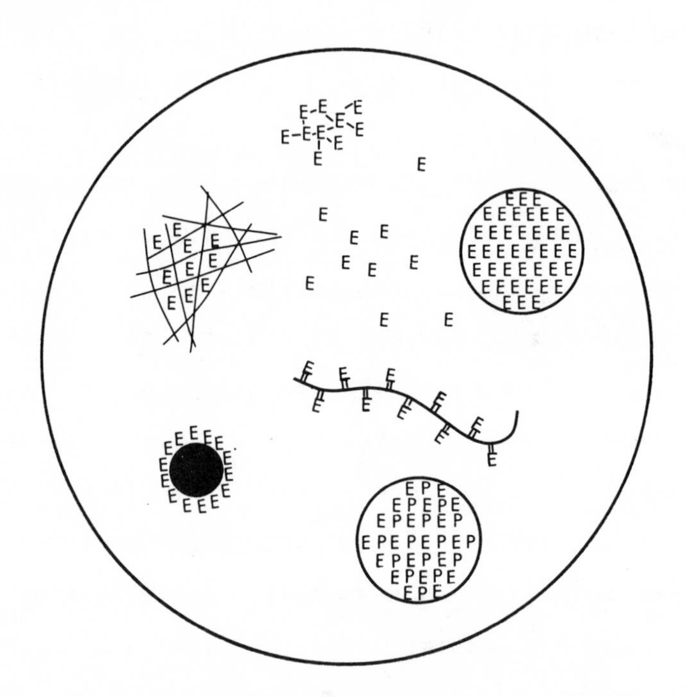

Fig. 4. Variation in content of microencapsulated enzyme - enzyme molecule (E) in solution or immobilized by various basic approaches enclosed within a single microcapsule.

A. Features of Biomedical Significance

The membranes of typical artificial cells (200 Å) are of the same order of thickness as biological cell membranes (4, 67). The membrane (i.e. nylon) can withstand a membrane tension of up to 2,520 dynes/cm (80), which is many times greater than that of biological cell membranes. The typical artificial cell membrane has a pore radius of about 16 Å (4, 81), which is comparable to that in dialysis membrane.

However, two important points differentiate semipermeable microcapsules from a standard dialysis membrane. Ten ml of 20-micron-diameter microcapsules has a total surface area of more than 2 m^2, which is more than twice the surface area present in a standard hemodialysis machine (4, 70). In addition, the membrane of the semipermeable microcapsule is at least 100 times thinner than the membrane of a standard dialyzer.

With these two factors in mind, the potential transport rate of 10 ml of 20-micron-diameter microcapsules is at least 200 times higher than that for a standard hemodialysis machine (Fig. 5), (4, 70, 82). As a result of this transport rate, the equilibration of permeant molecules is extremely fast. For example, the half-life ($t_{1/2}$) of equilibration is 5 seconds for urea and 35 seconds for larger molecules like sucrose (4, 81). This means that small molecules up to a molecular weight of 5000 can equilibrate rapidly across the membrane, whereas, macromolecules and cells cannot cross the membrane. As a result, microencapsulated enzymes do not leak out, but can act very efficiently on permeant substrates. This high permeability greatly differentiates microencapsulated enzymes from matrix-entrapped enzymes in solid spheres.

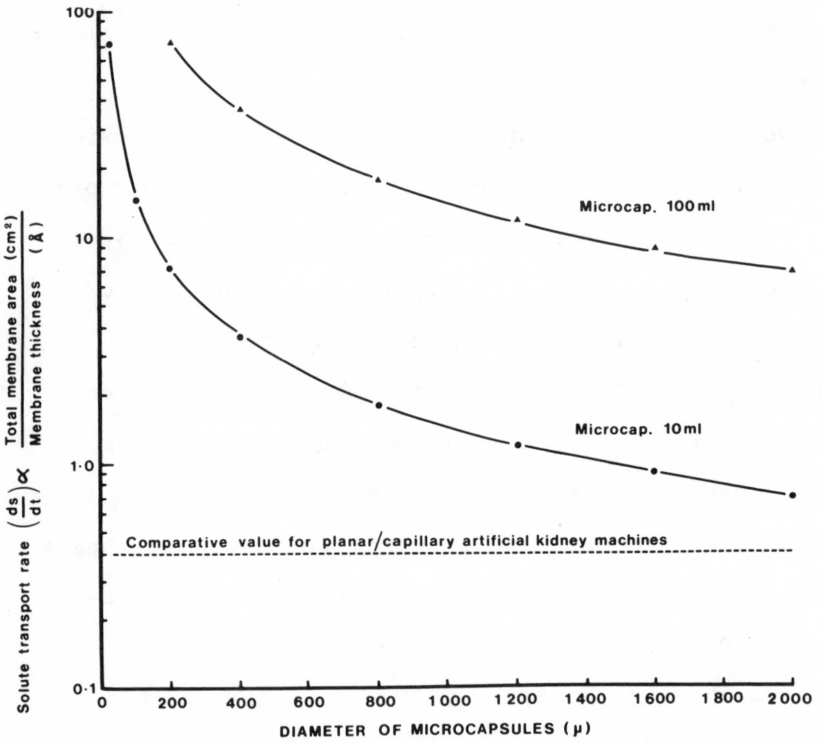

Fig. 5. Solute transport rate of microcapsules with ultrathin membranes compared to the solute transport rates of a complete standard artificial kidney machine.

The stability of microencapsulated enzymes can be greatly increased in two ways. One is to microencapsulate enzymes with a high concentration of other protein (e.g. 10 gm% hemoglobin or albumin) (4, 42, 66). Since the protein cannot leak out, it will remain at high concentration within the microcapsules, even if the microcapsules themselves are dispersed in very dilute suspension. Thus, unlike enzymes in a dilute solution, the microencapsulated enzyme, like biological cells, are always maintained in an intracellular environment with a high concentration of protein. When stored at $4^{\circ}C$, catalase in free solution lost 50% of its initial activity after 15 days. However, catalase microencapsulated with a high concentration of protein retained more than 90% of its initial activity for more than 100 days (42). In the same way, when stored at $4^{\circ}C$, dilute asparaginase solution lost 50% of its initial activity after 20 days; but when microencapsulated with a high concentration of protein, retained 100% of its activity after 100 days (83). Increased stability was also observed for urease and lactase (4).

Another way to enzyme stability is to microencapsulate enzymes which have already been immobilized by covalent binding, matrix-entrapment, or adsorption (3, 4). Yet another method is to microencapsulate the enzyme and then treat the enzyme-containing microcapsules with cross-linking agents, like glutaraldehyde, thereby cross-linking the enzymes with the microcapsules (42). Catalase, asparaginase, and lactase microencapsulated in this manner retained more than 90% of their activity after storage at 4°C for more than 100 days. Compared to enzyme in solution, the temperature stability of the microencapsulated enzyme system is also greatly increased.

A 100-micron-diameter nylon microcapsule contains a 0.02-micron-thick polymer membrane. This means that nearly the whole microcapsule consists of the enzyme system. There is, therefore, a negligible accumulation of polymer after *in vivo* introduction. The membrane material could be formed from bio-compatible or non-thrombogenic membranes (4, 84). This would allow the microencapsulated enzyme to come in contact with biological fluids without causing any adverse effect on the blood cells or protein fractions. The microencapsulated enzymes are no longer in contact with macromolecules (e.g., antibodies), nor with leucocytes, so immunological reaction problems are avoided (3, 4, 66). On the other hand, in certain cases where it is necessary for the enclosed enzymes to act on macromolecules, the membrane permeability can be modified to allow permeant macromolecules to enter (4).

Detailed analysis of the transport properties and enzyme kinetics of microencapsulated enzymes have been reported (4, 81-82, 85-90).

VI. ADMINISTRATION OF IMMOBILIZED ENZYMES

A. Introduction

Many of the basic functions of the human body are based on enzymatic reactions carried out in intracellular environments. Despite the basic importance of enzymes in body functions, enzyme therapy has not yet played an important role in the field of medicine. The reason for this is that, except for a few digestive enzymes that can be administered orally and one or two enzymes which might be useful for external wound debridement by local application, most enzymes would have to be administered into the body. This presents a number of problems (3, 4).

Enzymes are usually administered in the form of a solution, whereas in nature, they usually act in an intracellular environment and are therefore stabilized and protected. In addition, most enzymes now available in appreciable quantity are obtained from bacterial sources. This means that even if the enzyme is highly purified, it would still be treated by the human body as a foreign protein. As a result, their use may result in hypersensitivity reactions, and repeated injections might result in the production of sufficient antibodies rapidly remove and inactivate subsequently injected enzyme

In summary, the problems enzyme therapy using enzymes extracted from cells are: Instability and short duration of action in the body because the enzyme is rapidly removed as a foreign protein; possible allergic reaction; possible toxic contamination from the cell extract; and, development of immunological reaction and the production of antibodies to inactivate subsequently injected enzymes.

Can immobilized enzymes help solve these problems? The questions of stability and duration of action may be solved by any of the immobilized enzymes described, since many of the enzymes become much more stable when immobilized. However, not all immobilized enzymes can solve the problems of recognition as a foreign protein, the production of allergic reaction, the production of antibodies with repeated injection, and enzyme inactivation by antibodies. Thus, enzymes immobilized by covalent binding are not completely prevented from coming into direct contact with external antibodies or leucocytes. Enzymes prepared by molecular entrapment are prevented from coming into direct contact with the external environment. Unfortunately, the problem here is the large polymer-to-enzyme ratio. This large amount of polymer in the form of a solid sphere would greatly restrict diffusion and only the enzyme in the outer aspects of the solid gel particle can act efficiently (58). Thus, to have sufficient *in vivo* enzymatic action, a very large amount of matrix-entrapped enzyme would have to be injected.

Enzyme immobilization by microencapsulation may be the most appropriate approach for administering enzymes (3, 4, 66, 91), since it gives the enzyme an intracellular environment even after administration into the body. The enclosing membrane prevents the enzymes from coming in contact with external antibodies (leucocytes), and from leaking out to cause immunological and allergic toxic reaction. In addition, the polymer-to-enzyme ratio is such that a negligible amount of polymer is introduced into the body. The injection of 5 ml of 100-micron-diameter microcapsules carries with it polymer material (0.02 micron membrane thickness) which is only a minute fraction of the polymer suture material normally used in surgery. Since the ultrathin membrane of the microcapsule also permits rapid exchange of substrate and metabolites across the microcapsule membrane, small volumes of microcapsules will produce sufficient enzymatic action. The following experiment reports readily these points.

B. Basic Demonstration

The first experimental administration of immobilized enzymes was in the form of intravenously injected microcapsules containing red blood cells (4, 64, 66). The microcapsules were rapidly removed from the blood stream by the liver and spleen. The first reported successful demonstration of the ability of immobilized enzymes to act in the body was in 1963 using microencapsulated urease

(65-67, 91). Urease was used as a model system because its urea substrate is the major diffusable nitrogenous constitutent of the body fluid. In this first study, microencapsulated urease was injected intraperitoneally into dogs. The effect of the microencapsulated urease on the conversion of body urea into ammonia was followed by measuring the change in blood ammonia level. Only 0.25 ml/Kg of microencapsulated urease was injected into each dog. After injection, it was found that there was no leakage of urease from the microcapsules and that the microcapsules efficiently converted blood urea to blood ammonia. Long-term experiments showed that intraperitoneally injected microencapsulated urease had a half-life of three days. It is interesting that urease in dilute solution had a half-life of less than half a day, even when tested *in vitro*. Microencapsulated enzymes have also been used for extracorporeal hemoperfusion, whereby the microencapsulated enzyme is retained within an extracorporeal chamger (Fig. 6) (70) and comes in contact with blood flowing through the chamber. This way, the microencapsulated urease acts efficiently on blood-borne substrate without entering the body. These initial studies using microencapsulated urease led to further studies in experimental enzyme therapy.

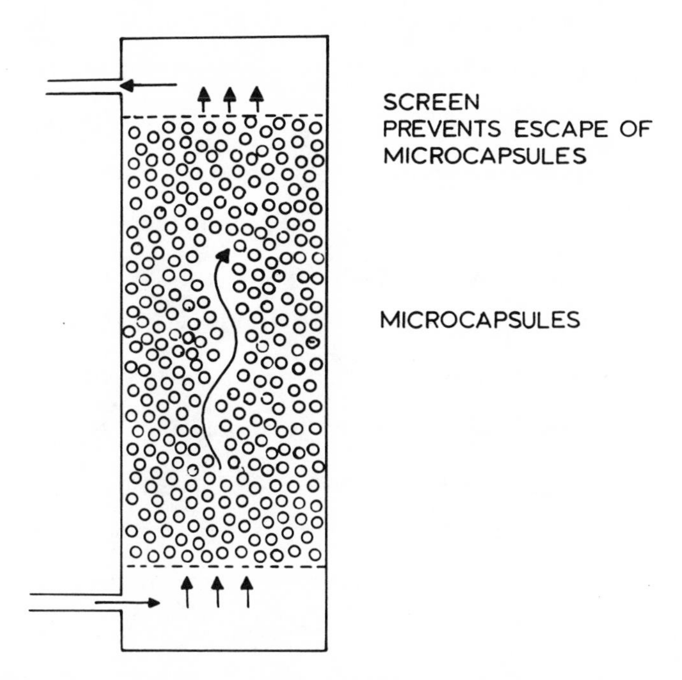

Fig. 6. Extracorporeal shunt chamber containing microcapsules for perfusion by body fluid.

C. Replacement For Hereditary Enzyme Defects

Hereditary enzyme defects may result in an accumulation of substrates to toxic level, or the deficiency of essential products. Although replacing the defective enzymes is an obvious theoretical solution to the problem, this is not done because of the earlier mentioned problems of enzyme administration. However, the advancements described in the preceding section were followed by experiments to determine the feasibility of using immobilized enzyme to correct inborn errors of metabolism (92-94).

Acatalasemia, a hereditary enzyme-deficiency disease, is also present in a strain of C_s^b mice (95). Catalase solution injected into these mice is removed rapidly from the body, and also produces immunological reaction and antibodies (95). The microencapsulated red blood cells hemolysate prepared here (64-67) also contained active catalase. Additional crystalline catalase was added to the hemolysate and was enclosed within the microcapsules. These were used for experimental enzyme replacement therapy in acatalasemic mice (92).

The initial study used four groups of mice - one group of controlled normal mice, and three groups of acatalasemic mice. Injection of sodium perborate into the normal mice produced no adverse affect, and within 20 minutes most of the injected perborate was removed by the catalase present in the normal animals. Injecting the same amount of perborate into the acatalasemic mice resulted in methemoglobin formation and respiratory distress, and large amounts of injected perborate were still present after 20 minutes. Animals protected with the same assayed amount of catalase solution or catalase microcapsules were apparently protected from perborate and the amount of perborate in the body after 20 minutes was significantly reduced (Fig. 7). This led to further study (94) which demonstrated that intraperitoneally injected catalase solution quickly entered the circulating blood and was then rapidly removed. Microencapsulated catalase injected intraperitoneally remained in the peritoneal cavity, where it continued to act efficiently on perborate in the body fluid. The microencapsulated catalase exhibited a much longer active life after injection than an identical amount of enzyme in solution. Furthermore, studies showed that repeated injection of catalase in solution resulted in higher antibody titers. Repeated injection of microencapsulated catalase, on the other hand, did not result in the production of antibodies.

In many cases, injecting catalase solution into immobilized acatalasemic mice resulted in anaphylactic shock and death. Furthermore, the injected catalase solution was cleared much more rapidly and was much less effective. Injecting microencapsulated catalase into animals immunized with catalase produced no anaphylactic shock, but continued to efficiently protect the animal from perborate, (94).

These studies demonstrated that some enzyme administration problems may be solved by using microencapsulated enzymes. This approach avoids immunological reaction, and in animals which have developed antibodies to the enzyme, microencapsulated enzyme still can protect the enzyme from the antibodies already present in the body.

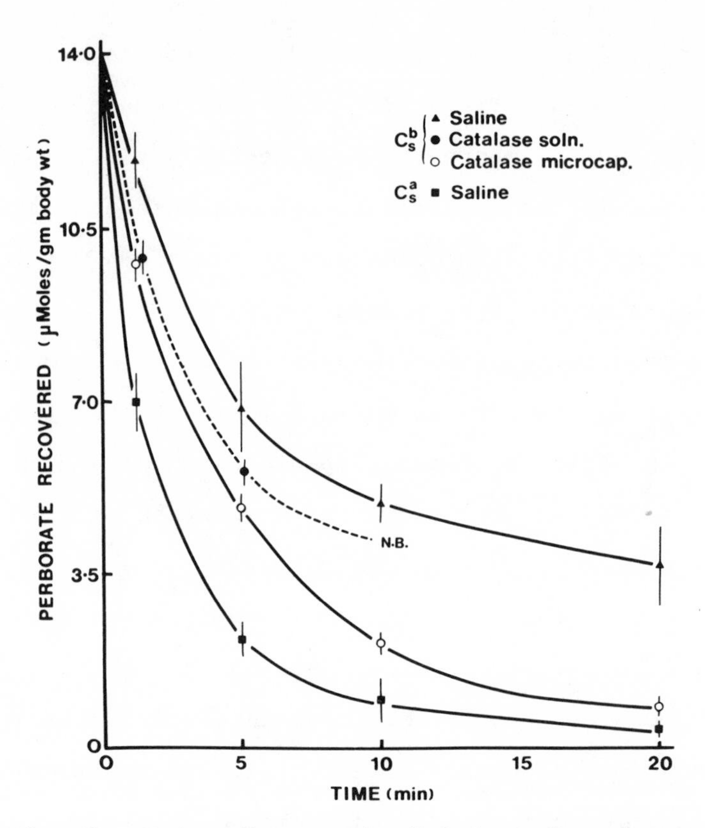

Fig. 7. Rate of removal of injected perborate in normal (C_s^a) mice and in immunized acatalasemic mice (C_s^b) (N.B. - death of animals from anaphylactic shock when injected with catalase solution).,

In the previous experiments, the microencapsulated enzymes acted only on a substrate present in the body fluid. However, there are other enzyme-deficiency conditions in which the accumulated metabolite can also be removed from body fluid. And, there are a number of other enzyme-deficiency conditions in which the accumulation of substrate is intracellular, and the counteracting enzyme must be introduced into certain types of cells.

Having demonstrated the use of immobilized enzymes to correct enzyme-deficiency diseases, further developmental work was needed to learn how microencapsulated or modifications of microencapsulated enzymes might be introduced within cells. For instance, it was found that properly microencapsulated enzyme with suitable surface properties could be removed selectively by the reticuloendothelial system (4). Other studies have also been carried out using the liposome system (77, 78). In these studies, entrapped enzymes deficient in certain lipid or glycogen storage diseases were injected intravenously and it was found that the liposomes could be located intracellularly in the liver. Autoradiographic examination of the liver showed that hepatocytes are involved in the uptake of liposomes. Most of the liposome radioactivity was recovered in the mitochondria-lysosomal fraction. Further analysis showed that the entrapped enzymes were located mostly in the lysosome. These studies suggest that further analysis is needed to determine the feasibility of using liposomes to treat enzyme deficiency diseases involving an intracellular accumulation of substrates, especially in the lysosome fraction. In other studies, red blood cells have been used to microencapsulate the deficient enzyme (79) by rapid hemolysis of the cells in the presence of these enzymes. Further study is being carried out (79) to see whether this type of immobilized enzyme may be used for enzyme replacement.

D. Suppression of Substrate-Dependent Tumor

Basic studies by Broome (96) and others demonstrated that L-asparaginase can suppress the growth of certain asparagine-dependent tumors. Sufficient L-asparaginase for clinical trial could only be obtained from bacterial sources. However, parentally injected bacterial L-asparaginase is removed rapidly as a foreign protein. Furthermore, antibodies produced as a result of repeated injections caused even more rapid removal and inactivation of injected L-asparaginase, making this treatment ineffective.

The first animal studies using immobilized asparaginase to suppress asparagine-dependent tumors was reported in 1969 (97). In these studies, the asparaginase was immobilized by microencapsulation and then injected intraperitoneally into C3H/HEJ mice (97, 98). Identically assayed amounts of enzyme in solution and in microencapsulated form were injected intraperitoneally. The results showed that, although the enzyme solution significantly suppressed the growth of 6C3HED lymphoscoma, the same assayed amount of microencapsulated asparaginase suppressed the growth of the tumor significantly longer (Fig. 8). Further study (83) demonstrated that this is due to the increased stability of the microencapsulated asparaginase as compared to the enzyme solution. After intraperitoneally injecting asparaginase solution, the plasma asparagine level was lowered to zero, maintained this level for only three days, and rose to normal in four days. With the same assayed amount of microencapsulated asparaginase the plasma asparagine level was maintained at zero for seven days, and required another four days to return to normal.

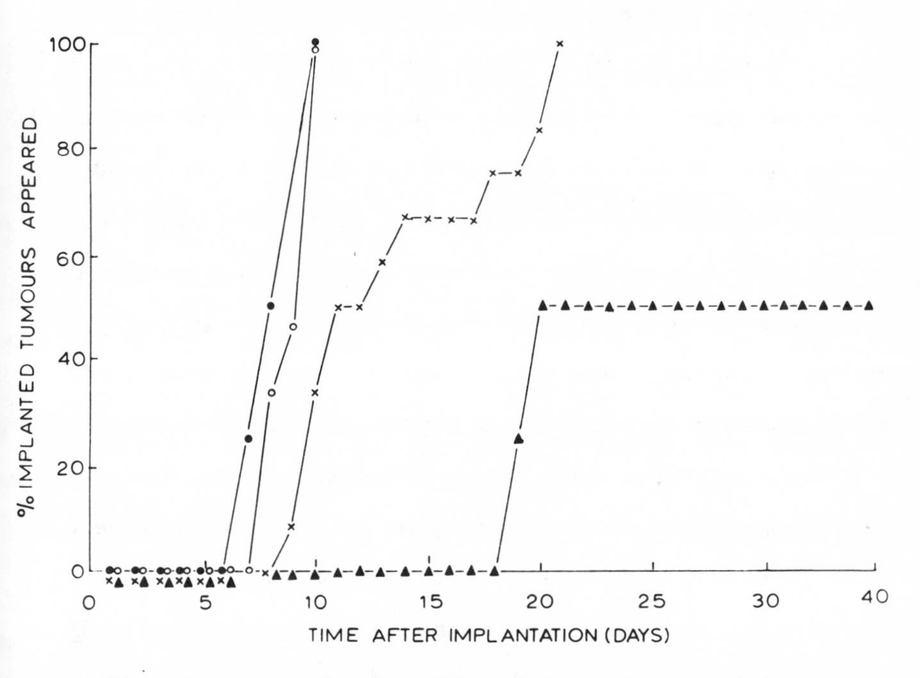

Fig. 8. Appearance of lymphosarcoma in mice receiving injection of control saline, control microcapsule, X asparaginase solution, and asparaginase microcapsules (4).

In vitro studies showed that microencapsulated asparaginase can retain its activity even longer when cross-linked with glutaraldehyde (83). Further studies (99) demonstrated that asparaginase in solution, injected intraperitoneally, rapidly entered the blood circulation stream from which it was rapidly removed. Microencapsulated asparaginase on the other hand, remained active in the peritoneal cavity, and continued to act on body asparagine.

In the same studies (99), it was also found that microencapsulated asparaginase was more effective in suppressing established lymphosarcoma. More recently, the preparation and *in vitro* studies of microencapsulated asparaginase have also been carried out by other groups (69, 100). Using other approaches, Hasselberger et al (102), tested immobilized asparaginase *in vitro* with a view toward eventual animal testing. Updike (63, 101) entrapped asparaginase within solid gel particles by the gel entrapment technique and injected these solid particles intraperitoneally. Intraperitoneal injections into rats lowered the plasma asparagine level to nearly zero and held this level for at least two days, but the ability to suppress lymphosarcoma was not studied. Weetall (103) covalently linked asparaginase to dacron vascular prothesis and *in vivo* stability tests showed that about 50% of its activity still remained after seven days. He also studied the covalent binding of asparaginase to inorganic carriers such as glass or methycrytate (104).

This type of covalent binding of asparaginase was tested in animals (105, 106). Blood was allowed to perfuse over the covalently bound asparaginase. This approach successfully removed asparagine from body fluid, although there was a slight leakage of asparaginase into the blood (106). Here, the enzyme is also exposed to circulating leucocytes and any antibodies which might be present. In another study (107), asparaginase was mixed with collagen, cast as a membrane, and then treated with glutaraldehyde. The membrane is then wound to form a modular reactor, and tested *in vitro*.

E. Artificial Kidney, Detoxification, Liver Support

The feasibility of using artificial cells to construct a compact artificial kidney was apparent in 1964 (66) and demonstrated in 1966 (70). Thirty-three ml of 100-micron-diameter artificial cells with a membrane thickness of 0.02 micron has a total surface area of 2 m^2. In the standard artificial kidney, the membrane thickness is at least five microns and the total surface area is 1 m^2. If 33 ml of 100-micron-diameter artificial cells are put in a shunt, then their total transport rate would be at 200 times faster than that of the standard artificial kidney (70, 82) (Fig. 5).

Study was conducted (70) in which 10 ml of 90-micron-diameter artificial cells were placed in an extracorporeal shunt chamber in such a way that blood circulating through the shunt chamber directly contacted the artificial cells retained within the shunt by screens (Fig. 6). In dogs with an average weight of 25 Kg, it was found that within 90 minutes of hemoperfusion across the shunt chamber of microencapsulated urease, the blood urea level of 18 mg% fell to 8 mg%. As was also demonstrated in the same study, ammonia can be removed by microencapsulated ammonia adsorbent (70). This feasibility was further supported by analysis from other laboratories (108).

Another possible use for microencapsulated urease was with ammonia adsorbent for oral administration (92, 109). These experiments showed that the microencapsulated urease removed enough urea to significantly lower the urea level in animals. Oral administration potential was supported by the work of other groups (110, 111). The possible use of immobilized enzymes for removing urea has also been modified by Gordan's group (112) for removing urea in the dialysate compartment of the standard artificial kidney.

The use of microencapsulated urease for treating chronic renal failure is still being developed and is not yet ready for clinical use. The other component of the artificial cells containing activated charcoal has already been used here to treat patients with chronic renal failure, acute intoxication, and liver failure

(113-119). In these applications, albumin has been immobilized onto the surface of the microencapsulated activated charcoal to make the surface blood compatible (Fig. 9) (113).

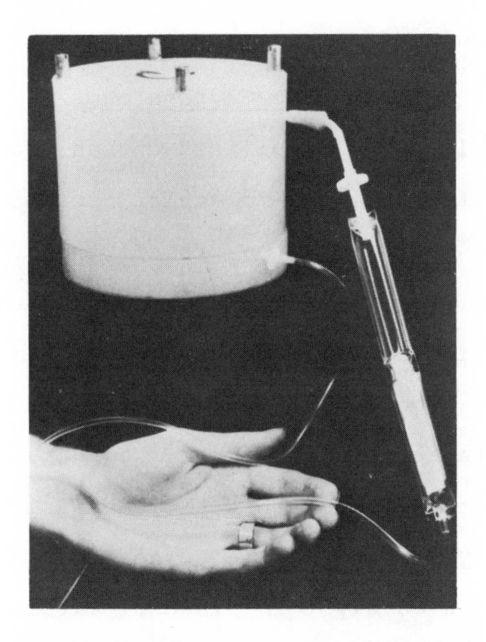

Fig. 9. Compact microcapsule artificial kidney prepared from microencapsulated activated charcoal. Surface of microcapsules made blood compatible by immobilization with albumin.

F. Red Cell Substitutes

The original reason for immobilizing red blood cell enzymes and hemoglobin in 1957 (64) was to study their potential use as red blood cell substitutes (4, 64, 66). Microencapsulated red blood cell hemolysate retained its ability to combine reversibly with oxygen and to catalyse the hydration of carbon dioxide, thus functioning like red blood cells. However, after intravenous infusion, they are removed rapidly from the blood stream circulation. Once the problem of survival in circulation can be solved, microencapsulated hemoglobin and carbonic anhydrase may hold promise as red blood cell substitutes.

G. Membrane Oxygenator

Broun et al (120) immobilized carbonic anhydrase in sheets of silastic by inter-molecular cross-linking with glutaraldehyde. This way, the transport of CO_2 across the membrane is 1.5 times higher when compared with unmodified silastic sheets. Updike's group immobilized catalase inside hemodialysis membrane. This way, the immobilized catalase can convert hydrogen peroxide into oxygen. However, he found that the catalase he used was not sufficiently stable, and he turned to using transitional metal as a catalyst in place of catalase. His results using transitional metal as catalyst appear to be very promising (121).

H. Substitutes For Islet Cell Insulin Secretion

If a small implantable system that can detect small changes in the blood glucose level is available, it could be connected to an automatic insulin infusion system and, through feedback mechanisms, it might be possible to control the glucose level in diabetic patients. Updike and Hicks (122) made use of immobilized enzymes to prepare a glucose sensor. They immobilized glucose oxidase in acrylamide gel by molecular entrapment. The gel containing the immobilized enzyme is coated as a thin film over a Clark polarographic oxygen electrode. As glucose and dissolved oxygen from tissue fluid diffuses into the immobilized enzyme film, glucose is oxidized by the glucose oxidase resulting in a decrease of dissolved oxygen, which is monitored by the oxygen electrode. This way, the changes in glucose concentration in the tissue fluid can be monitored.

This system works well in _in vitro_ situations, however, there are problems in using these for _in vivo_ operations, especially for long term use. These involve the instability of the glucose oxidase enzyme, even when immobilized in the gel (62). There are also problems involved in the drift and instability of the Clark type electrode in prolonged use. Furthermore, there would be problems of biocompatibility of the electrode after implantation, since formation of fibrous capsule around the electrode may alter the sensitivity of the electrode.

Wingard's group (123) carried out studies to immobilize glucose oxidase in a polyacrylamide gel and found enzyme prepared this way to have increased stability. William et al (124) demonstrated the feasibility of using an electrode containing immobilized glucose oxidase for blood glucose analysis by measuring the changes in current of the enzyme electrode compared to a reference electrode.

Bessman and Schultz (125) prepared an implantable glucose sensor consisting of a membrane-covered oxygen-sensitive electrolytic cell covered with a cloth matrix in which glucose oxidase is covalently bound and cross-linked. Depending on the level of glucose in the body, the electrode could detect different amounts of oxygen diffusing into the cell. Changes in current output are compared to an enzyme-free control cell. They found the deivce to be highly sensitive to glucose in the physiological range of 50 to 150 mg/100 ml, and standard curves obtained *in vitro* in flowing aqueous buffer solution and in flowing blood are nearly identical (125). Another approach is to use other forms of catalysts; e.g., transitional metals can act as catalysts for the oxygenation of glucose (126). These sensors have been implanted into the tissue of Rhesus monkeys and experiments showed correlation between the blood glucose level and sensor current changes (126).

I. Fuel Cells For Cardiac Pacemaker

It is feasible to use immobilized enzymes to generate electricity for fuel cells, like those in cardiac pacemakers for example. Drake (127) studied the use of immobilized glucose oxidase as a fuel cell, using covalently bound or embedded enzyme. The best result obtained was only 0.5 mA per square cm with about 0.5 volt polarisation in a solution of 0.5 M glucose. Wingard et al (123, 128-130) extended this approach further. Most recently, they tested a polyacrylamide gel matrix, a glutaraldehyde-enzyme matrix, and a charged glutaraldehyde-enzyme matrix for immobilizing glucose oxidase on the surface of a platinum electrode (130). They found that at a glucose concentration of 0.0055 M, currents in the range of 5 to 50 mA were observed, depending on the type of immobilization. A glutaraldehyde enzyme-charged matrix produced the highest current. Further studies are being carried out. Another group has studied implantable fuel cells for powering a miniature implantable pacemaker (131). They entrapped hyaluronidase into silicone rubber foam. The silicone rubber foam containing the enzyme is then placed near the membrane covering the anode side of the fuel cell. *In vitro* studies showed that some power could be produced from hyaluronic acid. However, further study is required before practical use of this approach.

J. Use of Immobilized Protein or Enzymes for Blood-Compatible Surfaces

Albumin has been immobilized on the surface of polymers by adsorption or cross-linking to form a blood-compatible surface (113, 132). This way the adverse effect of foreign surfaces on platelets of perfusing blood is markedly decreased. As a result, microencapsulated activated charcoal coated with albumin has been used successfully for treating patients with chronic renal fialure, acute intoxication, and liver failure (Fig. 9) (<u>113-119</u>).

Another approach using immobilized protein or enzymes is as immobilized proteolytic or fibrinolytic agents to prevent thrombus formation (<u>133, 134</u>). In the study on urokinase, the enzyme was immobilized by simple adsorption. It has also been immobilized by adsorption followed by cross-linking with glutaraldehyde. *In vitro* studies showed that urokinase immobilized by adsorption and cross-linking retained significant thrombolytic activity. *In vivo* studies in dogs demonstrate that the implanted surface in contact with blood retained its thrombolytic activity (<u>134</u>). Hoffman's group (<u>135</u>) also studied the covalent binding of albumin and streptokinase to radiation-grafted hydrogels on polymer surfaces.

K. Other Biomedical Applications

The example discussed so far demonstrate the more applied aspects of the biomedical applications of immobilized enzymes. They have also been used in basic biomedical research, especially for studying transport in enzyme membrane models which have been discussed earlier in relationship to the microenvironment of immobilized enzymes (<u>33, 48, 49</u>) and oxygenators (<u>120</u>). Additional references are also available (<u>136-138</u>). One large area of biomedical application is in analytical chemistry

(139-140). Another area which has important biomedical implications is the immobilization of co-enzymes and co-enzyme analogs (15). These and other related areas, including immobilized antigen and antibodies, solid-phase radioimmunoassay (RIA), and affinity chromatography, are presented in other chapters of this book.

REFERENCES

(1). P. V. Sundarum, E. K. Pye, T. M. S. Chang, V. H. Edwards, A. E. Humphrey, N. O. Kaplan, E. Katchalski, Y. Levin, M. D. Lilly, G. Manecke, K. Mosbach, J. Porath, H. Weetall, and L. B. Wingard, Jr., Biotech. and Bioeng. 14, 15, 1972.
(2). E. Brown and A. Racois, Bull. Soc. Chim. Fr., No. 12, 4613, 1971.
(3). T. M. S. Chang, Science Tools, 16, 33, 1969.
(4). T. M. S. Chang, Artificial Cells, Monograph, Charles C. Thomas, Publisher, Springfield, Illinois 1972.
(5). I. Chibata and T. Tosa, Kagaku to Seibutsu, 7, 147, 1969.
(6). E. M. Crook, Metabl. Regul. Enzyme, Action, Fed. Bur. Biochem. Soc. Meeting 6th, Sols. A., Ed., Academic Press, London, 297, 1970.
(7). R. D. Falb and G. A. Grode, Fed. Proc. 30, 1688, 1971.
(8). R. Goldman, L. Goldstein, and E. Katchalski, Biochemical Aspects of Reactions on Solid Surfaces, Stark, G. R. Ed., Academic Press, New York, 1, 1971.
(9). L. Goldstein, Methods, Enzymol., Vol. 19, Perlman, G. E. and Lorand, L., Eds., Academic Press, New York 935, 1970.
(10). J. Gryszkiewiez, Folia Biol. (Warsaw), 19, 119, 1971.
(11). G. Kaye, Process Biochem. 3, 36, 1968.
(12). M. Lilly, C. Money, W. Hornby, and E. M. Crook, Biochem. J., 95, 54, 1965.
(13). M. D. Lilly, S. P. O'Neill, and P. Dunnill, Biochem., 55, 985, 1973.
(14). G. Manecke, Biochem. J., 107, 2, 1968.
(15). G. J. H. Melrose, Rev. Pure and Appl. Chem. 21, 83, 1971.
(16). K. Mosbach, Sci. Am., 224, 26, 1971.
(17). J. Porath, R. Axen, and S. Renback, Nature 215, 1491, 1967.

(18). W. R. Bieth and K. Venkatasubramanian, Chemtech, 1973.
(19). H. H. Weetall, Res./Devel., 22, 18, 1971.
(20). M. K. Weibel and H. J. Bright, Biochem. J., 124, 801, 1971.
(21). L. B. Wingard, Jr. Enzyme Engineering, John Wiley & Son, N. Y. 1972.
(22). O. R. Zaborsky, Immobilized Enzymes, Monograph, CRC Press, Cleveland, Ohio, 1973.
(23). N. Grubhofer and L. Schleith, Naturwissenschaften. 40, 508, 1953.
(24). N. Grubhofer and L. Scheitch, Hoppe-Seyler's Z., Physiol. Chem., 297
(25). M. A. Mitz and L. J. Summaria, Nature, 189, 576, 1961.
(26). A. Bar-Eli and E. Katchalski, Nature 188, 856, 1960.
(27). G. Kay and E. M. Crook, Nature, 216, 514, 1967.
(28). R. Epton and T. H. Thomas, Aldeichimica Acta, 4, 61, 1971.
(29). H. H. Weetall, Science, 166, 615, 1969.
(30). Y. Levin, M. Pecht, L. Goldstein, and E. Katchalski, Biochemistry, 3, 1905, 1964.
(31). S. A. Barker, P. J. Somers, R. Epton, and J. V. LcLaren, Carbohyde, Res. 14, 287, 1970.
(32). J. K. Inman and H. M. Dintzis, Biochemistry, 8, 4074, 1969.
(33). L. Goldstein, M. Pecht, S. Blumberg, D. Atlas, and Y. Levin, Biochemistry, 9, 2322, 1970.
(34). S. A. Barker, A. N. Emery, and J. M. Novais, Process Biochem., 5, 11, 1971.
(35). R. Axen, P. Vretblad, and J. Porath, Acta Chem. Scand., 25, 1129, 1971.
(36). M. A. Stahmann and R. R. Becker, J. Am. Chem. Soc., 74, 2695, 1952.
(37). T. M. S. Chang, Science, 146, 524, 1964.
(38). F. Wold, Methods Enzymol. Vol. 11, Hirs, C. H. W., Ed., Academic Press, New York, 1967, 617.
(39). S. Avrameas, Immunochemistry, 6, 43, 1969.
(40). E. F. Jansen and A. C. Olson, Arch.Biochem. Biophys., 129, 221, 1969.
(41). F. A. Quiocho and F. M. Richards, Proc. Natl. Acad. Sci. USA, 52, 833, 1964.
(42). T. M. S.Chang, Biochem. Biophys. Res. Commun., 44, 1531, 1971.
(43). R. Haynes and K. A. Walsh, Biochem. Biophys. Res. Commun., 36, 235, 1969.
(44). R. Goldman, O. Kedem, I. H. Silman, S. R. Caplan, and E. Katchalski Science, 150, 758, 1965.
(45). G. Broun, E. Selegny, S. Avrameas, and D. Thomas, Biochim. Biophys. Acta, 185, 258, 1969.

(46). J. R. Wykes, P. Dunnill, and M. D. Lilly, Biochim. Biophys. Acta, 250, 522, 1971.
(47). S. P. O'Neill, J. R. Wykes, P. Dunnill, and M. D. Lilly, Biotech. Bioeng., 13, 319, 1971.
(48). E. Katchalski, I. Silman, and R. Goldman, Adv. Enzymol. Relat. Areas Mol. Biol., 34, 445, 1971.
(49). L. Goldstein, Biochem., 11, 22, 1972.
(50). R. Goldman, O. Kedem, I. H. Silman, S. R. Caplan, and E. Katchalski, Biochemistry, 7, 486, 1968.
(51). A. D. McLaren, J. Phys. Chem., 58, 129, 1954.
(52). A. D. McLaren, G. H. Peterson, and I. Barshad, Soil Sci. Soc. Am. Proc. 22, 239, 1958.
(53). J. P. Hummel and B. S. Anderson, Arch. Biochem. Biophys., 112, 443, 1965.
(54). T. Tosa, T. Mori, N. Fuse, and I. Chibata, Enzymologia, 31, 214, 1966.
(55). T. Tosa, T. Mori, N. Fuse, and I. Chibata, Agr. Biol. Chem., 33, 1047, 1969.
(56). A. Y. Nikolaev and S. R. Mardashev, Biokhimiya, 26, 641, 1961.
(57). P. Bernfeld and J. Wan, Science, 142, 678, 1963.
(58). P. Bernfeld, R. E. Bieber, and P. C. MacDonnell, Arch. Biochem. Biophys., 127, 779, 1968.
(59). T. Tosa, T. Mori, and I. Chibata, Hakko Kogaku Zasshi, 49, 522, 1971.
(60). H. D. Brown, A. B. Patel, and S. K. Chattopadhyay, J. Biomed. Mater. Res., 2, 231, 1968.
(61). E. K. Bauman, L. H. Goodson, G. G. Guilbault, and D. N. Kramer, Anal. Cheml, 37, 1378, 1965.
(62). G. P. Hicks and S. J. Updike, Anal. Chem., 38, 726, 1966.
(63). H. L. Nadler and S. J. Updike, Enzyme (In Press).
(64). T. M. S. Chang, Report of Research Project for B.Sc. Honours Physiology, McGill University, Montreal, 1957.
(65). T. M. S. Chang, F. C. MacIntosh, and S. G. Mason, Proc. Canad. Fed. Biol. Sci., 6, 16, 1963.
(66). T. M. S. Chang, Science, 146, 524, 1964.
(67). T. M. S. Chang, F. C. MacIntosh, and S. G. Mason, Canad. J. Physiol. Pharmacol., 44, 115, 1966.
(68). M. Shiba, S. Tomioka, M. Koishi, and T. Kondo, Chem. Pharm. Bull, 18, 803, 1970.
(69). T. Mori, T. Tosa, and I. Chibata, Biochim. Biophys. Acta, 321, 653, 1973.
(70). T. M. S. Chang, Trans. Amer. Soc. Artif. Intern. Organs, 12, 13, 1966.
(71). M. Kitajima, S. Miyano, and A. Kondo, J. Chem. Soc. Japan, Ind. Chem. Soc., 72, 493, 1969.
(72). SNAM Progetti S.p.A., Ger. 1,932,426, Jan. 2, 1970.
(73). S. W.May and N. N. Li, Biochem. Biophys. Res. Commun., 47, 5, 1972.

(74). P. Mueller, and D. O. Rudin, J. Theor. Biol, 18, 222, 1968.
(75). T. M. S. Chang, Fed. Proc., 28, 461, 1969.
(76). A. D. Bangham, M. M. Standish, and J. A. Watkin, J. Molec. Biol., 13, 238, 1965.
(77). G. Gregoriadis, and B. E. Ryman, Eur. J. Biochem., 24, 485, 1972.
(78). G. Gregoriadis and B. E. Ryman, Biochem. J., 129, 123, 1972.
(79). G. M. Ihler, R. H. Glew, and F. W. Schnure, Proc. Nat. Acad. Sci., 70, 2663, 1973.
(80). A. W. L. Jay and K. S. Sivertz, J. Biomed. Mat. Res. 3, 577, 1968.
(81). T. M. S.Chang and M. J. Poznansky, J. Biomed. Mat. Res. 2, 187, 1968.
(82). T. M. S. Chang, in Recent Developments in Separation Science, Ed. N. Li, Chemical Rubber Company Press, Cleveland, Ohio 203, 1972.
(83). T. M. S. Chang, Enzyme J. 14, 95, 1973.
(84). T. M. S. Chang, L. J. Johnson, and O.Ransome, Canad. J. Physiol. Pharmacol., 45, 70, 1967.
(85). A. O. Mogensen and W. R. Vieth, Biotech. Bioeng. 15, 467, 1973.
(86). P. V. Sundaram, Biochim. Biophy. Acta, 293, 1, 1973.
(87). R. C. Boguslaski and A. M. Janik, Biochim. Biophys. Acta, 250, 260, 1971.
(88). P. S. K. Choi and L. T. Fan, J. Appl. Chem. Biotechnol. 23, 531, 1973.
(89). A. O. Mogensen and W. R. Vieth, Biotech. Bioeng., 15, 467, 1973.
(90). J. C. W. Ostergaard and S. C. Martiny, Biotech. Bioeng. 15, 561, 1973.
(91). T. M. S. Chang, Biomed. Eng. J., 8, 334, 1973.
(92). T. M. S. Chang and M. J. Poznansky, Nature, 218, 243, 1968.
(93). T. M. S. Chang, J. Dent. Res. 2 Supp., 319, 1972.
(94). M. J. Poznansky and T. M. S. Chang, Bioch. et. Bioph. Acta. 334, 103, 1974.
(95). R. N. Feinstein, J. T. Braun, and J. B. Howard, J. Lab. Clin. Med., 68, 952, 1966.
(96). J. D. Broome, Nature, 191, 1114, 1961.
(97). T. M. S. Chang, Proc. Canad. Fed. Biol. Sci., 12, 62, 1969.
(98). T. M. S. Chang, Nature, 229, 117, 1971.
(99). E. Siu Chong and T. M. S. Chang, (In Press), Enzyme J.
(100). T. Mori, T. Sato, Y. Matui, T. Tosa, and I. Chibata, Biotech. Bioeng., 14, 663, 1972.
(101). S. Updike, C. Prieve, and J. Magnuson, Birth Defects, Original Article Series, 9, 77, 1973.

(102). F. X. Hasselberger, H. D. Brown, S. K. Chattopadhyay, A. D. Mather, R. O. Stasiw, A. B. Patel, and S. N. Pennington, Res., 30, 2736, 1970.
(103). H. H. Weetall, J. Biomed. Mater. Res., 4, 597, 1970.
(104). H. H. Weetall, Research/Develop. 22, 18, 1971.
(105). H. Hyden, Arzneim. Forsch., 21, 1671, 1971.
(106). D. Sampson, L. S. Hersh, D. Cooney, and G. P. Murphy, Trans. Amer. Soc. Artif. Int. Organs, 18, 54, 1972.
(107). K. Venkatasubramanian, W. R. Vieth, and F. R. Bernath, Enzyme Engineering, (in press), 1973.
(108). S. N. Levine and W. C. LaCourse, J. Biomed. Mat. Res., 1, 275, 1967.
(109). T. M. S. Chang and S. K. Loa, Physiologist, 13, 70, 1970.
(110). R. E. Sparks, O. Lindan, N. S. Mason, M. H. Litt, and P. M. Meier, Trans. Amer. Soc. Artif. Intern. Organs, 17, 229, 1971.
(111). D. L. Gardner, D. R. Falb, B. C. Kim, and D. C. Emmerling, Trans. Amer. Soc. Artif. Intern. Organs, 17, 239, 1971.
(112). A. Gordon, M. A. Greenbaum, L. B. Marantz, M. J. McArthur, and M. H. Maxwell, Trans. Amer. Soc. Artif. Intern. Organs, 15, 347, 1969.
(113). T. M. S. Chang, Cand. J. Physiol. Pharmacol., 47, 1043, 1969.
(114). T. M. S. Chang and N. Malave, Trans. Amer. Soc. Artif. Intern. Organs, 16, 141, 1970.
(115). T. M. S. Chang, A. Gonda, J. H. Dirks, and N. Malave, Trans. Amer. Soc. Artif. Intern Organs, 17, 246, 1971.
(116). T. M. S. Chang, A. Gonda, J. H. Dirks, J. F. Coffey, and T. Lee Burns, Trans. Amer. Soc. Artif. Int. Organs, 18, 465, 1972.
(117). T. M. S. Chang. Lancet, ii, 1371, 1972.
(118). T. M. S. Chang and M. Migchelsen, Trans. Amer. Soc. Artif. Int. Organs, 19, 314, 1973.
(119). T. M. S. Chang, J. F. Coffey, P. Barre, A. Gonda, J. H. Dirks, M. Levy and C. Lister, Canad. Med. Assoc. J., 108, 429, 1973.
(120). G. Broun, E. Selegny, C. T. Minh, and D. Thomas, FEBS Lett. 7, 223, 1970.
(121). S. J. Updike, M. C. N. Shults, and J. Magnuson, J. Appl. Phys., 34, 271, 1973.
(122). S. J. Updike and G. P. Hicks, Nature 214, 986, 1967.
(123). L. B. Wingard, C. C. Liu, and N. L. Nagda, Biotech. Bioeng. 13, 629, 1971.
(124). D. L. Williams, A. R. Doig, Jr., and A. Korosi, Anal. Chem., 42, 118, 1970.

(125). S. P. Bessman and R. D. Schultz, Trans. Amer. Soc. Artif. Int. Organs, 19, 361, 1973.
(126). K. W. Chang, S. Aisenberg, J. S. Solldner, and J. M. Hiebert, Trans. Amer. Soc. Artif. Int. Organs, 19, 352, 1973.
(127). R. F. Drake, U. S. Government Report, PB 177695, 1968.
(128). L. B. Wingard, and C. C. Liu, 8th ICMBE, 1969.
(129). L. B. Wingard and C. C. Liu, Proc. Intl. Symp. on Instrumentation for Heart Transplantation and Substitution, Prague, Czech., 99, 1969.
(130). E-J. J. Lahode, C. C. Liu, and L. B. Wingard, Proc. 7th Intersoc. Energy Conversion Conf., San Diego, Calif., Sept. 25-29, 1972.
(131). L. A. Doham, S. J. Yao, S. K. Wolfson, Jr., Trans. Amer. Soc. Artif. Int. Organs, 17, 411, 1971.
(132). D. J. Lyman, K. G. Klein, J. J. Brash, B. K. Gritzinger, and J. D. Andrade, Thromb. Diath, Haem, 42, 109, 1970.
(133). B. Kusserow, R. Larrow and J. Nichols, Trans. Amer. Soc. Artif. Int. Organs, 17, 1, 1971.
(134). B. K. Kusserow, R. W. Larrow, and J. E. Nichols, Trans. Amer. Soc. Artif. Int. Organs, 19, 8, 1973.
(135). A. S. Hoffman, G. Schmer, C. Harris, and W. G. Kraft, Trans. Amer. Soc. Artif. Int. Organs, 18, 10, 1972.
(136). D. Thomas and G. Broun, Biochim., 55, 975, 1973.
(137). S. R. Caplan, Biochim., 55, 967, 1973.
(138). R. Goldman, Biochim., 55, 953, 1973.
(139). G. G. Guilbault, J. Inter. Un. Pure and Appl. Chem., 25, 727, 1971.
(140). G. G. Guilbault, CRC Crit. Rev. Anal. Chem., 1, 377, 1970.

Chapter 7

ENZYME ELECTRODE PROBES

Dr. George G. Guilbault

Louisiana State University
New Orleans, Louisiana

I. INTRODUCTION

A. General

Undoubtedly, one of the most exciting new fields of potentiometry is the enzyme electrode. The past decade has witnessed the advent of almost 50 new "ion-selective electrodes" -- electrodes that work well for all types of cations and anions, but only when inorganic species are involved. Ion-selective electrodes for organics have, in effect, been totally non-selective, irregular in response, and not useful. For this reason, none are available commercially, with the exception of the acetylcholine electrode sold by Corning.

However, by combining an ion-selective electrode with an enzyme, that biological catalyst which acts specifically and sensitively with **almost all organic** and inorganic compounds, one obtains an enzyme electrode which can assay organic compounds.

The principle here is quite simple: Choose an enzyme which acts specifically or highly selectively with the compound to be assayed. This enzyme is immobilized or insolubilized, and is then mounted on a conventional ion-selective electrode which measures either a product of the reaction (e.g., NH_3 from urea or acid from penicillin) or a reactant (e.g., O_2 used in the oxidation of glucose or uric acid). As the substance to be assayed diffuses into the enzyme layer it produces products or consumes a reactant, and this is then measured by the ion-selective electrode. The potential produced is a log function of the concentration of the substance assayed. Thus, it is no longer necessary to prepare reagents, make a standard curve, effect a reaction and measure. A single, self-contained electrode placed into a solution allows the concentration to be read directly, with no difficulty and requiring only a buffer as the reagent. Furthermore, the same electrode can be used for several hundred to a thousand assays, decreasing the cost-per-test to fractions of a cent.

This chapter introduces the novice reader to the enzyme electrode concept, if a novice, and can make experts of readers with some experience in this area. He will be taught how to make and use an enzyme electrode, what he can expect from such an electrode, and what types of electrodes have been described.

Since the enzyme electrode is a union of an ion-selective electrode and an enzyme, these elements are each discussed in detail so the reader can recognize what can be expected from the total -- the enzyme electrode.

B. Ion-Selective Electrodes

One of the most commonly made measurements in the chemical laboratory is that of pH. A glass hydrogen ion-selective electrode, a reference electrode, and a pH meter combine to form an extremely useful analytical tool. The advantages of this measuring system, often taken for granted, are speed, sensitivity, cost and reliability. In addition, the sample is not destroyed or consumed in the process. The same advantages apply to other ion-selective electrodes which have become available in the past few years. Now, ion-selective electrodes sensitive to particular cations and anions can be purchased or constructed at moderate cost. The analytically useful range of these sensors is generally from 10^{-1} M to 10^{-5} M, although several sensors are useful at much lower concentrations. Since the response of ion-selective electrodes is logarithmic, the precision of measurements is constant over their dynamic range.

Potentiometry with ion-selective electrodes is finding wide application in chemical studies. Particularly useful applications are in the fields of clinical and environmental chemistry, where the large number of samples and the need for rapid analysis rule out many slower, more involved methods.

The hydrogen-selective or pH electrode is the best-known ion-selective electrode. Its discovery is credited to Cremer (1) and Haber and Klemensiewicz (2) who found that certain glasses respond to hydrogen ion activity. The response of the glass electrode was commonly believed to be the result of hydrogen ions migrating through the thin glass membrane. The studies carried out by Karreman and Eisenman (3) and the work of Nicolsky et al. (4) provided the insight necessary for developing new ion-selective electrodes. Presently, electrodes selective for many ions are available (Table 1). As newer methods of preparing ion-selective electrodes (I-SE) have been developed, several kits for preparating different electrodes using a common body or housing have been introduced. With the exception of the single-crystal F electrode and the glass electrodes, all electrodes can be easily prepared with a minimum of time and effort.

The field of ion-selective electrodes was reviewed by Pungor (5) and Rechnitz (6). More recent developments and applications of ion-selective electrodes may be found in the report of the Symposium held by the National Bureau of Standards at Gaithersburg, Maryland (7) or in the article by Thomas (8).

An ion-selective electrode may be defined as a device that develops an electrical potential proportional to the logarithm of the activity of an ion in solution. The term "specific" which is sometimes used to describe an electrode, indicates that the electrode responds only to a particular type of ion. Since no electrode is truly "specific" for one ion, the term ion-selective is more appropriate.

TABLE 1

List of Commercially Available Electrodes

Cations		Anions		Gases
Ammonium	G,S,L	Bromide	S	CO_2
Hydrogen	G	Chloride	S,L	NH_3
Cadmium	S	Cyanide	S	NO_x
Calcium	L	Fluoride	S	SO_2
Copper	S	Fluoroborate	L	H_2S
Divalent Ions	L	Hydroxide	G	HCN
Lead	S	Iodide	S	HF
Monovalent	G	Nitrate	L	O_2
Potassium	G,S,L	Perchlorate	L	
Silver	G,S	Sulfide	S	
Sodium	G,L	Thiocyanate	S	
Rubidium	L			

G = Glass; S= Solid; L-Liquid-Membranes. Gas Electrodes all available from Orion (Cambridge, Mass.) and CO_2 from Radiometer (Copenhagen, Denmark) or Instrumentation Labs (Mass.). Other electrodes are available from many companies such as Orion, Corning (Corning, N.Y.), Beckman (Fullerton, Calif.), or Radiometer.

The response of an ion-selective electrode to an ion, A, of activity α_A and charge z_A, is given by the Nernst equation:

$$E = \text{constant} \pm \frac{2.303\ RT}{zF} \log \left[\alpha_A + k_{A,B}\, \alpha_B\right]^{z_A/z_B}, \qquad (1)$$

in which E is the measured potential, R is the gas constant and is equal to 0.314 joules deg^{-1} $mole^{-1}$, T is the absolute temperature in ^{o}Kelvin, F is the Faraday constant equal to 96,487 coulombs $equiv^{-1}$, $k_{A,B}$ is the selectivity coefficient, B is any interfering ion of charge z_B and activity α_B. The sign of the second term on the right-hand side of Eq. 1 is positive for cations and negative for anions.

The selectivity coefficient is a numerical description of the preferential response of an ion-selective electrode to the major ion, A, in the presence of the interfering ion B. The lower the numerical value of $k_{A,B}$ for a particular ion-selective electrode, the greater the concentration of B that can be tolerated before it will cause measurement errors. Values of $k_{A,B}$ can be calculated from:

$$\pm \frac{E_2 - E_1}{2.303\ RT/zF} = \log k_{AB} + \frac{z_A}{z_B} - 1\ (\log\ \alpha_A), \qquad (2)$$

in which E_1 and E_2 are the potential measurements of separated solutions of the principle ion and interfering ion, respectively, at the same activity. Or, more realistically, k_{AB} may be obtained by making measurements of the potential of an ion-selective electrode in solutions of constant interferent activity, α_B, and changing primary ion activity, α_A. Then, k_{AB} may be determined by:

$$\alpha_A = k_{A,B}(\alpha_B)^{z_A/z_B} . \qquad (3)$$

The value of $_A$ is taken at the point where serious deviation from Nernstian response is noted (8).

Ion-selective electrodes can be divided into several categories according to the composition of their sensor membranes (Table 2).

Glass electrodes are available which are sensitive to H^+ (pH electrodes) and to cations in the order $Ag^+ > H^+ > K^+ > NH_4^+ \gg Na^+$, Li^+, Ca^{++}, Mg^{++} over a concentration range 10^{-1} to 10^{-5} M.

Solid-state electrodes are ion-selective electrodes in which the sensor is a thin layer of a single or mixed crystal or precipitate which is an ion conductor. There are two classes of solid-state electrodes: homogeneous and heterogeneous. Homogeneous or crystal membrane electrodes are electrodes in which the membrane is a pellet prepared from a precipitate, mixture of precipitates, or a single crystal. In the heterogeneous electrodes, a precipitate or mixture of precipitates is dispersed in an inert supporting matrix such as silicone rubber of polyvinyl chloride (PVC).

Table 2

Classification of Ion-Selective Electrodes

A. Solid Membranes. May be homogeneous or heterogeneous.

1. Homogeneous Membranes.

a. Glass Electrodes are ion-selective electrodes in which the sensing membrane is a thin piece of glass. The chemical composition of the glass determines the selectivity of the membrane. This group includes:

pH electrodes;

cation-selective electrodes.

b. Crystal Membrane Electrodes are ion-selective electrodes in which the membrane is a crystalline material prepared from either a single compound or a homogeneous mixture of compounds; e.g., F^- or halide electrodes.

2. Heterogeneous Membranes. These are formed when an active substance is mixed with an inert matrix such as silicone rubber to form the sensing membrane; e.g., halide electrodes.

B. Liquid Membranes. May comprise positively or negatively charged ions or neutral species.

Table 2 (continued)

1. Charged Species.

   a. Positively Charged -- Bulky cations (e.g. those of quaternary ammonium salts or salts of transition metal complexes such as derivatives of 1,10-phenantholine) dissolved in a suitable organic solvent and held in an inert support provide membranes sensitive to changes in the activity of anions; e.g., NO_3^- or ClO_4^- electrode.

   b. Negatively Charged -- Complexing agents (e.g., of the type $(RO)_2PO_2^-$) dissolved in a suitable organic solvent and held in an inert support provide membranes sensitive to changes in activity of cations; e.g., Ca^{++} electrode.

2. Neutral Species. These electrodes are based on solutions of molecular carriers of cations (e.g., antibiotics, macrocyclic compounds, or other sequestering agents) which can be used in membrane preparations that show sensitivity and selectivity to certain cations, e.g., NH_4^+ or K^+ electrode.

Table 2 (continued)

C. Gas Electrodes. These sensors use a gas-permeable and an ion-permeable membrane to separate the sample solution from a thin intermediate solution film that is held between the membranes. This intermediate solution interacts with the gaseous species to produce a change in the intermediate solution's ionic level. This change which is sensed by the ion-selective electrode is directly proportional to the partial pressure of the gaseous species in the sample. Such electrodes have been constructed for CO_2, NH_3, NO_x, SO_2, HCN and H_2S.

Liquid membrane electrodes, which can be either liquid or "solid", are prepared by dissolving an organic exchanger in an appropriate solvent. The solution is held in an inert matrix. Such exchangers used in preparing these electrodes may be a ligand association complex such as those formed by the transition metals with derivatives of 1,10-phenanthroline, quarternary ammonium salts, organic-phosphate complexes and antibiotics. In some cases, the exchanger and solvent are entrapped in an inert polymer matrix such as PVC or polymethyl methacrylate. They can also be coated on a platinum wire or graphite rod ("solid" liquid membrane electrodes).

There are special electrodes which employ coatings over the membranes of ion-selective electrodes. If the coating is a gas-permeable membrane, the electrode is sensitive to CO_2 or NH_3. The gas diffuses through the membrane and alters the pH of an internal filling solution. This pH change is measured with a glass electrode and is proportional to the concentration of gas entering the membrane. Gas electrodes are available for a number of substances as shown in Table 1.

When using an ion-selective electrode to make potentiometric measurements of the activity of a given ion in solution, it is important to remember that the device is affected by the activity of the ion. Therefore, species which may complex the ion of interest and lower its activity must be removed or masked. It is often necessary to use a buffer solution to control ionic strength and pH, and to prevent changes in the activity of the ion being measured by oxidation reduction or complexation. This is also true for enzyme electrodes (work on the glass-based enzyme electrodes is discussed later in this chapter).

There are numerous reports of applications and progress in the design and manufacture of ion-selective electrodes. The use of PVC in preparating ion-selective electrodes has been an important development. Electrodes manufactured with PVC cost much less, have essentially the same response characteristics and better selectivities, and can usually be used longer than previous electrode assemblies. Moody, Oke and Thomas constructed a calcium-selective electrode using a liquid ion exchanger incorporated in a PVC matrix (9). The optimum concentration of calcium exchanger used in the preparation was described by Griffiths, Moody and Thomas (10). The electrode constructed in this manner gave a near-Nernstian response (30 mV per pCa unit) over the range 2.6×10^{-2} to 6.0×10^{-5} M in a $CaCl_2$ solution. Davies, Moody and Thomas prepared nitrate ion-selective electrodes by incorporating commercially available liquid ion exchangers in PVC (11). These electrodes overcome the leakage problem associated with other liquid ion exchange assemblies. Furthermore, these "solid" liquid membrane electrodes have greater selectivity for the ion of interest than the corresponding "liquid" electrode.

Pick, Toth, Pungor, Vasak and Simon used valinomycin in a variety of neutral carriers to prepare ion-selective electrodes for potassium (12). The response time of this type of electrode is usually less than three seconds, and the useful range of the electrodes is 10^{-1} to 10^{-5} M. The electrode prepared from this material is exceptional in that there is very little or no drift in potential over a three-day period. As the working characteristics of electrodes are improved and new electrodes are introduced, many new applications of these devices can be expected. It is the author's opinion that all second generation "liquid" electrodes will be "solid", and the classical "liquid" electrode will disappear from the scientific scene. Any of these electrodes can be prepared by simply following the method of Moody and Thomas (9,11).

The family of electrodes prepared from silver sulfide or silver sulfide mixed with halogen salts of silver are well characterized (5,6). The precipitates are used either in pellet form or mixed in an inert supporting matrix such as silicone rubber. When silver sulfide is used alone, the electrode responds to sulfide and silver ions over the concentration range 10^0 to 10^{-7} M.

When a particular silver halide is mixed with silver sulfide to form a sensor membrane, the membrane behaves as though it were composed of the halide salt alone. Electrodes for iodide, bromide and chloride have been prepared in this manner. The iodide electrode also responds to cyanide. An electrode for thiocyanate can be prepared by mixing silver thiocyanate with silver sulfide to form a sensor membrane. The selectivity coefficients of the halide and pseudo-halide electrodes can be estimated by,

$$k_{A,B} = \frac{\text{Solubility product of AgA}}{\text{Solubility product of AgB}} . \qquad (4)$$

Another group of electrodes is prepared by mixing copper or cadmium or lead sulfide with silver sulfide. The electrodes constructed from these mixed sulfides are sensitive to Cu^{++}, Cd^{++} and Pb^{++}, respectively, over the concentration range 10^{0} to 10^{-7} M. Silver and mercury (II) are serious interferences. High levels of ferric ion also interfere, but this can be overcome by adding fluoride to complex the Fe(III).

Any of these solid electrodes can be prepared by pressing a thin wafer of freshly precipitated salt (e.g., Ag_2S for the sulfide electrode) using high pressure in an infra-red pellet press. The wafer is then sealed into a glass tube of appropriate diameter using epoxy cement. The electrode is completed by adding 0.1 M electrolyte and an inner reference electrode (e.g., Ag/AgCl).

Cation-selective electrodes for H^{+} and monovalent ions (e.g., Ag^{+}, Na^{+}, K^{+}, NH_4^{+}) are available from glass electrode companies such as Corning, Beckman or Orion. These electrodes are rugged, have response times of a few seconds and linear ranges of 10^{0} - 10^{-5} M.

The text edited by Eisenman (13) and the chapter of the N.B.S. publication written by Khuri (14) provide more information on glass electrodes. For use with enzymes, it is recommended that a flat combination (glass-calomel) electrode be used. Such electrodes are available from Radiometer (Copenhagen) and Ingold (Zurich, Switzerland).

Antibiotics and similar compounds have been used successfully to prepare cation-selective electrodes. The calcium ion-selective electrode developed by Ammann, Pretsch and Simon reportedly has superior selectivity for calcium over sodium and magnesium (15). Pioda, Wachter, Dohner and Simon have studied the properties of the antibiotics nonactin for use as a sensor membrane (16). An electrode prepared from this material in an inert matrix is sensitive to NH_4^+ with linear near-Nernstian response of 51 mV per decade change over the concentration range 10^{-1} to 10^{-4} M NH_4^+ (17). This electrode responds poorly to Li^+ ion activity, and the selectivity for NH_4^+ over K^+ and Na^+ was reported to be superior to glass electrodes sensitive to ammonium ion. The nonactin electrode can be used in an enzyme system, and will be discussed later. Valinomycin-based potassium ion-selective electrodes have been studied (18).

Gas electrodes, prepared by placing a gas-permeable membrane over housing containing a pH electrode (or Pt electrode in the case of O_2) and internal filling solution, are available for many species that either participate in or are produced from enzyme reactions, such as O_2, CO_2, and NH_3. The CO_2 electrodes from Instrumentation Laboratories, Radiometer and Orion, and the NH_3 electrode from Orion work well, but have longer response times than desirable, due to the presence of the gas membrane. If this membrane could be omitted, a much improved gas electrode -- and hence, enzyme electrode -- would result. This can be done very easily in the case of O_2 by simply using a flat Pt electrode (Beckman) polarized at a potential of -0.6 V vs. S.C.E. This electrode is covered with immobilized enzyme and works very well (19,20). For measurements of CO_2 and NH_3, it is recommended that the new air-gap electrode developed by Ruzicka and Hansen (21) be used. In this electrode the hydrophobic gas permeable membrane generally employed in gas electrodes is replaced by an air gap which separates the electrolyte layer from the sample solution. By avoiding the membrane construction and by utilizing a very thin, hydrophilic layer of electrolyte at the surface of the indicator electrode, a very high response speed is obtained. Furthermore, the lifetime of the sensor is substantially increased because the electrode does not come into direct physical contact with the sample solutions. Thus there are no interferences from surfactants, particulate matter, or organic solvents. Because the electrolyte layer can be easily renewed or even changed according to the requirements of a particular analysis, the same electrode can be used to measure a variety of gases. For a CO_2 electrode, for example, a layer of 0.01 M $NaHCO_3$+ surfactant (Victawet 12, Stauffer Chem. Co.) is placed on the glass electrode; and for NH_3, a layer of 0.01 M NH_4Cl + surfactant is used. The gas liberated from solution diffuses to the surface and interacts with the surface layer, effecting a change in pH that is measured:

$$NH_4^+ \rightleftharpoons NH_3 + H^+. \quad (5)$$

Finally, in addition to these ion-selective electrodes, another large class of electrodes -- the inert metal electrodes such as Pt or Au -- could be used to follow enzyme reactions involving electroactive products or reactants. These electrodes have been well characterized in the literature, and are available from companies such as Beckman Instrument Co.

In summary, many commercially available (or homemade) electrodes have been used and can be used to construct enzyme electrodes.

C. The Enzyme as Reagent

The second half of the enzyme electrode team is the enzyme -- that biological catalyst ever present in nature which enables the many complex chemical reactions, upon which depends the very existence of life as we know it, to take place at ordinary temperatures.

Enzymes work within complex living systems, and one of their outstanding properties is specificity. An enzyme is capable of catalyzing a particular reaction of a particular substrate regardless of the presence of other isomers of that substrate or similar substrates.

An example of the specificity of enzymes with respect to a particular substrate is provided by luciferase, which catalyzes the oxidation of luciferin to oxyluciferin (22). A rather complete study of many compounds similar in structure to luciferin, showed that the catalytic oxidation which produces the green luminescence occurs only with luciferine (Eq. 6). Substitution of an amino group for a hydroxyl group, or addition of another hydroxyl group to the luciferin molecule alters the enzymic action so that no green luminescence is produced. Another example of the specificity of enzymes is glucose oxidase, which catalyzes the oxidation of β-D-glucose to gluconic acid. A rather complete study of about 60 oxidizable sugars and their derivatives showed that only 2-deoxy-D-glucose is catalyzed at a rate comparable to that of β-D-glucose. The anomer a-D-glucose is oxidized catalytically less than one percent as rapidly as the β-anomer (23). Urease, which catalyzes the hydrolysis of urea, is even more specific.

$$\text{Luciferin} + O_2 \xrightarrow[\text{Mg}^{2+}\ \text{ATP}]{\text{LUCIFERASE}} \text{OXYLUCIFERIN (GREEN LUMINESCENCE)} \qquad (6)$$

Enzymes exhibit specificity with respect to particular reactions. For example, in attempting to determine glucose by oxidation in an uncatalyzed way, heating a solution of glucose and an oxidizing agent like ceric perchlorate, would result in uncontrollable side reactions that would yield other products in addition to gluconic acid. With glucose oxidase, however, catalysis at room temperature and a neutral pH is so effective that the rates of the other thermodynamically possible reactions are negligible.

This specificity of enzymes and their ability to catalyze reactions of substrates at low concentrations is of particular use in chemical analysis. Enzyme-catalyzed reactions have long been used analytically to determine substrates, activators, and inhibitors, and also enzymes themselves. Until recently, however, disadvantages associated with using enzymes have seriously limited their application. Frequently cited objections to using enzymes for analyticaly purposes have been their unavailability, instability, poor precision, and the labor involved in performing analyses. While these objections were valid in the past, numerous enzymes are now available in purified form, with high specific activity, and at reasonable prices, thus removing many of these objections.

More than 200 purified enzymes are now available from companies such as Boehringer-Mannheim (Mannheim,

Germany and New York City), Calbiochem (Los Angeles), Nutritionaly Biochemicals (Cleveland), Pierce (Rockford, Illinois), Sigma (St. Louis) and Worthington (Freehold, New Jersey). A listing of enzymes available through 1968 has been published by Guilbault (24). In addition, many hundreds of other enzymes can be isolated and purified for use in constructing enzyme electrodes.

Enzyme instability and the fact that enzymes are "consumed" when they are used in the soluble form are other difficulties. Continuous or semi-continuous routine analyses using soluble enzymes would require large amounts of these materials -- quantities greater than could be reasonably supplied, and which would represent prohibitive expenditures in many cases. However, if the enzyme could be immobilized (insolubilized) without loss of activity so that one sample could be used continuously for many days, considerable advantage would be realized. Immobilized enzyme can be used analytically in much the same way soluble enzyme is used; e.g., to determine the concentration of a substrate that is acted upon by the enzyme.

Several ways that enzyme can be immobilized to provide a product useful in an enzyme electrode are discussed later in Section II. Furthermore, a number of immobilized enzymes are available from companies such as Boehringer (catalase, chymotrypsin, glucose oxidase, papain, ribonuclease, trypsin, urease); Koch Light of Buckinghamshire, England, or Aldrich of Milwaukee, Wisc., (catalase, dextranase, glucose oxidase, β-glucosidase and uricase); Miles Labs of Elkhart, Ind.; and Corning Glass Works of Corning, New York, which offers to provide any enzyme insolubilized on glass beads.

In constructing an enzyme electrode one need only (1) pick an enzyme that reacts with the substance to be determined, (2) obtain the enzyme from commercial sources or isolate it oneself, (3) immobilize the enzyme by standard techniques (II,A) or buy it already immobilized, if possible, and (4) place the immobilized enzyme around the appropriate electrode to monitor the reaction that occurs. (Note: This will probably be the limiting factor in constructing an enzyme electrode, since steps 1 through 3 are always possible). These steps will be discussed more thoroughly in Section II.

D. Enzyme Electrodes--A Survey

Enzyme electrodes represent the most recent advance in analytical chemistry. They combine the selectivity and sensitivity of enzymatic methods of analysis with the speed and simplicity of ion-selective electrode measurements. The resulting device can quickly determine the concentration of a given compound in solution, and the method requires a minimum of sample preparation.

1. Soluble Enzymes

We will first consider using ion-selective electrodes to monitor enzyme activity, and then proceed to self-contained biochemical electrode systems in which the enzyme or substrate is placed on the sensor itself.

Christian (25) has presented a good review on electrochemical methods of monitoring enzymatic reactions that mentions the use of electrodes.

Probably the most common electrochemical method for assaying enzymes that produce or consume an acid is to use the most selective of all electrodes, the low-sodium-error glass electrode, to follow the pH change of the reaction mixture as a measure of the activity of the enzyme. This method is not generally employed directly since the activity of a given enzyme is affected by changes in pH. Instead, a "pH stat" method is used in which the pH of the assay mixture is maintained by adding an acid or base. The rate of the addition of reagent gives the reaction velocity.

The activity of an enzyme in a system in which oxygen is consumed can be determined using an oxygen electrode. The electrode is a gold cathode separated by an epoxy casting from a silver anode. The inner sensor body is housed in a plastic casing and comes in contact with the assay solution only through the membrane. When oxygen diffuses through the membrane, it is electrically reduced at the cathode by an applied potential of 0.8 volts. This reaction causes a current proportional to the partial pressure of oxygen in the sample to flow between the anode and cathode. The rate of oxygen uptake can be related to the activity of the enzyme or the concentration of substration in the assay mixture. Good correlation between glucose values in blood determined by measuring oxygen uptake and by standard chemical tests was found by Kadish and Hall (26) and by Makino and Koono (27).

Ion-selective electrodes have been used to determine the activity of rhodanase and cholinesterase. In the rhodanase system, a cyanide-selective electrode followed the decrease of cyanide ion during the reaction:

$$CN^- + S_2O_3^{2-} \xrightarrow{\text{rhodanase}} SCN^- + SO_3^{2-}, \qquad (7)$$

which is catalyzed by rhodanase (28). Results obtained by this method are comparable to those obtained by spectrophotometric procedures. The method was easily adapted to automated systems as shown by Guilbault et al (28). A solution method for assaying rhodanase using a CN elec-trode was described by Llenado and Rechnitz (29). The cholinesterase assay was performed using a sulfide-selective electrode to monitor the amount of thiocholine released under the influence of cholinesterase:

$$\text{Acetylthiocholine} + H_2O \xrightarrow{\text{Cholinesterase}} \text{thiocholine} + CH_3COOH. \quad (8)$$

The amount of thiocholine released is proportional to the activity of cholinesterase (30). Llenado and Rechnitz used systems very similar to those described earlier for the assay of β- glucosidase (31,32), rhodanase (31), and glucose oxidase (31). An ion-selective electrode for cyanide was used to follow the production of cyanide ion in the assay of β-glucosidase,

$$\text{Amygdalin} + H_2O \xrightarrow{\beta\text{-glucosidase}} \text{benzaldehyde} + 2\ \text{glucose} + HCN, \quad (9)$$

and the consumption of cyanide in the rhodanase system as was described previously (28). An iodide-selective electrode was used with the glucose oxidase assay to measure the decrease in iodide concentration resulting from oxidation of iodide to iodine by hydrogen peroxide. An autoanalyzer system was used to monitor the three enzymes.

$$\beta\text{-D-glucose} + H_2O + O_2 \xrightarrow[\text{oxidase}]{\text{glucose}} \text{D-gluconic acid} + H_2O_2; \quad (10)$$

$$H_2O_2 + 2H^+ + 2I^- \xrightarrow{\text{Mo(VI)}} 2H_2O + I_2. \quad (11)$$

This same reaction sequence for assaying glucose oxidase was subsequently turned around by Llenado and

Rechnitz (33) and used for assaying glucose. A soluble glucose oxidase was used, again in the AutoAnalyzer system. However, the use of the catalyzed iodide oxidation in conjunction with an iodide electrode in assaying glucose was shown by Guilbault et al (34) to be subject to many intereferences in blood. Reducing agents, such as ascorbic acid, tyrosine and uric acid interfere strongly, and glucose measurements in samples of blood containing them can only be made after vigorous pre-treatment of the sample.

A cyanide ion-selective electrode was used by Butknecht and Guilbault (35) to monitor the removal of cyanide from aqueous systems. The enzyme β-cyanoalanine synthase (injectase) was used in an immobilized form:

$$CN^- + \text{cysteine} \longrightarrow HS^- + \beta\text{-cyanoalanine} \tag{12}$$

One of the first uses of ion-selective electrodes in enzymic analysis was the report of Guilbault et al (36) who described the use of a monovalent cation glass electrode for following the urea-urease, glutamine-glutaminase, asparagine-asparaginase and D- and L-amino acids-amino acid oxidase reactions. In all cases, the NH_4^+ produced was measured:

$$\text{Substrate} \xrightarrow{\text{Enzyme}} NH_4^+. \tag{13}$$

In order to overcome the interference effects of Na^+ and K^+ on such a glass electrode, Montalvo (37) constructed an ammonium "ion-specific" electrode by coupling a hydrophobic ammonia permeable membrane to a Beckman cation electrode. In the electrode, the ammonia produced from the urea-urease reaction diffuses through the ammonia membrane, where it reacts with tris buffer to form NH_4^+ which is then sensed by the cation electrode. Response times of about 20 minutes were required for the enzyme reaction. Moreover, a constant flow of buffer must be pumped over the cation electrode, making the electrode totally impractical.

This same specificity for NH_4^+, and elimination of response to Na^+ and K^+ can be accomplished by using the Orion NH_3 electrode directly at pH 7 or 7.5 as described by Anfalt, Granelli and Jagner (38) (response time, two minutes) or with the air gap electrode as used by Hansen and Ruzicka (39) (response time, $<$ one minute at pH 11 after a two minute incubation time at lower pH). In this latter paper, the ammonia, catalytically released from urea in the presence of soluble urease, is volatilized and sensed by a glass electrode covered with a film of electrolyte. Since the electrode never actually touches the sample solution, the problems caused by the presence of proteins, blood cells and species which reportedly block pores in the membranes of other gas electrodes, are avoided.

Baum (40) has reported an electrometric method for determining cholinesterase (ChE) activity based on the use of a liquid-membrane ion-selective electrode which responds to the concentration of acetylcholine remaining during an enzymatic catalyzed hydrolysis:

$$\text{Acetylcholine} \xrightarrow{\text{ChE}} \text{Choline + acetic acid.} \quad (14)$$

The electrode sold by Corning Glass Works (Corning, New York) is perhaps the only selective non-enzymatic organic species-measuring device with a selectivity for acetylcholine over choline, which is similar in structure. As the acetylcholine reacts, a measured drop in potential is equated to the activity of enzyme.

2. Insolubilized Enzymes

a. Glucose Electrodes. The first report of an enzyme electrode was given by Clark and Lyons (41), who proposed that glucose could be determined amperometrically using soluble glucose oxidase held between Cuprophane membranes. The oxygen uptake was measured with an O_2 electrode:

$$\text{Glucose} + O_2 + H_2O \xrightarrow{\text{G.O.}} H_2O_2 + \text{gluconic acid} \quad (15)$$

The term "enzyme electrode" was introduced by Updike and Hicks (42), who coated an oxygen electrode with a layer of physically entrapped glucose oxidase in a polyacrylamide gel. The decrease in oxygen pressure was equivalent to the glucose content in blood and plasma. A response time of less than a minute was observed. Clark (43) proposed measuring the hydrogen peroxide produced in the enzymic reaction with a Pt electrode. An instrument based on this concept, using glucose oxidase held on a filter trap, is now sold by Yellowsprings Instrument Company (Yellowsprings, Ohio). Two platinum electrodes are used. One compensates for any electrooxidizable compounds in the sample, such as ascorbic acid. The second monitors the enzyme reaction.

Williams, Doig and Korosi (44) used quinone in place of oxygen as the hydrogen acceptor, and described enzyme electrodes for blood glucose:

$$\text{Glucose} + \text{quinone} + H_2O \xrightarrow{\text{G.O.}} \text{gluconic acid} + \text{hydroquinone}; \quad (16)$$

$$\text{Hydroquinone} \xrightarrow{\text{Pt}} \text{quinone} + 2H \ + 2e^- \quad (E = 0.4 \text{ V vs. S.C.E.}). \quad (17)$$

Using glucose oxidase trapped in a porous or jelled layer and covered with a dialysis membrane over a Pt electrode, glucose could be determined by monitoring the electrooxidation of quinone. About 3 to 10 minutes was required to obtain a steady-state current.

Guilbault and Lubrano (45,46) described a simple, stable, rapid-reading electrode for glucose. The electrode consists of a metallic sensing layer (Pt or Pt-glass (47)) covered by a thin film of immobilized glucose oxidase held in place by cellophane. When poised at the correct potential, the resulting current is proportional to the glucose concentration. The time of measurement with this amperometric approach is less than 12 seconds using a kinetic method. Stored at room temperature, the electrode is stable for more than a year with only a 0.1% reponse change from maximum per day. The enzyme electrode determination of blood sugar compares favorably with commonly used methods with respect to accuracy, precision, and stability, and the only reagent needed for assay is a buffer solution.

Nagy, von Storp and Guilbault (34) described a self-contained electrode for glucose based on an iodide membrane sensor:

$$\text{Glucose} + O_2 \xrightarrow[\text{Oxidase}]{\text{Glucose}} \text{gluconic acid} + H_2O_2 \quad (18)$$

$$H_2O_2 + 2I^- + 2H^+ \xrightarrow{\text{Perodixase}} 2H_2O + I_2 \quad (19)$$

The highly selective iodide sensor monitors the local decrease in iodide activity at the electrode surface. The glucose assay was performed in a flow stream and at a stationary electrode. Pre-treatment of the blood sample was required to remove interfering reducing agents, such as ascorbic acid, tyrosine, and uric acid.

Nilsson, Akerlund and Mosbach (48) described the use of conventional hydrogen ion glass electrodes for preparing enzyme-pH electrodes by either entrapping the enzymes within polyacrylamide gels around the glass electrode or as a liquid layer trapped within a cellophane membrane. In an assay of glucose based on a measurement of the gluconic acid produced, the pH response was almost linear from 10^{-1} to 10^{-3} M with a ΔpH of about 0.85/decade. Electrodes of this type were also constructed for urea and penicillin. The ionic strength and pH were controlled using a weak (10^{-3} M) phosphate buffer, pH 6.9 and 0.1 M Na_2SO_4.

b. Urea Electrodes. Guilbault and Montalvo (49,50) prepared several electrodes for urea by physically entrapping urease in a polyacrylamide gel held over the surface of a monovalent cation electrode by cellophane film:

$$\text{Urea} + 2H_2O \xrightarrow{\text{Urease}} 2NH_4^+ + 2HCO_3^- . \quad (20)$$

The urea diffuses into the urease layer where it is converted to ammonium ions, which are sensed by the cation electrode. The electrode could be used for up to three weeks with no loss of activity, and responded to urea in the concentration range 5×10^{-5} to 1.6×10^{-1} M with a response time of about 35 seconds.

Because sodium and potassium ions interfered with the measurement, Guilbault and Hrabankova (51) used an uncoated NH_4^+ electrode, as the reference electrode to the urease-coated NH_4^+ electrode, and added ion-exchange resin in attempts to develop a urea electrode useful for assaying blood and urine. Good precision and accuracy were obtained.

In attempting to improve the selectivity of the urea determination, Guilbault and Nagy (52) used a silicone rubber-based nonactin ammonium ion-selective electrode as the sensor for the NH ions liberated in the urease reaction. The selectivity coefficients of this electrode were 6.5 for NH_4^+/K^+; 7.50 for NH_4/Na^+; and much higher for other cations. The electrode's reaction layer was made of urease enzyme chemically immobilized on polyacrylic gel. A still further improvement was described by Guilbault, Nagy and Kuan (53) using a three-electrode system, which allowed dilution to a constant interference level. Analysis of blood serum showed good agreement with spectrophotometric methods, and the enzyme electrode was stable for four months at 4^oC.

Still further improvement in the selectivity of this type of electrode was obtained by Anfalt, Grameli and Jagner (38) who polymerized urease directly onto the surface of an Orion ammonia gas electrode probe by means of glutaraldehyde. Sufficient NH_3 was produced in the enzyme reaction layer even at pH's as low as 7 or 8 to allow direct assay of urea in the presence of large amounts of Na^+ and K^+. A response time of two to four minutes was observed.

Guilbault and Tang (54) described a still better, total interference-free, direct-reading electrode for urea, using the air gap electrode of Hansen and Ruzicka (21). A thin layer of urease chemically bound to polyacrylic acid was used at a solution pH of 8.5, where good enzyme activity was still obtained, yet where sufficient NH_3 is liberated to yield a sensitive measurement with the air-gap NH_3 electrode. The urea diffuses into the gel; the NH_3 produced diffuses out of solution to the surface of the air gap electrode where it is measured. A linear range of 3×10^{-2} to 5×10^{-5} M was obtained with a slope of 0.75 pH unit/decade. The electrode could be used continuously for blood serum analysis for up to one month (at least 500 samples) with an accuracy precision of less than 2%. The response time is two to four minutes at pH 8.5, and the electrode is washed under a water tap for 5 to 10 seconds after each measurement. Absolutely no interference from any level of substances commonly present in blood was observed (Na^+, K^+, NH_4^+, ascorbic acid, etc.).

A urea electrode using physically entrapped urease and a glass electrode to measure the pH change in solution was described by Mosbach et al (48). The response time of the electrode to urea was about 7 to 10 minutes and was linear from 5×10^{-5} to 10^{-2} M with a change of about 0.8 pH unit/decade. The electrode could be kept at room temperature for two to three weeks. The ionic strength and pH were controlled using a weak (10^{-3} M) tris buffer and 0.1 M NaCl.

Still another urea electrode possibility is the use of a CO_2 sensor to measure the second product of the urea-urease reaction, HCO_3^-. Guilbault and Shu (55) evaluated the use of the CO_2 sensor and found that a urea electrode, prepared by coupling a CO_2 electrode with a layer of urease covered with a dialysis net, had a linear range of 10^{-4} to 10^{-1} M, a response time of about one to three minutes, and a slight response to only acetic acid. Na^+ and K^+ ions had no interference.

c. Amino Acid Electrodes. The CO_2 sensor was evaluated by Guilbault and Shu for response to tyrosine when coupled with tyrosine decarboxylase held in an immobilized form by a dialysis membrane (55). A linear range of 2.5 x 10^{-4} to 10^{-2} M was observed with a slightly faster response time than with the urea electrode mentioned earlier. A slope of 55 mV/decade was obtained, compared to 57 mV/decade for the urea electrode.

Enzyme electrodes for determining l-aminoacids were developed by Guilbault and Hrabankova (56), who placed an immobilized layer of L-amino acid oxidase over a monovalent cation electrode to detect the ammonium ion formed in the enzyme-catalyzed oxidation of the amino acid. These electrodes are stable for about two weeks, and have a one to two minute response times.

Two different kinds of enzyme electrodes were prepared by Guilbault and Nagy for determining L-phenylalamine (57). One of the electrodes used a dual enzyme reaction layer -- L-amino acid oxidase with horseradish peroxidase -- in a polyacrylamide gel over an iodide-selective electrode. The electrode responds to decreasing iodide activity at the electrode surface due to the enzymatic reaction and subsequent oxidation of iodide.

$$\text{L-Phenylalanine} \xrightarrow{\text{L-aminoacid oxidase}} H_2O_2. \qquad (21)$$

$$H_2O_2 + 2H^+ + I^- \xrightarrow[\text{peroxidase}]{\text{horseradish}} I_2 + H_2O. \qquad (22)$$

The other electrode used was a silicone rubber-based, nonactin-type ammonium ion-selective electrode covered with L-amino acid oxidase in a polyacrylic gel. The same principle of substrate diffusion into the gel layer, enzymatic reaction, and detection of the released ammonium applied to this system. Linear calibration plots were also obtained for L-leucine and L-methionine in the range 10^{-4} to 10^{-3} M.

Electrodes specific for D-amino acids, which are oxidatively catalyzed by D-amino acid oxidase were reported by Guilbault and Hrabankova (58). The NH_4^+ ion produced is monitored with a cation electrode:

$$\text{D-amino acid} + O_2 \xrightarrow{\text{Oxidase}} NH_4^+ + HCO_3^- . \quad (23)$$

The stability of these electrodes could be maintained for 21 days stored in a buffered flavine adenine dinucleotide (FAD) solution, because the FAD is weakly bound to the active site of the enzyme and is needed for its activity. Electrode probes suitable for the assaying of D-phenylalanine, D-alanine, D-valine, D-methionine, D-leucine, D-norleucine, and D-isoleucine were developed. An electrode for asparagine was also developed using asparaginase as the catalyst (58). No co-factor was necessary.

Guilbault and Shu (59) described an enzyme electrode for glutamine, prepared by entrapping glutaminase on a nylon net between a layer of cellophane and a cation electrode. The electrode responds to glutamine over the concentration range 10^{-1} to 10^{-4} M with a response time of only one to two minutes.

Guilbault and Lubrano (60) prepared an electrode for L-amino acids by coupling chemically bound L-amino acid oxidase to a Pt electrode to sense the peroxide produced in the enzyme reaction:

$$\text{L-aminoacid} + O_2 + H_2O \longrightarrow \text{R-COCOOH} + NH_3 + H_2O_2 . \quad (24)$$

The time of measurement using a kinetic measurement of the rate of increase in current per unit time is less than 12 seconds, and the only reagent required is a phosphate buffer. The L-amino acids cystein, leucine, tyrosine, phenylalanine, tryptophan and methionine were assayed.

d. Alcohol Electrodes. Alcohol oxidase catalyzes the oxidation of lower primary aliphatic alcohols.

$$RCH_2OH + O_2 \xrightarrow[\text{oxidase}]{\text{alcohol}} RCHO + H_2O_2. \qquad (25)$$

The hydrogen peroxide produced in these reactions may be determined amperometrically with a platinum electrode as described earlier for determining glucose. Guilbault and Lubrano (61) used the alcohol oxidase obtained from Basidiomycete to determine the ethanol concentration of 1-ml samples over the range 0 to 10 mg/100 ml, with an average relative error of 3.2% in the 0.5 to 7.5 mg/100 ml range. This procedure should be adequate for clinical determinations of blood ethanol, since normal blood from individuals who have not ingested ethanol ranges from 40 to 50 mg/100 ml. Methanol seriously interferes with the procedure because the alcohol oxidase is more active for methanol than ethanol. However, the concentration of methanol in blood is negligible compared to that of ethanol.

e. Uric Acid Electrode. A self-contained rapid reading electrode for uric acid was described by Nanjo and Guilbault (19). The electrode was prepared by placing a layer of glutaraldehyde-bound uricase over the tip of a Beckman Pt electrode. The enzyme was then covered for support with a thin layer of dialysis membrane. The decrease in the level of dissolved oxygen in solution due to the enzymic reaction, given by

$$\text{Uric Acid} + O_2 \xrightarrow{\text{uricase}} \text{allantoin } H_2O_2 + H_2O, \quad (26)$$

was maintained at an applied potential of -0.6 V vs. S.C.E. The current observed was proportional to the level of uric acid at concentrations of 10^{-5} to 10^{-1} M. By measuring the initial rate of change in current, an assay can be performed in less than 30 seconds.

Further studies indicated the electrode could be used to assay glucose, amino acids, alcohols and cholesterol (20).

It was found that the peroxide produced in the reaction could not be monitored at + 0.6 V vs. S.C.E. as described in the method of Guilbault and Lubrano for glucose (45,46), amino acids (60) and alcohols (61), because the polarographic curves for uric acid and hydrogen peroxide are too close to be separated at any pH useful for the enzyme reaction, and because an allantoin-peroxide complex is the product of uric acid oxidation, not free peroxide. Additionally, the oxygen uptake method was found to be more sensitive, enabling the assay of lower substrate concentrations.

The use of a Pt electrode rather than a Clark-type oxygen electrodes eliminates the problems associated with gas membrane electrodes, i.e., slow response and blockage of the membrane by substances present in blood.

f. Lactic Acid Electrode. Williams, Doig and Korosi (44) used ferricyanide as a hydrogen acceptor for lactic acid, and described an enzyme electrode for lactate based on the reaction:

$$\text{Lactate} + 2\text{Fe(CH)}_6^{3-} \xrightarrow{\text{LDH}} \text{Pyruvate} + 2\text{Fe(CN)}_6^{4-}; \quad (27)$$

$$2\text{Fe(CN)}_6^{4-} \xrightarrow{\text{Pt}} 2\text{Fe(CN)}_6^{3-} + 2e^- + H^+. \quad (28)$$

In monitoring the electro-oxidation of the ferrocyanide produced at +0, 4 V vs. S.C.E. at a Pt electrode covered with a porous or jelled layer of lactate dehydrogenase and a dialysis membrane, a current proportional to the concentration of lactic acid was observed. About 3 to 10 minutes were required for measurement. Because of the low Km value of this enzyme (Km = 1.2×10^{-3} M), it was necessary to dilute the sample with buffered ferricyanide. A linear plot was obtained over the range 10^{-4} to 10^{-3} M.

g. Amygdalin Electrode. An electrode specific for amygdalin and based on a solid-state cyanide electrode was reported by Rechnitz and Llenado (62,63). The enzyme β-glucosidase immobilized in acrylamide gel, was used:

$$\text{Amygdalin} \xrightarrow{\beta\text{-glucosidase}} \text{HCN} + 2\,C_6H_{12}O_6 + \text{Benzaldehyde}. \quad (29)$$

A linear range of 5×10^{-3} to 10^{-5} M was reported, with a slope of about 40 mV/decade and response times of about 10 minutes at concentrations of 10^{-4} and 10^{-5} M amygdalin and 30 minutes at 10^{-4} to 10^{-5} M. The electrode rapidly lost activity, indicating an incomplete physical entrapment had been effected.

One reason for this long response time and poor stability was that the high pH used (10.4) is a pH at which the enzyme has low activity and is denatured. This was recognized by Mascini and Liberti (64) who improved the response time and other electrode characteristics by working at a pH of 7. The electrode was prepared by spreading the enzyme directly onto the membrane surface and covering it with a thin dialysis membrane. Since the enzyme was not immobilized, a stability of less than a week is obtained, but the response time was only about one to two minutes at 10^{-1} to 10^{-3} M and six minutes at 10^{-4} M. Furthermore, a linear calibration was obtained from 10^{-1} to 10^{-4} M with a slope of 53 mV/decade (compared to about 40 mV/decade at pH 10).

h. Penicillin Electrode. The first attempt at designing a penicillin electrode was made by Papariello, Mukherji and Shearer (65). The electrode was prepared by immobilizing penicillin β-lactamase (penicillinase) in a thin membrane of polyacrylamide gel molded around, and in intimate contact with, a glass (H^+) electrode. The increase in hydrogen ion from the penicilloic acid liberated from penicillin is measured:

$$\text{Penicillin} \xrightarrow{\text{Penicillinase}} \text{Penicilloic Acid.} \qquad (30)$$

The response time of the electrode was very fast (<30 sec) and had a slope of 52 mV/decade over the range 5×10^{-2} to 10^{-4} M for sodium ampicillin. The reproducibility of the electrode was very poor, probably because no attempt was made to control the ionic strength and pH.

Mosbach et al (47) prepared a penicillin electrode in similar fashion, using polyacrylamide-entrapped penicillinase on a glass (H^+) electrode, yet controlled the ionic strength and pH by using a weak 0.005 M phosphate buffer, pH 6.8, and 0.1 M NaCl. Good results were obtained in comparison to the results of Papariello et al. The calibration plot was linear from 10^{-2} to 10^{-3} M with a ΔpH of 1.4, and as little as 5×10^{-4} M sodium penicillin could be determined. The electrode could be stored for three weeks, and the average deviation was $\pm$ 2% with a response time of two to four minutes.

3. Theoretical Study

An excellent theoretical study of the kinetic behavior of enzymes immobilized in artificial membranes was made by Blaedel, Kissel and Boguslaski (66). Experimental support of the theoretical equations were obtained with urease in three systems: a membrane-covered sensor; a membrane separating two solutions; and a membrane immersed in solution.

E. Substrate Electrodes for the Assay of Enzymes

1. Urease Electrode

Attempts have been made to determine the activity of enzymes using "immobilized" substrates. Such electrodes have two limitations that the earlier described enzyme electrodes do not have: (1) the substrate, unlike the enzyme, is used up in an assay, hence, the amount of substrate available will be a limiting factor; (2) the enzyme, since it is not consumed, will continually act on the substrate to continually produce product, hence, the analysis must involve a kinetic rather than an equilibrium method.

Montalvo (67,68) designed an electrode for urease by continually passing a layer of soluble urea between the tip of a NH_4^+ cation electrode and a dialysis membrane. Urea diffuses through the membrane and is hydrolyzed by urease in the dilute aqueous solutions. The ammonium ion diffuses back through the membrane to the cation electrode, where it is sensed. Although an interesting approach, it is one that lacks practicality.

2. Cholinesterase Electrode

In another study, an enzyme-sensing electrode system for serum cholinesterase was prepared by coupling a pH-sensing electrode to a thin polymer membrane with a low-molecular-weight cut-off (69). The electrode system utilized two thin-layer solutions to form a microelectrochemical cell. One layer contained the serum to be assayed, the second contained the acetylcholine substrate which had been stabilized to balance the non-enzymatic decay of substrate by using a high-molecular-weight buffer. A pseudo-linear curve was obtained from 10-70 units/ml of cholinesterase, and assays could be performed in 1.5 to 4.5 minutes.

A more practical approach to a cholinesterase substrate electrode was proposed by Gibson and Guilbault (70), who prepared an insoluble Reinickate salt of acetylcholine and placed it on the tip of a pH electrode covered with a nylon net permeable to enzyme for support. The enzyme diffuses into the substrate layer producing acid which is then sensed by the pH electrode:

$$\text{Acetyl-choline Reineckate} \xrightarrow{\text{ChE}} \text{acetic acid} + \text{choline-Reineckate.} \quad (31)$$

The author believes that such substrate electrodes have a limited future, becoming generally accepted only if they can be preparable in simple, self-contained systems, such as those developed for the enzyme electrodes.

II. EXPERIMENTAL

Section I covered basic enzyme electrode principles and surveyed the electrodes made and used up to this time as reported in the literature. From Section I, it appears that highly specific enzyme electrodes with a high degree of accuracy and sensitivity can be made by using appropriate enzymes. Applications of such electrodes to the clinical, the pharmaceutical, industrial and pollution laboratories is quite feasible and sound. The electrodes have the further advantages of simplicity because they require only one solution -- a buffer -- and economy, being useable for as many as 500 to 1,000 determiniations per electrode.

This section covers the experimental parameters involved in constructing enzyme electrodes, and the observed characteristics and limitations of these electrodes.

A. Immobilization Methods

1. General

Cost is a factor in enzymic analysis because continuous or semicontinuous analysis require large amounts of unrecoverable enzymes. Immobilized enzymes can be used continuously, and expensive enzymes can be recovered, thus resolving the problem of cost. Also, because most solubilized enzymes are very unstable, they must often be stored at low temperatures, and then they may only retain activity for a short period of time. In cytochemical systems, most if not all of the enzymes are attached to cell surfaces or entrapped within cell membranes. Immobilization puts enzymes in a more natural environment with the result that they usually are most stable and efficient.

Within the last six years, a new technology based on enzyme immobilization has emerged. Five methods have been used in preparating water-insoluble derivatives of enzymes: Microencapsulation within thin-wall spheres; adsorption on inert carriers; covalent crosslinking by bifunctional reagents into macroscopic particles; physical entrapment in gel lattices; and covalent binding to water-insoluble matrices. Some enzyme immobilization methods and procedures for immobilizing enzymes for use in electrodes are covered in this section. More technical details are available in several review articles (71-82).

2. Immobilization Techniques

a. Microencapsulation. Microencapsulation within thin-wall spheres, the newest approach to enzyme insolubilization, was first introduced only several years ago by Chang (83-88). The thin wall of the sphere is semipermeable so enzyme is physically prevented from diffusing out of the microcapsule while reactants and products readily permeate the encapsulating membrane.

Microencapsulation is accomplished by depositing polymer around emulsified aqueous droplets, either by interfacial coacervation or interfacial polycondensation. Though such a technique has not yet been applied to preparing electrodes, it could be. Since this is discussed elsewhere in this book,it will not be considered further here.

The main disadvantage of using spheres is that many of the interfacial polymerization procedures cause enzyme deactivation. Rony (89) has circumvented this problem by first producing hollow fibers instead of spheres and then placing the enzyme in the fiber and sealing the ends. Hollow fibers have the added advantages of being easy to fill with many types of catalyst, allowing recovery of the trapped enzyme, and being mass-produced and, therefore, commercially available.

b. Adsorption. Adsorbing enzymes on insoluble supports results from ionic, polar, hydrogen bonding, hydrophobic or π-electron interactions. In early works, adsorption to inorganic carriers was used as a technique for characterizing enzymes. Since then, characterizing the adsorption phenomenon has been the major concern of most studies. Early works on protein adsorption have been reviewed by Zittle (90) and Silman and Katchalski (80). A few of the adsorbents that have been used are glass (90), quartz (91), charcoal (92), silica gel (93), alumina (94), ion-exchange resins (95,96), and bentonite (97).

The major advantage of this method of immobilization is its extreme simplicity. Attachment is by simple exposure, and conditions are mild. But the disadvantage is a very serious one. Adsorbed enzymes are easily desorbed by factors depending in pH, solvent, substrate, and temperature effects.

c. Covalent Crosslinking by Bifunctional Reagents

Bifunctional reagents have been used to insolubilize enzymes and other proteins by intermolecular crosslinking, with the concomitant formation of macroscopic particles (77,78,80). Bifunctional reagents can be divided into two classes: "homo" bifunctional, and "hetero" bifunctional, depending on whether the reagent possesses two identical or different functional groups.

A few "homo" bifunctional reagents are glutaraldehyde (98-101), bisdiazobenzidine-2, 2'-disulfonic acid (102,103), 4,4'-difluoro-3,3'-

$$\begin{array}{c} CHO \\ | \\ (CH_2)_3 \\ | \\ CHO \end{array} \quad \text{Glutaraldehyde}$$

dinitrodiphenyl sulfone (104), diphenyl-4,4'-dithiocyanate-2,2'-disulfonic acid (105), 1,5-difluoro-2,4-dinitrobenzene (106), and phenol-2,4-disulfonyl chloride (107). A few "hetero" bifunctional reagents are toluene-2-isocyanate-4-isothiocyanate (108), trichloro-s-triazine (109, 110), and 3-methoxydiphenyl methane-4,4'-diisocyanate (108). The advantages of this method are its simplicity and the chemical binding of the enzyme, which enables control of the physical properties and particle size of the final product. The major disadvantage is that many enzymes are sensitive to the coupling reagents and lose activity in the process. This method is thus limited in its applicability, but has been used to manufacture enzyme electrodes (19,20,38).

d. Inclusion in Gel Lattices. An enzyme can be occluded with a gel matrix by carrying out a polymerization in an aqueous solution containing the enzyme. The polymer used most often consists of acrylamide as monomer and N,N'-methylenebisacrylamide as cross-linking agent. The gel can be formed as a thin film where it is to be used, or it can be formed in a block and mechanically dispersed into the desired particle size to be used in solution, in a column, etc. Polyacrylamide gels were first prepared by Bernfeld and Wan (111) and later used by many others (112-117). Other types of gels that have been used are starch gel stabilized onto polyurethane foam (118-120) or silastic (121,122).

Inclusion of enzymes in a gel has the advantage of mild reaction conditions without significant enzyme alteration. Therefore, it can theoretically be applied to any enzyme.

However, several disadvantages should be mentioned. Because there is a broad distribution of pore size in polyacrylamide-type polymers, there is a continuous leakage of the occluded enzyme. Free radicals generated during the polymerizing process may cause the enzyme to lose some or most ot its activity. Also, there probably will be diffusion control of the substrates and products, and the reaction is limited to small substrates which can diffuse readily into the gel (i.e., glucose, urea, amino acids).

e. Covalent Binding to Water-Insoluble Matrices. Within the last six years, a new technology, based on immobilizing enzymes by covalent coupling to insoluble carriers, has emerged. Because covalent coupling places enzymes in a more natural environment, they usually function more efficiently, have increased stability, and have the added advantage of immobilization that is not reversed by pH, ionic strength, substrate, solvents, or temperature. It is little wonder that covalent binding of enzymes to insoluble carriers is now the most widely used method of enzyme insolubilization.

Carriers are chosen by their properties of solubility, functional groups, mechanical stability, surface area, swelling, and hydrophobic or hydrophilic nature. Essentially, three types of carriers have been employed: Inorganics, natural polymers, and synthetic polymers. A few of the more common carriers that have been used are given in Table 3. These carriers are usually activated by transformation into various derivatives.

Binding is carried out by way of functional groups on the enzyme which are not essential for its catalytic activity. Even though a particular amino acid residue may be necessary for catalytic activity, it is usually present in nonessential positions. Random chemical reaction will denature only a portion of the molecules present. Because binding must be carried out under conditions which do not cause denaturation, it is necessary to understand the effects of chemical modification of enzymes on their activity. Several reviews have been written on this subject (142-144).

Table 3

Common Carriers Used for Covalent Binding of Enzymes

Carrier	Reference
Porous glass	123, 124
Polyacrylamide	125, 126
Polyacrylic acid	127, 128
Polyaspartic acid	129
Polyglutamic acid	127
Polystyrene	130
Nylon	131, 132
Cellulose	133, 134
Sepharose	135, 136
Sephadex	136, 137
Ethylene-maleic anhydride copolymer	129, 138
Agarose	139, 140
Sepharose	135, 141
Carboxymethyl Cellulose	129, 133

The amino acid residues suitable for covalent binding are: α- and ε-amino groups; the phenol ring of tyrosine; β- and γ- carboxyl groups; the sulfhydryl group of cysteine; the hydroxyl group of serine; and the imidazole group of histidine. Of these, the most widely used are the first three. There are a large number of methods of covalently coupling enzymes to water-insoluble carriers. Several of the more common methods are shown in Fig. 1. Table 4 shows the amino acid residues that are reactive in several common immobilizing methods.

FIGURE 1

COMMON COUPLING REACTIONS

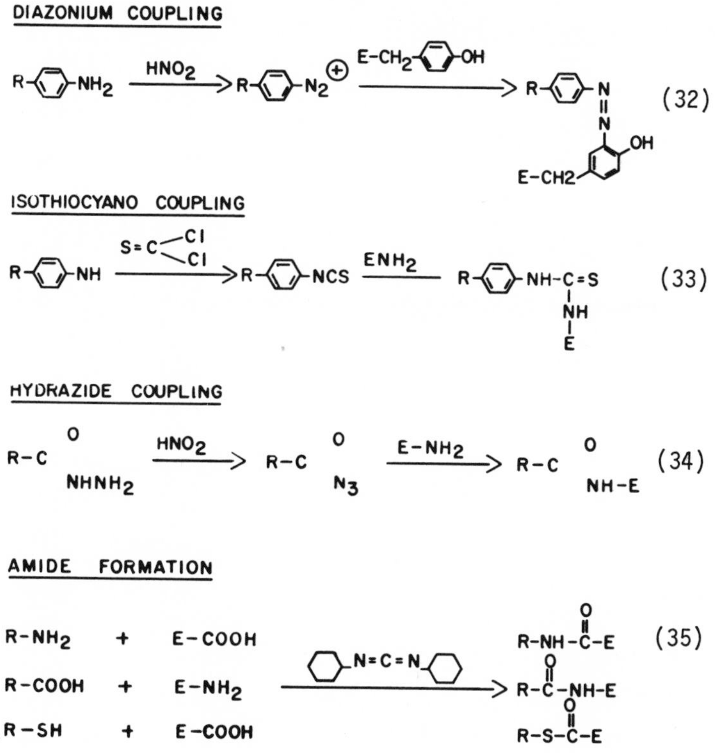

DIAZONIUM COUPLING

$$R-C_6H_4-NH_2 \xrightarrow{HNO_2} R-C_6H_4-N_2^{\oplus} \xrightarrow{E-CH_2-C_6H_4-OH} R-C_6H_4-N=N-C_6H_3(OH)-CH_2-E \quad (32)$$

ISOTHIOCYANO COUPLING

$$R-C_6H_4-NH \xrightarrow{S=C\langle^{Cl}_{Cl}} R-C_6H_4-NCS \xrightarrow{ENH_2} R-C_6H_4-NH-\underset{\underset{E}{|}}{\underset{NH}{|}}{C}=S \quad (33)$$

HYDRAZIDE COUPLING

$$R-C\langle^{O}_{NHNH_2} \xrightarrow{HNO_2} R-C\langle^{O}_{N_3} \xrightarrow{E-NH_2} R-C\langle^{O}_{NH-E} \quad (34)$$

AMIDE FORMATION

$$\begin{array}{lll} R-NH_2 + E-COOH & & R-NH-\overset{O}{\overset{\|}{C}}-E \\ R-COOH + E-NH_2 & \xrightarrow{C_6H_{11}-N=C=N-C_6H_{11}} & R-\overset{O}{\overset{\|}{C}}-NH-E \\ R-SH + E-COOH & & R-S-\overset{O}{\overset{\|}{C}}-E \end{array} \quad (35)$$

FIG. 1 (continued)

DIMETHYLACETYL COUPLING

$$R\text{-}CH(OCH_3)_2 \xrightarrow{H^+} R\text{-}CH_2\text{-}CHO \xrightarrow{E\text{-}NH_2} R\text{-}CH_2\text{-}\underset{}{\overset{OH}{\overset{|}{C}}}H\text{-}NH\text{-}E \qquad (36)$$

DISULFIDE COUPLING

$$R\text{-}CH_2SH + E\text{-}CH_2SH \underset{reduce}{\overset{oxidize}{\rightleftharpoons}} R\text{-}CH_2\text{-}S\text{-}S\text{-}CH_2\text{-}E \qquad (37)$$

MALEIC ANHYDRIDE COUPLING

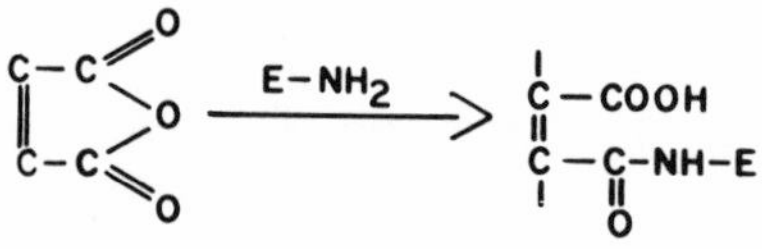

$$\text{(maleic anhydride ring: C=C with C(=O)–O–C(=O))} \xrightarrow{E\text{-}NH_2} \begin{array}{l} -\overset{|}{C}-COOH \\ \ \ \| \\ -\underset{|}{C}-\underset{\|}{C}\text{-}NH\text{-}E \\ \qquad O \end{array} \qquad (38)$$

THIOLACTONE COUPLING

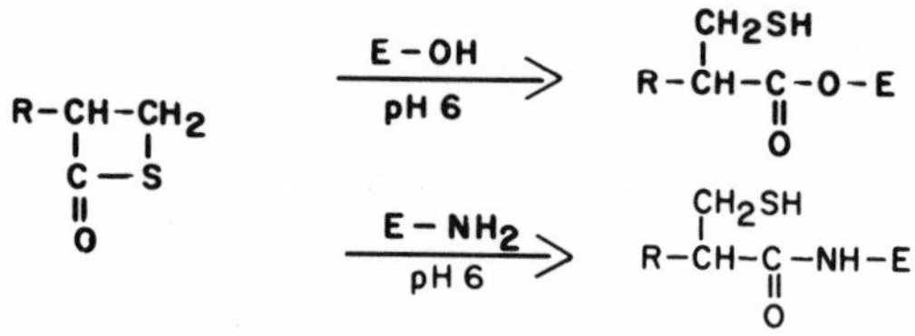

$$\begin{array}{l} R\text{-}CH\text{-}CH_2 \\ \quad | \qquad | \\ \quad C - S \\ \quad \| \\ \quad O \end{array} \begin{array}{l} \xrightarrow[pH\ 6]{E\text{-}OH} R\text{-}\overset{CH_2SH}{\overset{|}{C}}H\text{-}\underset{\underset{O}{\|}}{C}\text{-}O\text{-}E \\ \\ \xrightarrow[pH\ 6]{E\text{-}NH_2} R\text{-}\overset{CH_2SH}{\overset{|}{C}}H\text{-}\underset{\underset{O}{\|}}{C}\text{-}NH\text{-}E \end{array} \qquad (39)$$

DICHLOROTRIAZINE COUPLING

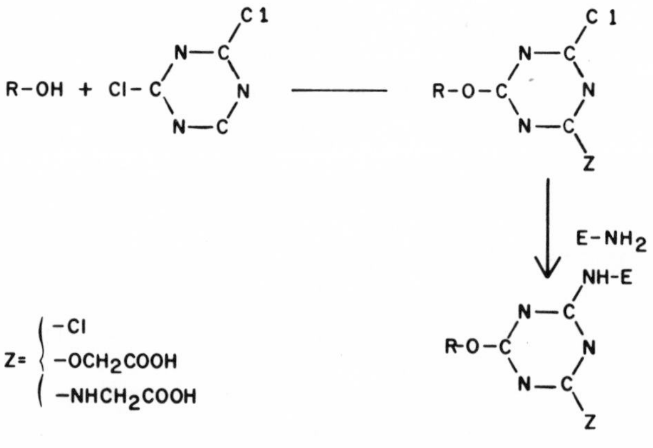

$$R\text{-}OH + Cl\text{-}C_3N_3(Cl) \longrightarrow R\text{-}O\text{-}C_3N_3(Cl)(Z) \xrightarrow{E\text{-}NH_2} R\text{-}O\text{-}C_3N_3(NH\text{-}E)(Z) \qquad (40)$$

$$Z = \begin{cases} -Cl \\ -OCH_2COOH \\ -NHCH_2COOH \end{cases}$$

Table 4

Reactive Amino Acid Residues in Common Covalent Insolubilizing Methods (72)

Insolubilizing Reagent	Reactive amino acid residues								Reference
Chloro-sym -trianzinl derivative	Lys								109, 110
Diazo derivative	Lys	His	Tyr	Arg	Cys				146
	Lys	His	Tyr						147
	Lys		Tyr	Arg					148
		His	Tyr						149, 150, 151
Iso-Thiocyanato derivative	Lys			Arg					152, 153
N-Ethyl-5-phenylisoxazolium-3'-sulphonate	Lys					Asp	Glu		154
Diimide	Lys		Tyr		Cys	Asp	Glu		155, 156
Acid azide	Lys		Tyr		Cys			Ser	157
Maleic acid or maleic anhydride copolymer	Lys								158
Glutaraldehyde	Lys								98, 159-162
	Lys	His	Tyr		Cys				163
Cyclic iminocarbonate	Lys								164

3. Properties of Immobilized Enzymes

Many immobilized enzyme properties are different from those of their solubilized counterparts. We have already mentioned kinetic effects brought about by diffusion control and heterogeneous catalysis. The most important properties are an increased long-term stability, and temperature stability. In general, stability can be increased by the proper choice of coupling procedure and insoluble carrier. Another property that has been observed is a change in reactivity. Immobilization may alter kinetic constants due to a change in activation energy, pH profile, Michaelis constants, or even specificity.

If the matrix is ionic, the microenvironment of the active site may be altered by the electrostatic field. The pH within the matrix will be different than that in the external solution. If substrates are also charged, the apparent Michaelis constant may be altered. Covalent coupling itself may bring about these effects as well as a change in specificity due to an alteration of the enzyme' net charge, nearest neighbor effects on the active site region, perturbations of intramolecular interactions, and conformational changes.

Covalent binding of enzymes to electrically neutral carriers, by way of non-active site residues, has been shown to have no effect on their kinetic behavior towards low-molecular-weight substrates (80,102,165). Detailed discussions of the effects of immobilization on enzyme properties can be found elsewhere (72, 78, 138, 166).

4. Experimental

a. Preparation of Polyacrylamide Entrapped Enzymes. Dissolve N,N'-methylenebisacrylamide (Eastman Chemical Co.) (1.15 g) in phosphate buffer (0.1 M, pH 6.8, 40 ml) by heating to 60°C. Cool the solution to 35°C and add acrylamide monomer (Eastman Chemical Co.) (6.06 g). After mixing, filter the solution into a 50-ml volumetric flask containing riboflavin (Eastman Chemical Co.) (5.5 mg) and potassium persulfate (5.5 mg) and dilute to the mark. The resulting gel solution is stable for months, if stored in the dark. Prepare the immobilized enzyme by mixing 0.10 g of enzyme with 1.0 ml of gel solution. Remove oxygen by purging with N_2. Polymerize by irradiating with a 150W photoflood lamp for one hour. Freeze-dry the preparation to form a powder or prepared slices for use.

b. Preparation of Covalently Bound Enzyme.

(1) Via Acryl Azide Derivative of Polyacrylamide. This preparation is a modification of one given by Inman and Dintzis (125).

Preparation of Polyacrylamide Beads. Acrylamide (21 g), N, N'-methylenebisacrylamide (0.55 g), tris-*hydroxymethyl) aminomethane (Sigma Chemical Co.) (13.6 g), N,N,N',N'-tetramethylethylenediamine (Eastman Chemical Co.) (0.086 ml), and HCl (1 M, 18 ml) are dissolved in 300 ml of distilled water. Ammonium persulfate (0.22 g) is then added as a catalyst. Free radical polymerization is effected by stirring magnetically under a nitrogen atmosphere and irradiating with a 150w projector spotlight (Westinghouse). After polymerization is complete, the stirring bar is removed and the polyacrylamide is broken into large pieces. The polymer is then made into small spherical beads by blending at high speed in 300 ml of water for five minutes. The resulting bead suspension is stored under refrigeration.

Preparation of Hydrazide Derivative. Into a siliconized, three-neck, round-bottom flask is placed 300 ml of the polyacrylamide bead suspension. The flask and a beaker containing anhydrous hydrazine (Matheson, Coleman and Bell) (30 ml) are immersed in a 50°C constant-temperature oil bath. After about 45 minutes, the hydrazine is added to the gel, the flask is stoppered, and the mixture is stirred magnetically for 12 hours at 50°C. The gel is then centrifuged and the supernate discarded. The gel is washed with 0.1 M NaCl by stirring magnetically for several minutes, centrifuging, and discarding the supernate. The washing procedure is repeated until the supernates are essentially free of hydrazine as indicated by a pale violet color after five minutes when tested by mixing 5 ml with a few drops of a 3% solution of sodium 2,4,5-trinitrobenzenesulfonate (Eastman Chemical Co.) and 1 ml of saturated sodium tetraborate (Mallinckrodt). The hydrazide derivative is then stored under refrigeration.

Coupling of Enzyme. The hydrazide derivative (100 mg) is placed in a plastic centrifuge tube and is washed with 0.3 M HCl. It is then suspended in HCl (0.3 M, 15 ml), cooled to 0°C, and sodium nitrite (1 M, 1 ml), also at 0°C, is added. After stirring magnetically for two minutes in an ice bath, the azide derivative formed is rapidly washed with phosphate buffer (0.1 M, pH 6.8) at 0°C by centrifugation and decantation until the pH of the supernate is close to 6.8. The azide gel is then suspended in phosphate buffer (0.1 M, pH 6.8, 10 ml) containing 100 mg of enzyme to be bound. After the mixture is stirred magnetically for one hour at 0°C, glycine (10 ml, 0.5 M) is added to couple with unreacted azide groups. After stirring for an additional hour at 0°C, the enzyme' gel is washed several times with phosphate buffer (0.1 M, pH 6.8) and stored under refrigeration.

The general reaction scheme for this preparation is given by the equations:

$$CH_2{=}CH\text{-}C({=}O)NH_2 + (CH_2{=}CHCONH)_2CH_2 \xrightarrow{(NH_4)_2S_2O_8} \vdash C({=}O)NH_2 \quad ; \tag{41}$$

$$\vdash C(=O)NH_2 + H_2NNH_2 \xrightarrow[\text{12 HRS}]{50^\circ C} \vdash C(=O)NHNH_2 \quad ; \tag{42}$$

$$\vdash C(=O)NHNH_2 + HNO_2 \xrightarrow[\text{2 MIN.}]{0^\circ C} \vdash C(=O)N_3 \quad ; \tag{43}$$

$$\vdash C(=O)N_3 + NH_2\text{-}\boxed{\text{ENZYME}} \xrightarrow{0^\circ C} \vdash C(=O)NH\text{-}\boxed{\text{ENZYME}} \tag{44}$$

(2) Via Diazonium Derivative of Polyacrylic Acid.

Polymerization of Acrylic Acid. Approximately 50 ml of reagent grade acrylic acid (Aldrich Chemical Co.) are dissolved in 20 ml hexane and placed in a round-bottom flask. A few milligrams of ammonium persulfate are added as a free radical initiator, and the system is kept in a dry nitrogen atmosphere. The flask is heated with a heating mantle until rapid polymerization is observed. The mantle is then quickly removed and the flask is allowed to cool to room temperature.

Preparation of Copolymer. The polymer is broken into small particles and neutralized with sodium hydroxide. The sodium salt is evaporated to dryness in a rotary evaporator and ground to a fine powder. The powder (approx 3.6 g) is suspended in 6 ml of hexane and is cooled to approximately 4°C. The acid is converted to the acyl chloride by the addition of $SOCl_2$ (2.8 ml) with stirring in an ice bath for one hour, removing generated gases by suction. The acyl chloride polymer is washed with ether and dried under vacuum. Then p-nitroaniline (0.5 g) and ether (6 ml) are added, and the mixture is allowed to stir overnight. The product formed is filtered, washed with ether, and air dried.

Coupling of Enzyme. The p-nitroaniline derivative (150 mg) is dissolved in 10 ml of distilled water, and the solution is adjusted to pH 5 with dilute acetic acid. Ethylenediamine is added slowly with stirring until a fine white precipitate is observed. The precipitate is then washed three times with distilled water and suspended in 5 ml distilled water. The polymer is then reduced by adding TiCl , and the product is washed several times with distilled water by centrifugation and decantation. The reduced derivative is converted to a diazonium salt by adding nitrous acid (0.5 M, 10 ml) at approximately 4°C with stirring for two minutes. The diazonium salt intermediate is flushed with cold distilled water and is rapidly washed several times with phosphate buffer (0.1 M, pH 6.8) by centrifugation and decantation. It is then mixed with a phosphate buffer solution (0.1 M, pH 6.8) containing 100 mg of pure enzyme to be bound at approximately 4°C for one hour. The resulting gel is washed several times with buffer and stored under refrigeration.

The general reaction scheme for this preparation is given by the equations:

$$CH_2{=}CHC(=O)OH \xrightarrow[(NH_4)_2S_2O_8]{80-110°C} \vdash C(=O)OH \quad ; \tag{45}$$

$$\vdash C(=O)OH \xrightarrow{SOCl_2} \vdash C(=O)Cl \quad ; \tag{46}$$

$$\vdash C(=O)Cl + H_2N{-}C_6H_4{-}NO_2 \longrightarrow \vdash C(=O)NH{-}C_6H_4{-}NO_2 \quad ; \tag{47}$$

$$\vdash C(=O)NH{-}C_6H_4{-}NO_2 \xrightarrow{TiCl_3} \vdash C(=O)NH{-}C_6H_4{-}NH_2 \quad ; \tag{48}$$

$$\vdash C(=O)NH{-}C_6H_4{-}NH_2 \xrightarrow[0\,°C]{NaNO_2/HCl} \vdash C(=O)NH{-}C_6H_4{-}N^{\oplus}Cl^{\ominus} \quad ; \tag{49}$$

$$\vdash C(=O)-NH-C_6H_4-N_2^{\oplus}\ Cl^{\ominus} \xrightarrow{HO-C_6H_4-[ENZYME]} \vdash C(=O)-NH-C_6H_4-N{=}N-C_6H_3(OH)-[ENZYME] \quad (50)$$

(3) Via Glutaraldehyde Crosslinking. To a solution containing phosphate buffer (0.1 M, pH 6.8, 2.7 ml) and bovine serum albumin (Nutritional Biochemicals Corp.) (17.5%, 1.5 ml) is added pure enzyme to be bound (50 mg). The resulting solution is rapidly mixed with glutaraldehyde (Sigma Chemical Co.) (2.5%, 1.8 ml) for several seconds and is then immediately applied to the electrode surface. The general reaction for this preparation is given by the equation:

$$CHO-CH_2-CH_2-CH_2-CHO + NH_2-[ENZYME] \longrightarrow [ENZYME]-NH-CH(OH)-CH_2-CH_2-CH_2-CH(OH)-[ALBUMIN] \quad (51)$$

(4) Using ENZACRYL AA. Enzacryl AA (Aldrich Chemical Co.) (200 mg) is placed into a 50-ml plastic centrifuge tube and allowed to stir magnetically overnight with HCl (2 M, 20 ml). It was then cooled to $0^{o}C$ in an ice bath and an ice-cold sodium nitrate solution (4%, 8 ml) is added. The mixture is stirred for 15 minutes, then washed four times with phosphate buffer (0.1 M, pH 7.8) at $0^{o}C$ by centrifugation and decantation. To the diazonium salt formed is added a phosphate buffer solution (0.1 M, pH 6.8, 2 ml) containing high-purity enzyme (80 mg). The mixture is allowed to stir at $0^{o}C$ for up to 48 hours. The bound enxyme is then washed several times with phosphate buffer (0.1 M, pH 6.8) and stored under refrigeration.

The general reaction scheme for this preparation is given by Eqs. 46 and 47.

B. Construction of Enzyme Electrodes

In Section I, subsection C, it was mentioned that there are four steps to follow in constructing an enzyme electrode. Each of these factors will now be covered in more detail:

1. Step 1. Pick an enzyme that reacts with the substance to be determined. From standard reference books on enzymology, such as Biochemists Handbook (C. Long, ed., Van Nostrand, Princeton, U.S.A.) find an enzyme system suitable for your determination. In one case, this will

involve using the primary function of the enzyme, i.e., the main substrate-enzyme reaction. For example, for a penicillin electrode, penicillinase would be used; for a glutamine electrode, glutaminase; for an L-glutamic acid electrode, L-glutamate dehydrogenase. In other cases, this might necessitate using an enzyme that acts on the compound of interest as a secondary substrate; i.e., urease for N-methyl urea, or malic dehydrogenase for acetic acid (170). Of course, this latter case will introduce more interferences and less selectivity into the assay. If you cannot find a suitable system yourself, seek assistance from biochemists.

Note: In some cases, several enzymes may act on the substrate of interest via different reactions. For example, L-tyrosine could be determined using L-tyrosine decarboxylase and measuring the CO liberated (55), or by using L-amino acid oxidase and a Pt electrode (60), or an NH electrode (56,58). The latter enzyme, although less selective, can be obtained commercially in high purity; the former is available in low purity and would have to be purified before use. Hence, the scientific capabilities of one's laboratory might dictate the choice of enzyme.

2. Step 2. Once the enzyme for an application is found, check commercial suppliers' catalogs (Sigma, Boehringer, Calbiochem, Worthington, etc.) to check its availability

and purity. The latter may or may not present a problem. Many enzymes, such as jack bean urease or glucose oxidase from the food industry (General Mills), are stable in an impure state, and can be used satisfactorily in a pseudo "immobilized form" (i.e., as a liquid covered with a dialysis membrane) for up to a week. In other cases, like any of the decarboxylases available from Sigma, the impure enzyme has too low an activity to be useful in the low-purity state without further purification. In the latter case, the enzyme must be purified. Though not difficult, this involves further work and maybe even assistance from others.

In still other cases, one might find that the enzyme one wants to use is not available commercially. In this case, there are two possibilities: ask a large biochemical supply house if it will isolate and purify the enzyme you want (many will, if the price is right), or look up the enzyme in the literature or standard biochemistry-enzymology reference books, determine the isolation and purification methods used, and do it yourself. We have followed this second course in many cases with excellent results. In most cases, the techniques are simple enough to be carried out by a person with reasonable scientific training. Frequently, the results are well worth the effort.

3. Step 3. It is well to keep in mind that the better an enzyme is immobilized, the more stable it will be, the longer it will be useful, and the more assays will be possible from one batch. Let's now consider the various possibilities, and the characteristics of the product.

a. Commercially available immobilized enzyme. This is the ideal case and is the first choice, if possible. A number of enzymes are available in immobilized form. Most of these are fine products, and as good or better than most scientists can do themselves. The author has personally used the products of Boehringer and Aldrich with good success. Available enzymes that are likely to be of most use to the reader include urease (Boehringer), glucose oxidase (Boehringer or Aldrich), ribonuclease (Boehringer), and uricase (Aldrich). Furthermore, under certain conditions, Corning has offered to sell almost any enzyme bound to glass, and the reader is invited to write for details.

b. Soluble "immobilized" enzyme. The second choice available, which is the easiest for the novice if the first choice is not possible, is to simply use the soluble enzyme in constructing the electrode. A thick paste is made of the enzyme powder with a little water (1 to 1 μl); the paste is spread over the surface of the electrode, and this layer is covered with a 20 to 25 μm-thick cellophane dialysis membrane (Will Scientific, Inc., or

Arthur H. Thomas, U.S.A.). Such soluble enzyme electrodes are stable for up to about a week, if kept in a 5-10°C refrigerator between use. Electrodes using the more crude enzymes such as urease or glucose oxidase mentioned earlier, might be stable for longer periods of time.

c. Physically entrapped enzymes. For ease of preparation, this is the next choice. Many enzymes have been physically bound in polyacrylamide gels, which are cross-linked polymers with the enzyme held inside. A typical preparation is mentioned in Section II, Subsection A, and in 4b of this section. Similar preparations can be effected with a minimum of effort by almost anyone. As carefully pointed out by Guilbault and Montalvo (50), the stability of the final product depends on the care taken and the control of experimental conditions and can be as long as three to four weeks or about 50 to 100 determinations.

d. Chemically bound enzymes. Although these are the most difficult to prepare, preparation can be effected by anyone who has had a year's course in organic chemistry. The products are most stable, can be used for 200 to 1,000 assays, and can be stored at room temperature for more than a year between assays (46). The best preparation method is determined by the individual enzyme, as will be discussed in other parts of this book. In the author's experience, the polyacrylacid diazo coupling (46,60) and the glutaraldehyde methods (19,20) have yielded extremely satisfactory results. The covalent binding to polyacrylamide crosslinked polymer, used by Boehringer in its commercial preparations, is also quite satisfactory.

Reactive intermediate for one- or two-step direct coupling of enzymes are available from Corning Glass Works (Corning, New York), Aldrich, Milwaukee, Wisconsin) and Koch Light (England). These are recommended to anyone interested in making chemically bound enzymes. The glutaraldehyde method is also quite simple to effect (glutaraldehyde is available from Sigma, St. Louis).

4. Step 4. Place the enzyme around the appropriate electrode. To develop an electrode for the substrate of interest, one must have a base sensor electrode that responds to reactant A or B in the equation

$$A + B \xrightarrow{\text{Enzyme}} C + D, \qquad (52)$$

or to one of the products, C or D. The sensor can be a gas electrode (to measure all O_2-consuming reactions, NH_3, or CO_2-liberating enzymes), a glass electrode (to follow H^+ changes in reactions that liberate acid, or NH_4^+-producing enzymes), a Pt electrode (to follow all enzyme reactions involving electroactive species or O_2), or some other ion-selective electrode (i.e., the CN^- electrode for amygdalin; a NH_4^+ antibiotic electrode for deaminase enzymes; an I^- electrode for oxidative enzymes coupled with the $I^- — I_2$ indicator reactions; the $S^=$

electrode for cholinesterase substrates, etc.). In most cases, the limiting factor in designing an enzyme electrode will be the availability of a sensor that can monitor the reaction. Of course, there are other possibilities for monitoring enzyme reactions. For example, a thermistor covered with enzyme could measure the temperature change resulting from the enzyme reaction being monitored (171). Considerable research is being done in this area, but no satisfactory systems have been developed so far.

Assuming that a sensor is available, and that the enzyme has been obtained and immobilized, we will now describe the preparation of typical enzyme electrodes (Fig. 2).

a. Type A -- Dialysis membrane electrode. Place 10 to 15 units of soluble, physically entrapped, or chemically bound enzyme (after immobilization of the enzyme the preparation should be freeze dried to form a powder) on a circular piece -- about twice the diameter of the electrode sensor -- of cellophane dialysis membrane (20 to 25 m thick, available from either Will Scientific or Arthur H. Thomas, U. S. A.) Wrap the cellophane around the electrode, taking care that the powder is spread over the surface of the electrode in a thin, even layer. (This might be coveniently done by placing a thick paste of the enzyme in water on the tip of the flat electrode while it is held upside down (Fig. 2B) and coating the enzyme on the surface with a spatula.) Place a rubber "O" ring (with a diameter that fits the electrode body snugly) around the cellophane (Fig. 2B) and gently push it onto the electrode body so the cellophane-enzyme layer on the bottom of the electrode is held tight and flat (Fig. 2B). Place the electrode in a buffer solution for a few hours or overnight. This allows buffer to penetrate the enzyme layer and lets entrapped air escape. Store the electrode in buffer between use (46,53,54,60, 61,64).

b. Type B electrode -- physical entrapment. Holding the electrode sensor upside down (Fig. 2A), cover it with a thin nylon net (about 90 μm thick - sheer nylon stocking obtained from any ladies shop is satisfactory), and secure it with a rubber "O" ring as described in the preceding subsection. The net serves as a support for the enzyme gel solution. Prepare the enzyme gel solution by mixing 0.1 g of enzyme (purity about 10-50 units/mg) with 1.0 ml of gel solution - 1.15 g of N,N'-methylene-bisacrylamide (Eastment Organics, Rochester), 6.06 g of acrylamide monomer (Eastman) 5.5 g of potassium persulfate, and 5.5 mg of riboflavin in 50 ml of water. Gently pour the enzyme gel solution onto the nylon net in a thin film, making sure all the pores of the net are saturated; one ml of this solution should be enough for several electrodes. Place the electrode in a water-jacketed cell at 0.5^{o}C and remove oxygen, which inhibits the polymerization, by purging with N_2 before and during polymerization. Complete the polymerization by irradiating with a 150W Westinghouse projector spotlight for one hour, after which time the enzyme layer should be dry and hard. Place a piece of dialysis membrane over the outside of the nylon net for further protection, and secure with a second rubber "O" ring. Soak the electrode in buffer solution overnight and store in buffer between use (42,46,48-52, 56-59,62,63,65).

c. Type C -- Direct polymerization onto the membrane. This can be effected by a direct attachment of the enzyme to the surface of the electrode by the Corning technique discussed elsewhere in this book, if the electrode is glass, or by direct chemical attachment on the electrode surface, as was done by Anfalt, Granelli and Jagner (38) in the case of the Orion NH electrode. In the latter study, membranes were prepared by dropping 0.1 ml of soluble urease solution (0.5 μ) onto the surface of the gas diffusion membrane. The membrane was set aside for 12 hours at 4° C to allow evaporation of the solvent. Glutaraldehyde solution was then added dropwise (2.5% in phosphate buffer, pH 6.2). The membrane was set aside for another 1.5 hours at 4°C, and was then rinsed carefully with water to remove free enzyme and buffer. Note that some activity was lost over a 20-day period, indicating insufficient enzyme was used (38). At least 5 to 10 units of urease would have been better in this case, again in 0.1 ml of solution.

Of the three types of electrodes described above, Type A is preferred for ease of preparation (once the bound enzyme is obtained) and for long-term stability (if the enzyme is chemically bound). The Type B electrodes are not difficult to prepare, but are time-consuming, requiring one hour of polymerization time per electrode, and have a maximum stability of about 3 to 4 weeks or 50 to 100 assays. The Type C electrode with direct attachment to the electrode is not favored because it esenstially commits the electrode sensor to the enzyme attached. With the Type A or B electrodes, the enzyme layer is easily replaced when it is no longer useful, so the sensor can be reused until its lifetime is exhausted.

C. Response Charcteristics of Electrodes

Once the electrode has been made, some of the factors that affect electrode response and stability should be considered. Table 5 lists the enzyme electrodes prepared to date, in which the enzyme is held "immobilized" in the vicinity of the electrode. The table also presents the response characteristics: response time, linear range, wash time, stability, amount of enzyme used, and the type of immobilization method. This listing is a guide to the various electrodes prepared, and enables comparison of the various types.

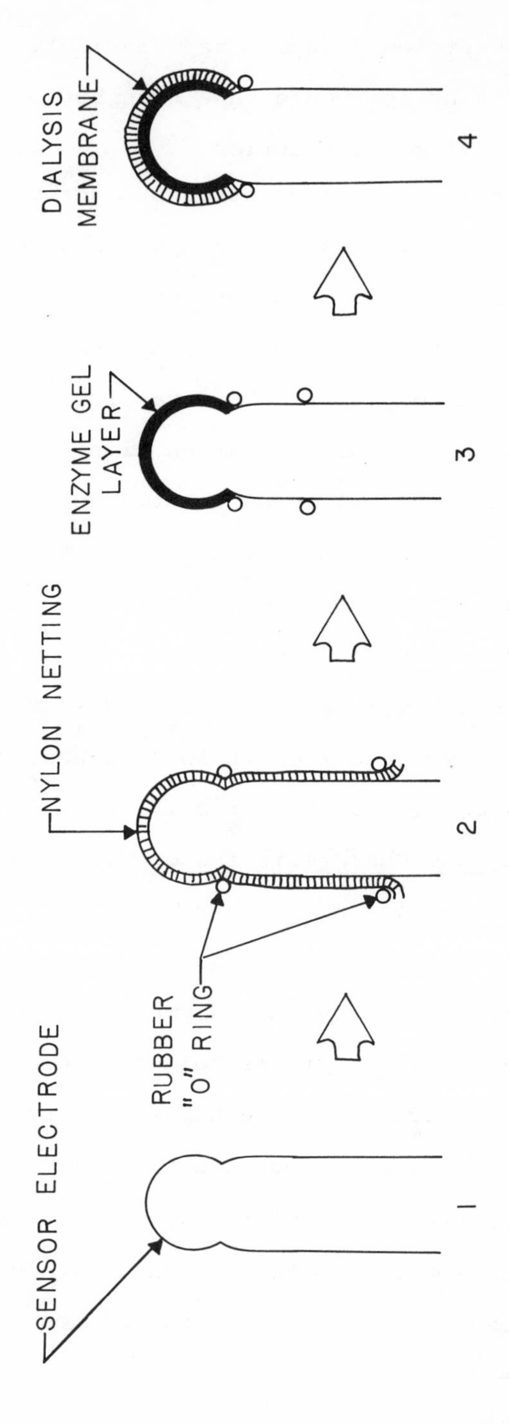

Fig. 2. Preparation of Enzyme Electrodes. A. Physically Entrapped Enzyme Electrodes.

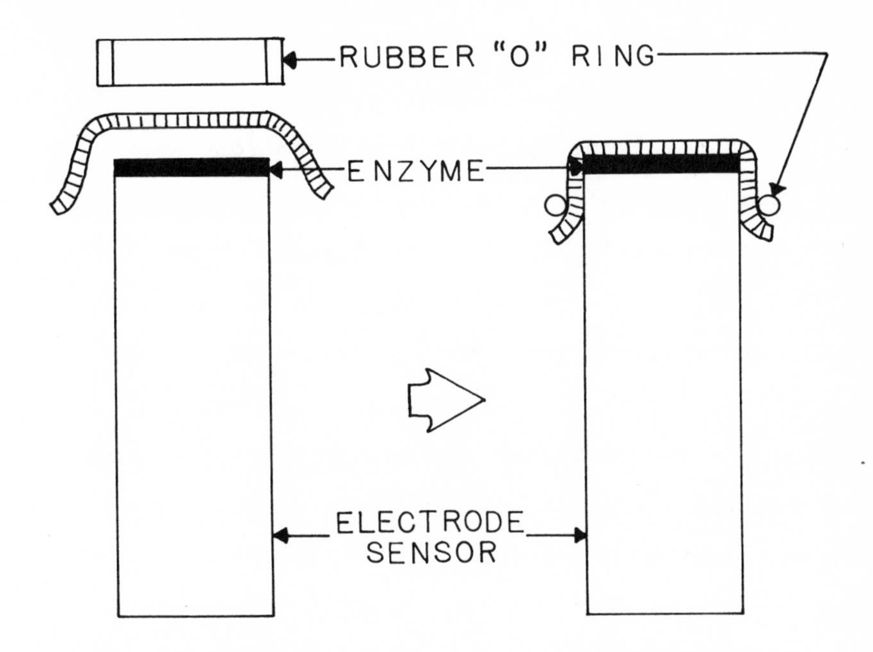

Fig. 2. Preparation of Enzyme Electrodes. B. "Soluble" or Chemically Bound Enzyme Electrodes.

1. Stability

The stability of an enzyme electrode is difficult to define, since an enzyme can lose some of its activity resulting in a shift of the calibration curve downwards. Yet, if the slope remains constant, as is frequently the case, the electrode is still useful, needing only calibration daily. This is seldom a problem, since those who use electrodes reset the pH or potential of their electrode using serum (i.e., Monitrol, Dade, Miami). Another problem in defining stability is that many workers measure the potential of their electrodes over long periods of time and report this data as the stability. However, the electrode may have been used once a day or week, 10 time a day, or 100 times a day. Naturally, the greater the use, the shorter the overall lifetime. As a general rule, a "soluble" electrode is useful for about one week or 25 to 50 assays if the electrode is kept refrigerated between uses. The physically entrapped "polyacrylamide" electrodes are good for about three weeks or 50 to 200 assays, depending crucially on the degree of care exercised in preparing the polymer. If they are not used very much, the chemical enzyme can be kept indefinitely even at room temperature (see Table 2 - as long as 14 months for glucose oxidase or uricase). One can expect to get about 200 to 1,000 assays per chemical enzyme electrode, again depending on how good a synthesis is effected.

Although the electrode can be stored at room temperature, it is recommended that all electrodes be kept in a refrigerator and covered with a dialysis membrane to prevent the action of bacteria which tend to feed on the enzyme, destroying its activity. The dialysis membrane (Mol Wt exclusion about 1,500) prevents the enzyme from getting out and bacteria from getting in. The stability of the physically entrapped enzyme varies greatly with experimental conditions, and a thorough study of these factors was made by Guilbault and Montalvo (50).

To determine the effect physical immobilization parameters have on the stability of the urea electrode, a series of enzyme electrodes was prepared, varying one immobilization parameter while maintaining all of the other parameters constant. To determine the stability of the immobilized urease coating on the surface of the cation electrode, the steady-state potential was obtained for a given urea substrate concentration at periodic time intervals. If the steady-state potential is constant within a certain time period, no loss of immobilized enzyme activity has occurred. All stability data reported were obtained with an electrode stored between measurements at 25°C in Tris buffer.

The maximum stability that could be achieved with the Type I enzyme electrode was obtained with the following immobilization parameters: photopolymerizing for one hour at 28°C with a No. 1 150W photoflood lamp; a gel-layer thickness of 350 μ; and an enzyme concentration in the gel of 175 mg/cm^{-3}gel. The slope of the stability curve, ΔmV/Δt, shows that the measured stability depends on the substrate concentration used in the stability measurements. When the urea concentration is high enough that the steady-state response is independent of the substrate concentration, ΔmV/Δt was 0.2 mV/day over a 14-day period. At lower substrate concentrations, 1 x 10^{-3} $\underline{M}$ urea for example, the steady-state response is first order in urea concentration, and a much smaller loss in activity -- 0.05 mV/day over a 14-day period -- was obtained. Since 1 x 10^{-3} $\underline{M}$ urea represents the upper limit of substrate concentration that can be measured with the enzyme electrode, the steady-state response falls by only 0.7 mV during 14 days operation at 25°C. After 14 days, the activity loss was much greater for both substrate concentrations.

Table 5

Various Enzyme Electrodes and their Characteristics

Type	Enzyme	Sensor	Immobilization[d]	Stability	Response Time	Amount of Enzyme	Range[g] (M)	Wash Time[h]	Reference
1. Urea	Urease	Cation	Physical	3 weeks	30 sec - 1 min.	25 units	10^{-2}-5 x 10^{-5}	2 min.	49,50
		Cation	Physical	2 weeks	1-2 min.	75 units	10^{-2}-10^{-4}	2 min.	52
		Cation	Chemical	>4 months	1-2 min.	10 units	10^{-2}-10^{-4}	2 min.	53
		pH	Physical	3 weeks	5-10 min.	~100 units	5 x 10^{-3} - 5 x 10^{-5}	10 min.	48
		Gas (NH_3)	Chemical	>4 months	2-4 min.	10 units	5 x 10^{-2} - 5 x 10^{-5}	20 sec.	54
		Gas (NH_3)	Chemical	20 days	1-4 min.	0.5 units	10^{-2}-10^{-4}	2-3 min.	38
		Gas (CO_2)	Physical	3 weeks	1-2 min.	25 units	10^{-2}-5 x 10^{-5}	2 min.	55
2. Glucose	Glucose Oxidase	pH	Soluble	1 week	5-10 min.	~100 units	10^{-1}-10^{-3}	10 min.	48
		Pt (H_2O_2)	Physical	6 months	12 sec - kinetic[f]	10 units	2 x 10^{-2}-10^{-4}	1 min.	45,46
		Pt (H_2O_2)	Chemical	>14 months	1 min - Steady State	10 units	2 x 10^{-2}-10^{-4}	1 min.	46
		Pt (H_2O_2)	Soluble	--	1-2 min.	~ 10 units	10^{-2}-10^{-4}	1 min.	41,43
		Pt (H_2O_2)	Soluble	<1 week[e]	3-10 min.	10 units	2 x 10^{-2}-10^{-3}	5 min.	44
		Pt (O_2)	Chemical	>4 months	1 min.	10 units	10^{-1}-10^{-5}	0.5-1 min.	20
		I^-	Chemical	>1 month	2-8 min.	10 units	10^{-3}-10^{-4}	2 min.	34
		Gas (O_2)	Physical	3 weeks	2-5 min.	20 units	10^{-2}-10^{-4}	2 min.	42

Table 5 (Continued)

Type	Enzyme	Sensor	Immobilization[d]	Stability	Response Time	Amount of Enzyme	Range[g] (M)	Wash Time[h]	Reference
3. L-Amino Acids (General)[a]	L-AA Oxidase	Pt	Chemical	4-6 months	12 sec - kinetic[f]	10 units	10^{-3}-10^{-5}	0.5-1 min.	60
		Cation	Physical	2 weeks	1-2 min.	10 units	10^{-2}-10^{-4}	2 min.	56
		NH_4^+	Chemical	>1 month	1-3 min.	10 units	10^{-2}-10^{-4}	2 min.	57
		I^-	Chemical	>1 month	1-3 min.	10 units	10^{-3}-10^{-4}	2 min.	57
L-Tyrosine	Decarboxylase	Gas (CO_2)	Physical	3 weeks	1-2 min.	25 units	10^{-1}-10^{-4}	2 min.	55
L-Glutamine	Glutaminase	Cation	Soluble	2 days[e]	1 min.	50 units	10^{-1}-10^{-4}	2 min.	59
L-Glutamic acid	Dehydrogenase	Cation	Soluble	2 days[e]	1 min.	50 units	10^{-1}-10^{-4}	2 min.	59
L-Asparagine	Asparaginase	Cation	Physical	1 month	1 min.	50 units	10^{-2} -5 x 10^{-5}	2 min.	58
4. D-Amino Acids (General)[b]	D-AA Oxidase	Cation	Physical	1 month	1 min.	50 units	10^{-2}-5 x 10^{-5}	2 min.	58
5. Lactic Acid	Dehydrogenase	Pt	Soluble	<1 week[e]	3-10 min.	2 units	2 x 10^{-3} - 10^{-4}	5 min.	44
6. Alcohols[c]	Oxidase	Pt(H_2O_2)	Soluble	1 week	12 sec.- kinetic[f]	10 units	0.5-100 mg%	0.5-1 min.	61
		Pt(H_2O_2)	Soluble	1 day[e]	1 min.	~1 unit	0.5-50 mg%	--	167
		Pt(O_2)	Chemical	>4 months	30 sec.	10 units	0.5-100 mg%	1 min.	20

Table 5 (Continued)

Type	Enzyme	Sensor	Immobilization[d]	Stability	Response Time	Amount of Enzyme	Range[g] (M)	Wash Time[h]	Reference
7. Penicillin	Penicillinase	pH	Physical	1-2 weeks[e]	0.5-2 min.	~400 units	10^{-2}-10^{-4}	3-5 min.	65
			Soluble	3 weeks	2 min	~1,000 units	10^{-2}-10^{-4}	10 min.	48
8. Uric Acid	Uricase	$Pt(O_2)$	Chemical	>4 months	30 sec.	~ 10 units	10^{-2}-10^{-4}	1 min.	19
9. Amygdalin	β-Glucosidase	CN^-	Physical	3 days[e]	10-20 min.	100 units	10^{-2}-10^{-5}	2 min.	62,63
			Soluble	1 week	1-3 min.	10 units	10^{-1}-10^{-5}	2 min.	64

[a] Electrode responds to L-cysteine, L-leucine, L-tyrosine, L-tryptophan, L-phenylalaine, L-methionine.

[b] Electrode responds to D-phenylalaine, D-alanine, D-valine, D-methionine, D-leucine, D-norleucine, D-isoleucine.

[c] Electrode responds to methanol, ethanol, allyl alcohol.

[d] Physical refers to polyacrylamide gel entrapment in all cases; chemical is attachment to glutaraldehyde with albumin, to polyacrylic acid or to acrylamide chemically followed by physical entrapment.

[e] Preparation lacks stability as evidenced by constant daily decrease in signal.

[f] Kinetic - rate of change in current measured after 12 seconds; steady state - current reaches a maximum in one minute.

[g] Analytically useful range - either linear or with reasonable change if curvature is observed.

[h] Time required for signal to return to base line before reuse.

To study the effect of immobilized urease activity on enzyme gel stability, Type I enzyme electrodes were prepared with activity of enzyme from 375 to 3500 Sumner units/gram of enzyme. No appreciable change in stability occurred with this relatively large change in enzyme activity. On the other hand, highly purified urease is known to be very unstable in solution. A similar trend in stability would be expected with immobilized urease.

Greater stability with the Type I enzyme electrode was always obtained when the gel solution was less than two days old. Gel solutions were stored without added polymerization catalysts when the storage period was greater than two days. The solutions were always stored in the dark at room temperature. The stability of the urease Type I electrode was studied as a function of enzyme gel-layer thickness in the range 30 to 350 μ . The stability increased with increased enzyme gel layer thickness, but response time also increased.

Several experiments were run to quantitatively determine the effect of photopolymerization light intensity and photopolymerization time on Type I enzyme electrode stability. When the high-intensity photoflood lamp is substituted with a 60W domestic lamp, the loss in activity rises from 0.2 to 4.2 mV/day for 8.33 x 10^{-2} $\underline{M}$ urea. A similar loss in activity for Type I electrode was obtained when only the photopolymerization time was reduced from one hour to 15 minutes.

To study the effect photopolymerization, temperature, and gel layer water content have on Type I electrode stability during photopolymerization, a series of enzyme electrodes was prepared with photopolymerization temperature ranging from 4°C to 43°C. The water content of the gel layer over the electrode surface was also varied when the photopolymerization temperature was changed. This is because the rate of evaporation of water for the thin enzyme gel layer varies directly with temperature. When the photopolymerization temperature and water content of the gel were varied to study Type I electrode stability, the other immobilization parameters were adjusted to give maximum stability. The stability, measured with 8.33×10^{-2} M urea, showed a loss of only 0.2 mV/day at 28°C photopolymerization temperature.

Upon lowering the immobilization or photopolymerization temperature to 6°C, the loss in electrode activity is much higher, 3.7 mV/day. At 6°C, the rate of water evaporation from the gel layer is sufficiently rapid so that when the polymerization is complete, the electrode is dry to the touch. The enzyme electrode is now more stable because a less porous polymer is formed. At higher polymerization temperature, such as 43°C, the resulting electrode is again less stable than when the polymerization temperature is 28°C. Therefore, maximum stability is obtained with Type I enzyme electrode when photopolymerized at 25°C to 28°C.

To determine the effect a cellophane film has on enzyme electrode stability, a Type II electrode was made by placing a thin film of cellophane over the enzyme gel layer. The cellophane was permeable to the urea substrate, but not to the high-molecular-weight enzyme. Polymerization parameters were the same as those used to obtain the maximum stability for the Type I (cellophaneless) electrode. Enzyme electrode Type II stability, measured with either 8.33×10^{-2} or 1×10^{-3} $\underline{M}$ urea, showed no measurable loss in activity for 21 days (electrode stored between measurements in Tris buffer at $25^{o}C$). After 21 days, the electrode began to lose activity. The increased stability of Type II electrode over Type I electrode is apparently due to the cellophane, which prevents any enzyme from leaching out of the enzyme gel layer. The stability of Type III electrode was identical with that of Type II.

Another factor that affects the apparent stability of all electrodes, especially the "soluble" and physically entrapped electrodes, is the enzyme content of reaction layer. As will be shown later, a certain amount of enzyme is required to yield a Nernstian calibration curve. Many times it is advantageous to add more enzyme, say twice as much, so more enzymic activity can be lost while still maintaining a linear Nernstian plot.

Still another factor affecting the stability of an electrode is the choice of operating conditions. An example of this is the comparison of the results obtained by Llenado and Rechnitz (62,63) and those of Mascini and Liberti (64) for the amygdalin electrode. Amygdalin is cleaved by β-glucosidase to give CN^- ions, which are sensed by a CN^- ion-selective electrode. Since this electrode responds best to free CN^- ions, obtainable only at pH's > 10, Lleando and Rechnitz used this pH in operating their electrode. Even though the enzyme was physically bound, it continually lost activity and had a lifetime of only a few days. Since it is known that almost all enzymes will lose activity at pH's < 3 and > 9, this undoubtedly was one contributant to the poor stability. Mascini and Liberti used only a soluble enzyme at pH 7, and found not only better stability (one week, which is all that can be expected of a soluble enzyme) but also faster response times. Another reason for the poor stability of Llenado and Rechnitz (63) is the "sausage" polymerization these authors tried, in which they made large pieces of physically entrapped enzyme and then cut slices for each assay. From our own and others' experiences with such a technique, the sausage obtained is like a roast beef placed in an oven for 30 minutes -- it is well done on the outside and raw on the inside. The reader is advised not to attempt such large-scale entrapment, but should prepare individual small batches of polyacrylamide enzyme gels.

A thorough comparison of the stability of the three types of "immobilized" enzyme electrodes was made by Guilbault and Lubrano (46) for glucose oxidase. The long-term stabilities of all three types of electrodes were studied by testing the response of each type of electrode to 5×10^{-3} M glucose in phosphate buffer, pH 6.6, at least once a week for several months. When not in use, the electrodes were stored in phosphate buffer at 25°C. The results, shown in Fig. 3, show that the long-term stability decreases in the order: chemically bound — physically bound — solubilized. Not only did the Type 1 electrode response decrease drastically with time, it also decreased with each determination of glucose. This is a serious problem, and, as a result, the Type 1 electrode is of little use analytically, except with frequent calibration and use of a large excess of enzyme. This problem is not encountered with the Type 2 and 3 electrodes using immobilized enzyme. The activity of these electrodes actually increased for the first 20 to 40 days before beginning to decrease. This performance is probably due to the establishment of diffusion channels in the matrix over a period of time with concomitant increase in apparent activity until the channel formation ceases and only denaturation is observed. Or, it could be due to changes in the conformation of the fraction of enzyme immobilized in a non-active conformation to the more stable and preferred conformation. Immobilization in an unfavorable conformation can be due to pH, temperature or stirring effects during the immobilization process. The decrease in response is due to a decrease in activity of the enzyme layer because of slow denaturation, and possibly to slow, irreversible inhibition. The Type 2 and 3 electrodes eventually reach a stability change of -0.25 and -0.08% of maximum response per day, respectively. The physically bound enzyme lost half its activity in seven months, but the chemically bound enzyme lost only 30% of its activity in 400 days (13 months). Of course, this stability would have been much less if the electrodes had been subjected to considerable daily use. Actually about 200 assays for the Type 2 electrode and almost 1,000 for the Type 3 enzyme electrode are possible.

Still another factor affecting the stability of some enzyme electrodes is the leaching of a loosely bound cofactor, which is needed for the enzymic activity, from the active site. Such was found by Guilbault and Hrabankova (58) in the case of D-amino acid oxidase in a polyacrylamide membrane. The bond between protein and coenzyme (flavine adenine dinucleotide, FAD) is very weak in D-amino acid oxidase, and FAS is easily removed by dialysis against buffer without FAD. Without FAD in the solution used to store the electrode, all activity is lost in one day. Storing the electrode between use in a 4 x 10^{-4} M solution of FAD in tris buffer, pH 8.0, resulted in a three-week stability, with little loss in activity.

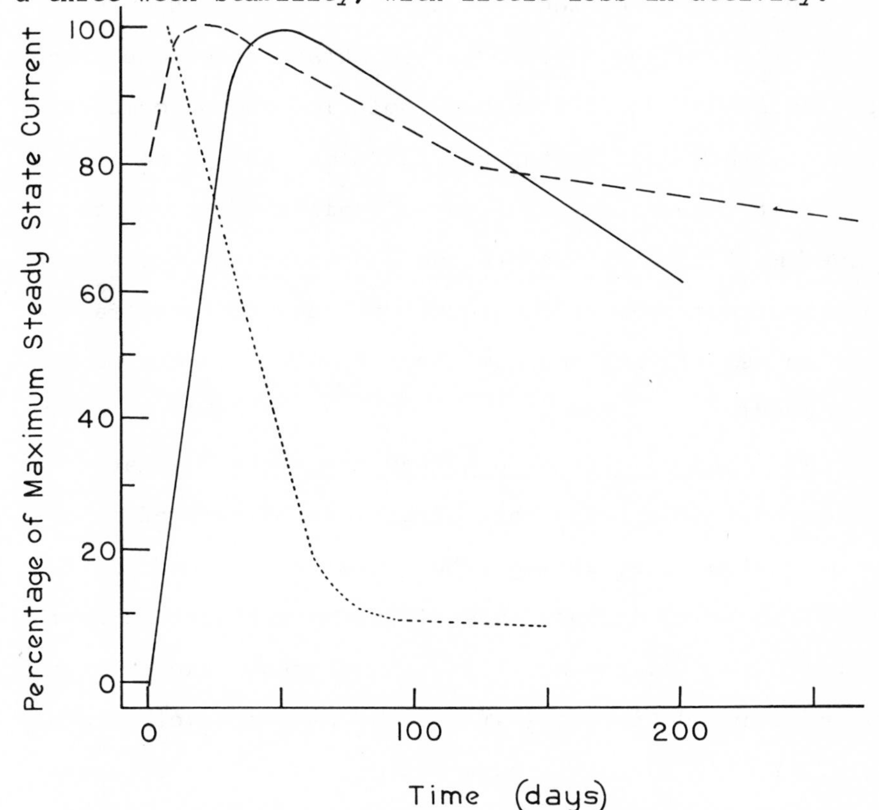

Fig. 3. Long-term stability of glucose electrodes by the Steady-State Method. --- Type 3 Electrode. ___ Type 2 Electrode, ... Type 1 Electrode. (Ref. 46 by permission of authors).

Finally, the stability of the enzyme electrode depends on the stability of the base sensor. In most cases, this is not the limiting factor in stability. The sensor has longer stability than the immobilized enzyme.

This factor should be considered however, in using some of the shorter life-span electrodes, such as the liquid membrane electrodes.

2. Response Time

There are many factors that affect the response speed of an enzyme electrode. To obtain a response, the substrate must diffuse through solution to the membrane surface, then diffuse through the membrane and react with enzyme at the active site, and, finally, the products formed must then diffuse to the electrode surface where they are measured. Let us consider each of these factors in detail, and see how the response time can be optimized.

a. Rate of Diffusion of the Substrate. A mathematic model describing this effect can be derived as was done by Blaedel et al (66), but in simplest, practical terms, the rate of substrate diffusion will depend on the stirring rate of the solution, as was shown experimentally by Mascini and Liberti (64) for the amygdalin electrode (Fig. 4). In an unstirred solution, substrate accesses the membrane surface slowly, so that long response times are observed. At high stirring rates, the substrate is quickly diffused to the membrane surface where it can react. The difference can be a reduction in response time of from 10 minutes to 1 or 1 minutes, or less. With rapid stirring for the urea electrode (49,50), a response time of less than 30 seconds was achieved. Also of importance is the relationship of stirring rate to the equilibrium potential observed. As shown in Fig. 4, the electrical potential shifts as a function of stirring rate and is due to the changing amount of substrate brought to the electrode surface and the degree of its reactivity. Hence, for fast response time and steady reproducibility values, fast, constant stirring is recommended (i.e., use the same stirrer speed for all readings).

<u>b. Reaction with Enzyme in Membrane.</u> According to the Michaelis-Menten equation,

$$v = \frac{k_3[E][S]}{Km = [S]}, \qquad (53)$$

the rate of reaction depend on the concentration of substrate and on enzyme activity and those factors that affect it, such as pH, temperature, and inhibitors. The equilibrium potential obtained, however, should be dependent only on the substrate concentration and the temperature (since this term appears in the Nernst equation). The response rate will also depend on the thickness of the membrane layer in which reaction occurs, and on the size of the dialysis membrane used to cover the enzyme layer, if one is used. Let us consider each of these factors separately.

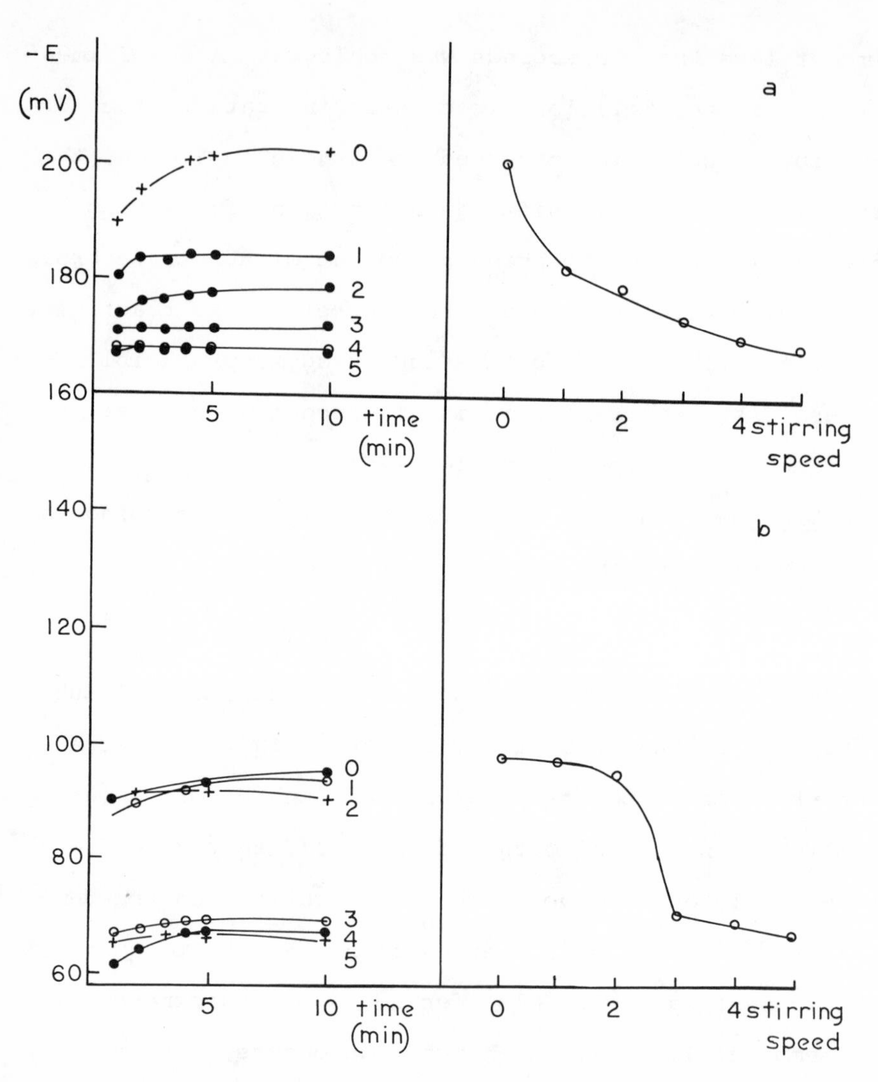

Fig. 4. Stirring Effect. Response time and equilibrium value with different stirring speed. The number 0 corresponds to unstirred solutions and numbers 1-5 to increasing stirring speed and are arbitrary numbers. (Amygdalin) = (a) 10^{-2} M, (b) 10^{-4} M; enzyme amount = 1 mg. (Ref. 64 by permission of authors.)

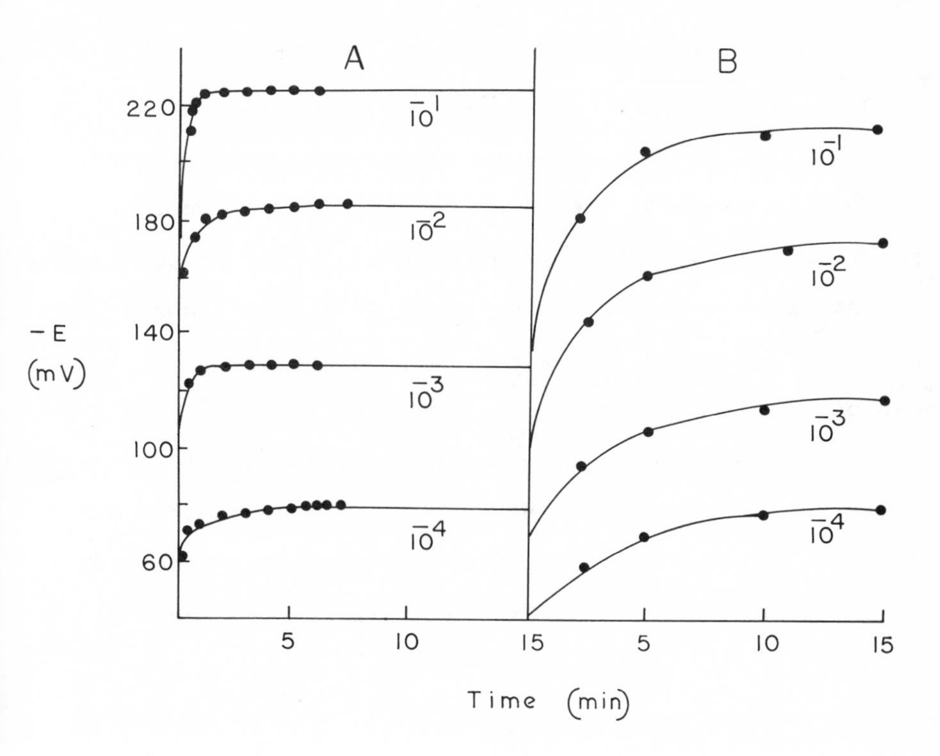

Fig. 5. Amygdalin response-time curves for an electrode containing 1 mg of β-glucosidase immobilized by a dialysis paper. (A) at pH 7 (B) at pH 10 (Ref. 64 by permission of authors.

(1) Effect of substrate. Two typical examples of the effect substrate concentration has on response rate as shown in Figs. 5 and 6. Figure 5 shows the response of a β-glucosidase membrane electrode to amygdalin at various concentrations (64). Figure 6 shows the response of a glucose-oxidase membrane electrode to glucose (46). In both case, the reaction rate increases (as indicated by the increased inflection of the E-time of i-time curve)

as the substrate concentration increases, and a faster response time is observed, i.e., one minute for 10^{-1} M amygdalin; five minutes for 10^{-4} M amygdalin. As an alternative to waiting until an equilibrium potential or current is reached, the rate of change in the current or potential can be measured and equated to the concentration of substrate. This was done by Guilbault and Lubrano for a glucose electrode (46) (Fig. 6), obtaining a result for glucose in 12 seconds.

(2) Effect of Enzyme Concentration. Enzyme activity in the gel will have two effects on an enzyme electrode. First, it will ensure that a Nernstian calibration plot is obtained, as discussed in the following subsection. Second, it will affect the electrode's response speed. However, this is a tricky effect, inasmuch as an increase in the amount of enzyme also affects the thickness of the membrane. This is demonstrated in Fig. 7, taken from the results of Mascini and Liberti (64), for the amygdalin electrode. As the amount of enzyme is increased from 0.1 to 2.5 mg of β-glucosidase, a shorter time is observed. Yet, when 5 mg of enzyme was used, the response time became quite longer. This latter effect is due to a further thickening of the membrane layer by using more weight of enzyme, resulting in an increase in the time required for the substrate to diffuse through the membrane. If one weight of enzyme had been chosen, and the activity of enzyme increased at constant mass, an initial steady in-

crease in response rate would be observed, followed by a gradual stabilizing of response time. For best results, therefore, it is recommended that as active an enzyme, as possible, be used in as thin a membrane as possible to ensure rapid kinetics.

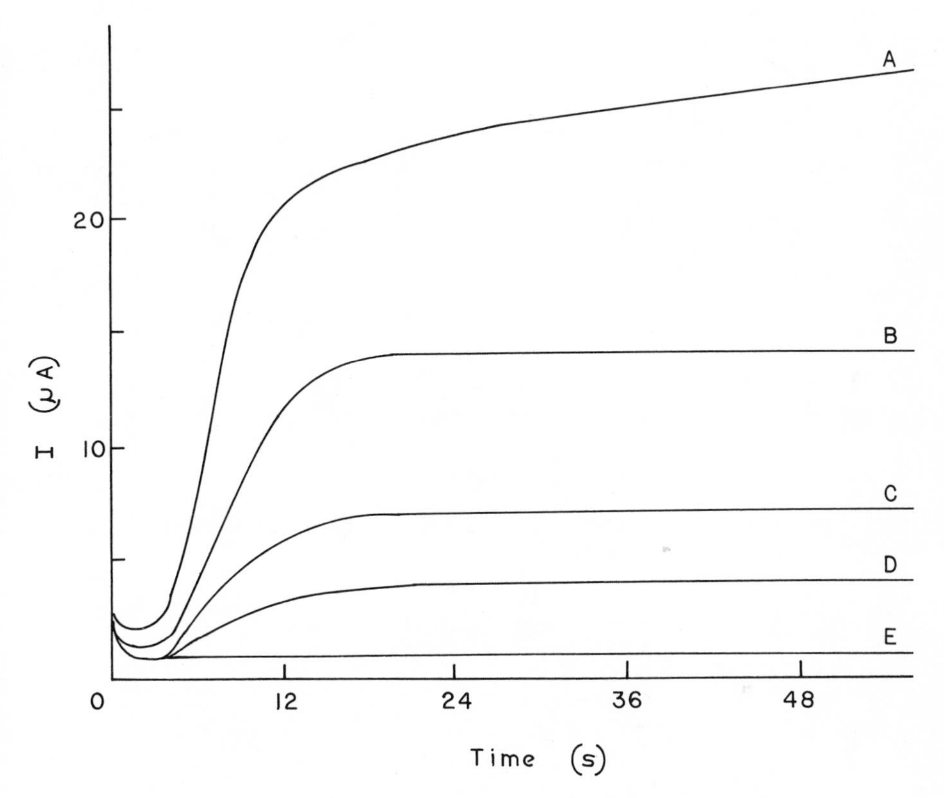

Fig. 6. Family of current-time curves for the glucose electrode poised at 0.6 V. Glucose solutions are in phosphate buffer, pH 6.0; ionic strength 0.1. A = 2.0 · 10^{-2} M; B = 1.0 · 10^{-2} M; C = 5.0 · 10^{-3} M; D = 2.5 · 10^{-3} M; E = 5.0 · 10^{-4} M. (From Ref. 46 by permission of authors).

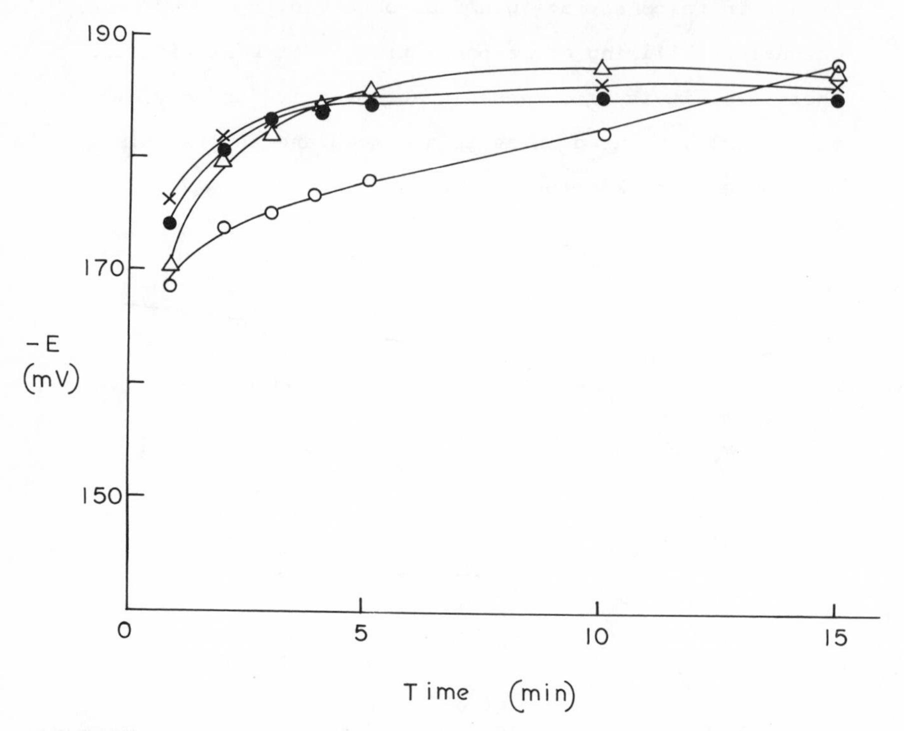

Fig. 7. Response-time curves with different amounts of enzyme. pH 7; (Amygdalin) = 10^{-2} M.

(●) 0.1 mg enzyme
(×) 1 mg enzyme
(▲) 2.5 mg enzyme
(○) 5 mg enzyme.

(Reference 64 by permission of authors).

(3) Effect of pH. Every enzyme will have a maximum pH at which it is most active, and a certain pH range within which it demonstrates reactivity. The immobilized enzyme will have a different pH range from that of the soluble enzyme because of its environment, as discussed earlier. The pH range for immobilized glucose oxidase is about 5.8 to 8.0 (46) (solution enzyme 5 to 7); β-glucosidase is about 5 to 8 (64). Hence, for fastest response, one should work at pH optimum. This is not always possible, however, because the sensor electrode might not respond optimally at the pH of the enzyme reaction. Therefore, a compromise is generally reached between these two factors. However, one should be careful not to be trapped into forcing the enzyme system to conform with the requirements of the sensor, as was done by Llenado and Rechnitz with the amygdalin electrode (62,63). These authors tried a pH of 10, which was shown to be optimum for the CN electrode sensor. Longer response times were obtained at pH 10 (Fig. 5) than at pH 7, because the enzyme has very little activity at the higher pH. Furthermore, the enzyme rapidly loses activity at high pH levels ($>$ 9 to 10) and Llenado and Rechnitz found their immobilized enzyme electrode very unstable, its response curve moving downward every day. Working at pH 7, Mascini and Liberti (64) found their soluble enzyme electrode useful for a week. Similar effects are noted in other

studies Guilbault and Shu (59), for example, found the optimum response time of a glutamine electrode when they decreased pH 6 from 5, the optimum for the enzyme reaction.

Further examples of making the electrode conform to the enzyme system, instead of vice-versa, are found in the reports of Anfalt et al (38) and Guilbault and Tarp (54) on the NH_3 sensor for the product of the urea-urease reaction. Although at the 7 to 8.5 optimum pH for this enzyme reaction, there is very little free NH_3 present to be sensed by the gas type electrode, and one would predict poor results for urea assay, both groups found that the sensitivity of the sensor was more than sufficient for each measurement at these low pH levels. This is partly due to the buildup of larger amounts of product in the reaction layer than in solution. As a result, increased sensitivity is obtained for the sensor, which is closest to the enzyme layer.

(4) Effect of Temperature. One would predict a dual temperature effect: an increase in reaction rate resulting in a faster response time; and a shift in equilibrium potential by virtue of the temperature coefficient in the Nernst and Van't Hoff equations.

This was demonstrated for the glucose electrode, by Guilbault and Lubrano (46) who studied the effect of temperature from 10 to 50°C on the electrode response.

Linear plots of the log rate and log total current vs. 1/T were observed as predicted by the Van't Hoff ($\ln K = \ln C - (\Delta H/RT)$, K being the equilibrium constant), and the Arrhenius ($\ln k - \ln A - (Ea/RT)$, k being the rate constant) equations. Practically, this means that the temperature of the enzyme electrode should be carefully controlled for best results, although the effect of temperature is most pronounced on reaction rate measurement. Similar effects were noted by Guilbault and Lubrano for amino acid oxidase (60).

Guilbault and Hrabankova (58) found that the response of the D-amino acid oxidase electrode D-methionine showed only very small effects at increasing temperatures (25 to 40°C), although the theoretical Nernstian slope is 61.74 mV/decade at 37°C. Similarly, Papariello et al (65) found that although the response time of his penicillin electrode was somewhat more rapid at 37°C than at 25°C, no great improvement was observed.

Hence, the enzyme electrode user is advised to control the temperature if he is making kinetic measurements. In making equilibrium measurements, however, simply use room temperature or about 25°C for convenience.

(5) Thickness of the Membrane. The time required to reach a steady state potential or current reading depends strongly on gel-layer thickness. This is due to the effect on substrate diffusion rate through the mem-

brane to the electrode sensor where it is measured. Guilbault and Montalvo (50) observed that the time interval for 98% of the steady-state response was about 26 seconds with a 60 μ-thick net of urease, and about 59 seconds with a 350 μ net for 8.33×10^{-2} M urea and an enzyme concentration of 175 mg/cc of gel. Similarly, Anfalt et al (38), using a urea electrode with glutaraldehyde-bound enzyme, and Mascini and Liberti (64), using a β-glucosidase amygdalin electrode, observed a response time increase in going from thin to thick membranes.

For best results, therefore, use of as thin a membrane as possible is recommended for best results. This can be achieved using high-activity enzyme.

(6) Effect of Dialysis Membrane. In most cases, it is advantageous to use a dialysis membrane to cover the electrode. This membrane protects the enzyme and prolongs electrode stability. Guilbault and Montalvo (50) noted that the cellophane coatings had little effect on the response time of the urea electrode.

A thorough study of the effect of dialysis membrane thickness on the response rate was made by Mascini and Liberti (64) for the amygdalin electrode. Their results, shown in Fig. 8, indicate that the response time and the equilibrium values are altered by varying membrane thickness. A thin, 20-μm membrane (Arthur H. Thomas) or a

25-μm membrane (Will Scientific) will have essentially no effect on response time, and are recommended for use.

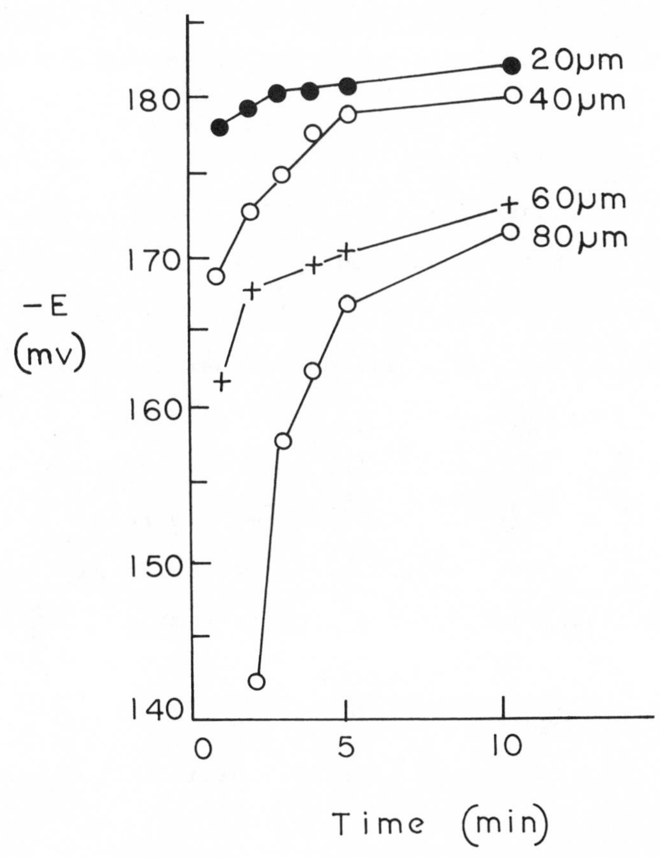

Fig. 8. Effect of thickness of dialysis paper on response time and on equilibrium values. (Amygdalin) = 10^{-2} M; enzyme amount = 1 mg. (Reference 64 by permission of authors.)

C. Measurement of Products at the Electrode Surface. The third factor affecting response speed is the electrode sensor itself, and how fast it gives a potential or current proportional to the amount of product or reactant sensed.

In most cases, where rapid stirring is used to minimize factor a, and a thin membrane of highly active enzyme is used under optimum conditions to minimize factor b, the determining factor will be the sensor's response time.

Guilbault and Montalvo (50) observed a response time of 26 seconds with a 60 -thick enzyme layer of urease in a urea electrode, compared to a response time of 23 seconds for an uncoated cation electrode. With their urea probe, Anfalt, Granelli and Jagner (38) observed response times of one-half to one minute, almost the same as those for the uncoated NH_3 gas electrode. Mascini and Liberti (64) noted a one-minute response time for their CN^--coated amygdalin electrode at high substrate concentration (10^{-1} M), quite similar to the uncoated CN^- electrode.

At low substrate concentrations (10^{-2} to 10^{-5}) membrane thickness becomes the rate-limiting factor for response time.

3. Effect of Concentration of Enzyme in Membrane

The enzyme concentration in the membrane will have two effects on electrode response. The first is on response time, as discussed earlier. The second is on the shape of the response curve. This is best indicated in Fig. 9, which is taken from the urea electrode work of Guilbault and Montalvo (50).

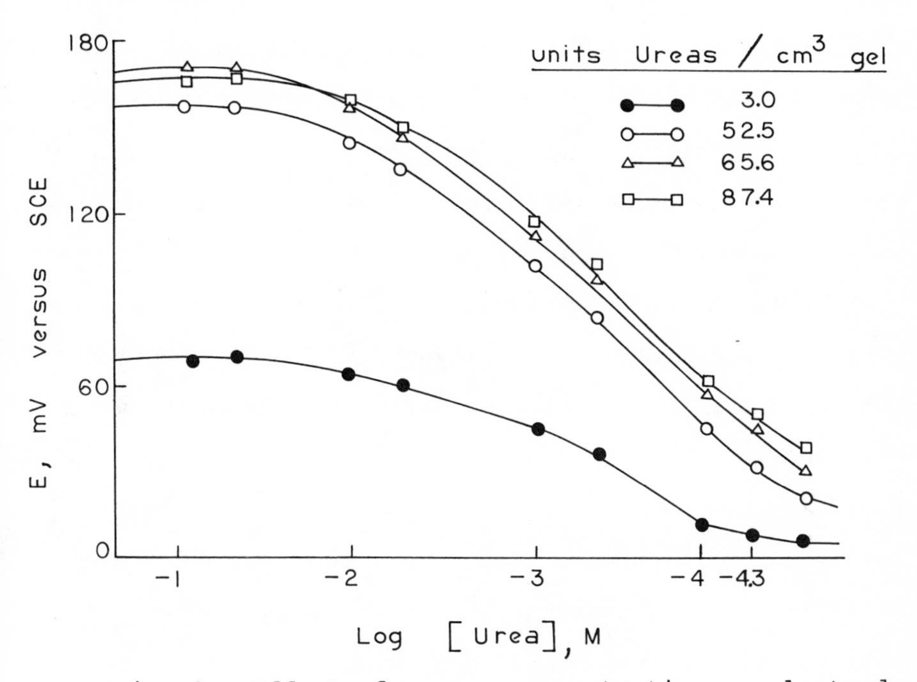

Fig. 9. Effect of enzyme concentration on electrode response; 250 μ netting. (Reference 50 by permission of authors.)

To study the effect enzyme concentration has on enzyme gel layer activity, gels were prepared with enzyme concentrations ranging from 3 to 110 units of urease/cm^3. The steady-state response of each enzyme-coated electrode, when dipped in urea solutions from 5×10^{-5} to 1.6×10^{-1} M, was measured. The results are shown in Fig. 9. The slope of each curve increases with the amount of enzyme in the electrode's gel layer until, with larger enzyme concentrations, only a small increase in gel membrane activity is obtained. Figure 10 is a plot of the steady-state response for 8.33×10^{-2} M urea against the amount of urease in the gel membrane. There is a rapid increase in response or activity up to 7.5 units of urease/cm^3 of gel. Above 7.5 units of urease/cm^3 of gel, large increases in enzyme concentration yield only small increase in enzyme gel membrane activity on the cation electrode. Optimum enzyme concentration in the gel, considering only the economy of enzyme, is obtained at about 7.5 units of urease/cm^3 of gel.

A series of urease electrodes was prepared with the same urease concentration (65.6 units of urease/cm^3 of gel) but with different gel compositions to determine if gel layer activity depends on the gel composition. Variation of the gel percent from 5 to 17.6 at constant monomer, with a 350-μ gel layer over the cation electrode, gave less than a two per cent difference in response with 8.33×10^{-2} M urea. Also, varying the percent of cross-linking material from 5 to 19 at constant gel concentration produced a very small difference in response.

With a urease concentration of 65.6 units/cm^3 of gel, the steady-state response to 8.33×10^{-2} M urea decreased by only two percent when the gel layer thickness was decreased from 350 to 60 μ. However, the response time decreased significantly.

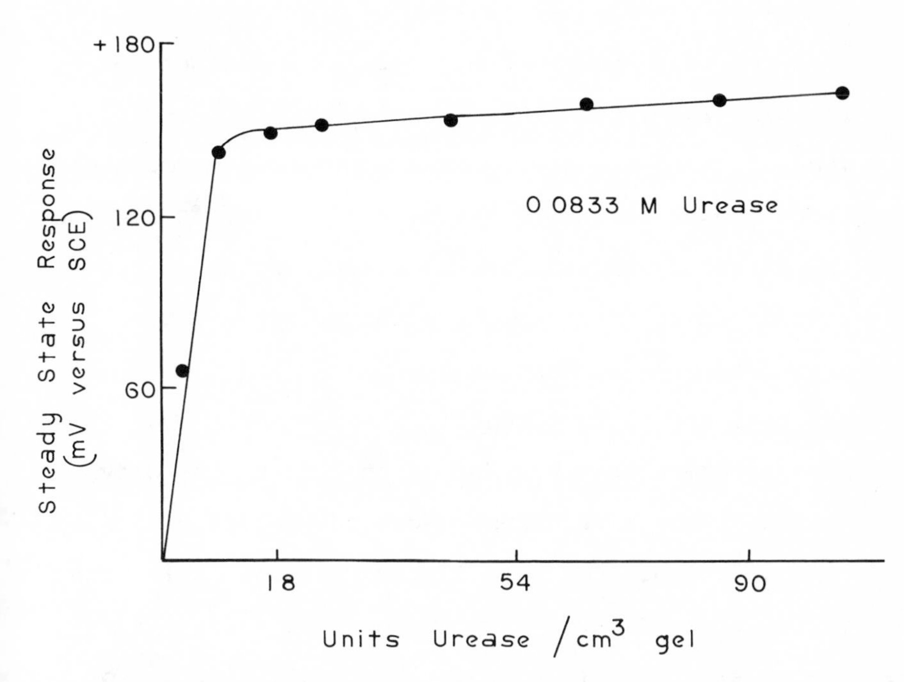

Fig. 10. Dependence of gel-layer activity on enzyme concentration, 350 μ netting. (Reference 50 by permission of author).

Similar results are obtained with every enzyme electrode. The amount of enzyme to be used depends on the individual electrode, and the curve, similar to that in Fig. 10, that is observed. However, as a rule of thumb, 10 to 20 units per membrane will generally be sufficient to give an excellent response curve. In the case of the more unstable "soluble" and physically entrapped enzyme electrodes, an excess of enzyme should be used (i.e., 50 units/cm^2 of gel as shown in Fig. 10), so enzyme loss will not affect observed potentials.

Likewise, purified enzymes should be used (at least 1 unit/mg) to hold membrane thickness at a minimum for fast response rates.

Table 5 lists the approximate amount of enzyme used by various groups in making enzyme electrodes. Since few of these carefully considered the effect the amount of enzyme used had on their response curves, this table should be used only as a rough guide.

4. Wash Time of the Electrode

Because there is a buildup of product in the enzyme membrane, every type of enzyme electrode requires washing after use to restore the base line potential. This wash time varies from 20 seconds in the case of urease and the air-gap electrode (54) to as long as 10 minutes for urease with a pH electrode sensor (48). Other approximate wash times are indicated in Table 5. Wash time increases with increasing enzyme membrane thickness, as expected, and with the nature of enzyme and the base sensor. This latter effect is due to the change on the enzyme, which is negative in the case of urease, with a resulting attraction for the positive NH_4^+ ions formed. The glass, likewise, has an attraction for these + ions, resulting in the longer wash times required.

The electrode can be washed in an automatic electrode washer as described by Montalvo and Guilbault (168) by simply placing the electrode in water, or by rinsing it under a distilled water tap. The type of washing will depend on the type of electrode. Soluble and physically entrapped enzymes must be washed gently since the enzymes can easily be washed out of their matrix. The more sturdy, chemically bound enzymes can be washed in tap water (54). The latter, of course, yields a much quicker return to base line. Oxygen-based electrodes (19,20) will show a quick return to base line merely by being placed in a fresh buffer solution. Other electrodes, such as the Pt-based amperometric enzyme electrodes (45,46,60,61) have to be pretreated before first use by applying a potential (+0.6 V vs. S.C.E.) until the anodic current decays to a low value. After each run, the electrode is washed in a stirring phosphate buffer solution until the current decays to a low value (0.5 to 1 min) indicating the removal of unreacted substrate and reaction products.

5. Range of Substrate Determinable

As indicated in Table 5, all enzyme electrodes sense substrate in the range of 10^{-2} to 10^{-4} *M*, with some electrodes useful to as high as 10^{-1} *M* or as low as 10^{-5} *M*. In all cases, curves similar to those in Fig. 9 are obtained, approximately Nernstian in the linear range with a slope of close to 59.1 mV/decade. All curves level off at high substrate concentration, as predicted by the Michaelis-Menten equation, which states that the reaction becomes independent of substrate at high substrate concentration. A leveling off of the curve at low substrate concentration is also observed. This is due to the detection limit of the electrode sensor used.

D. Effect of Interferences

Any enzyme electrode will only be as good as its selectivity. Hence, a consideration of possible interferences is now in order. Interferences fall into two categories: interferences in the electrode sensor; and interferences with the enzyme itself.

1. Interferences in the Electrode Sensor

If possible, the electrode used to sense the products of, or reactants in, the enzyme reaction should be one that will not react with other substances present in the sample being assayed.

The urea electrode probes originally described by Guilbault and Montalvo (49,50) could not be used for assaying urea in blood or urine because the cation glass sensor used to measure the NH_4^+ produced also responds to Na^+ and K^+ present in blood or urine.

In an attempt to eliminate this effect, a cell using a glass electrode (Beckman Electrode 39137 or 39047) as the reference unit was tried by Guilbault and Hrabankova (51). Calibration curves for urea were found to be the same when the cell with the SCE reference electrode was used. Also, the interferences of monovalent cations in solution are smaller in this case, because both electrodes are sensitive to these ions. However, concentrations of Na^+ and K^+ higher than 10^{-4} M, considerably decreased the electrode response. This effect could be explained as a decrease of activity coefficients in the presence of other ions, or as a decrease of enzymic reaction rate caused by a higher ionic strength.

Combining the uncoated glass reference electrode cell with a cation exchanger, the determination of urea in blood and urine is possible with a deviation of less than 3%.

A still further improvement was described by Guilbault and Nagy (52), in which a solid antibiotic nonactin electrode was used as the sensor. This electrode has a selectivity of NH_4^+/K^+ of 6.5/1 and NH_4^+/Na^+ of 7.5 x $10^2/1$, thus partially eliminating any response to these ions by the sensor. By using a three-electrode system (a chemically bound urease over a nonactin solid electrode vs. an uncoated nonactin electrode and a calomel electrode as reference) and diluting to a constant interference level of K^+ (Na^+ does not interfere because of the high selectivity coefficient of the sensor), urea in blood was assayed with an accuracy of 2 to 3% or better. In this procedure, a standard calibration plot of E vs urea was prepared in the presence of KCl, at its highest level present in blood. Before each run, the blood sample to be assayed was brought to the same potential as that observed in the KCl-buffer solution by adding more KCl to the sample, using the uncoated electrode vs. S.C.E. The potential of the urease sensor vs. S.C.E. was then read, and the urea determined.

Anfalt, Granelli and Jagner (38) used an Orion NH_3 electrode, which is a glass electrode with a layer of NH_4Cl and a gas-permeable membrane over the outside, to sense the NH_3 produced from the urea-urease reaction. In theory, the use of the gas NH_3 electrode, which senses only NH_3 and not K^+, Na^+ or other ions or substrates, should provide the desired sensor selectivity. However, at the optimum pH for the urea-urease reaction (7 to 8.5) the free NH_3 present is quite small compared to NH_4^+ ions. Yet, since the enzyme layer is placed directly onto the gas electrode, there was sufficient NH_3 produced at the electrode surface to give nice linear calibration plots at pH's of 7, 7.4, 8, and 9 with slopes close in 59.1 mV/decade (actually, the slope at pH 7 was 69.5 mV/decade, higher than Nernstian).

In a similar type study, Guilbault and Tarp (54) showed that the air-gap electrode for NH3 developed by Ruzicka and Hansen (21) could be used to provide a totally specific urea assay directly at pH 8.5. The air-gap electrode is designed on a principle similar to that of the gas electrodes, but minus the gas membrane, which usually causes such practical problems as long response times and clogging of the membrane pores by proteins and other substances in blood. This tremendously decreases the use life. In the air gap, NH_3, generated in solution by the enzyme reaction at pH 8.5, diffuses out of solution to the surface of the glass electrode, which is coated with a layer of 10^{-3} M NH_4Cl and surfactant. A change in pH is observed due to the reaction

$$NH_3 + H^+ \rightleftharpoons NH_4^+, \qquad (54)$$

which is monitored and is proportional to the log urea concentration. By calibrating the response curve with Monitrol II (freeze dried blood serum from Dade, Miami, Florida) the urea content of 30 blood samples from Riggs Hospital in Copenhagen were assayed with an accuracy of about 2%. The method requires only two to four minutes per assay, and using a chemically bound urease enables more than 500 assays per electrode.

In the Pt electrode devices described for measuring oxidative enzyme systems, two approaches have been taken. One is to measure the peroxide produced by monitoring the total current change (43,45,46), or the rate of change in the current (45,46) at +0.6V vs. S.C.E. The other is to measure the uptake of oxygen by the enzymic reaction (19,20) by measuring the reduction of O_2 at -0.6 V vs. S.C.E. In either system, compounds that are either oxidized or reduced at a Pt electrode at ± 0.6 V vs. S.C.E. would interfere. Clark (43) eliminates this problem by using a second uncoated Pt electrode, held at the same potential as the enzyme electrode, to measure any compounds present in blood. His system, as modified, forms the basis of an instrument for measuring glucose and sold commercially by Yellowsprings Instrument Company (Yellowsprings, Ohio, U.S.A.). However, the normal values of such interferences (ascorbic acid or uric acid are examples of oxidizable compounds) is generally $< 5\%$ of the glucose signal. Another way to eliminate the effect is to measure very quickly ($\sim$ 21 sec) the rate of change in the current. This was done by Guilbault and Lubrano (45,46), who found that assays of more than 200 blood samples could be performed with an accuracy and precision of better than 2%. Since all other compounds present in blood that consume O_2 can be subtracted out before measurement with the enzyme sensor (19,20), still better selectivity is obtained using the measurement of O_2 at -0.6V.

In other electrodes, such as the CN^- solid precipitate electrode used by Llenado and Rechnitz (62,63) and Mascini and Liberti (64) for amygdalin, any ions capable of forming insoluble silver salts will interfere because they form precipitate on the electrode surface. Also, substances capable of reducing silver ion will interfere, as will heavy metal ions and transition metal ions capable of forming cyanide complexes. In using an iodide sensor to measure glucose, Nagy, von Storp and Guilbault (34) found several types of interferences which make this type of measurement approach very limited: Interferences at the iodide electrode (thiocyanate, sulfide, CN^-, and Ag^+), and interferences from oxidizable compounds present in blood such as uric acid, tyrosine, ascorbic acid and Fe(II) which compete in the oxidation of iodide to iodine in the peroxide-peroxidase system. These compounds had to be removed by sample pretreatment.

Finally, in the glass electrode systems for pH measurement described by Papariello et al (65) and Mosbach and co-workers (48), any acidic or basic components present would interfere in the measurement. However, by adjusting to a definite pH before initiating the enzyme reaction, and assuming only the enzyme reaction will give rise to a pH change, these effects can be minimized or eliminated.

2. Interferences with the Enzyme Reaction

Such interferences fall into two classes: Substrates that can catalyze the reaction in addition to the compound to be measured; and substances that either activate or inhibit the enzyme.

With some enzymes, such as urease, the only substrate that reacts at a reasonable rate is urea -- hence, the urease-coated electrode is specific for urea (49,50). Uricase, likewise, acts almost specifically on uric acid (19). Others, like penicillinase (65), react with a number of substrates: ampicillin, naficillin, Penicillin G, Penicillin V, cyclibillin and dicloxacillin, all of which can be determined with a penicillinase electrode.

Similarly, D-amino acid oxidase (58) and L-amino acid oxidase (56,60) are less selective in their responses: The former in an electrode gives good response to D-phenylalanine, D-alanine, D-valine, D-methioine, D-leucine, D-norleucine, and D-isoleucine; the latter to L-leucine, L-tyrosine, L-phenylalanine, L-tryptophan, and L-methionine. Alcohol oxidase (61,167) responds to methanol, ethanol and allyl alcohol. Hence, in using electrodes of these enzymes, a separation must be used if two or more substrates are present, or the total must be determined. In the case of L-amino acid assay, the use of decarboxylative enzymes (55) which act specifically on different amino acids is an attractive alternative. Enzyme electrodes of this type are known for L-tyrosine, L-phenylalanine, L-tryptophan, and others.

Glucose oxidase acts on a number of sugars (34): glucose and 2-deoxy glucose are the main substrates, but celluboise and maltose also react, probably due to the presence of other hydrolytic enzymes in the glucose oxidase preparation.

The activity of the enzyme can be adversely affected by the presence of certain compounds, called inhibitors. Generally, these are heavy metal ions, such as Ag^{+}, Hg^{2+}, and Cu^{++}, and sulfhydryl-reacting organic compounds, such as p-chloromercuribenzoate and phenyl mercury(II) acetate (due to their reaction with the free S-H groups present at the active site of many enzymes, especially the oxidase) (46). One important point to realize, however, is that the immobilized enzyme is much less susceptible to inhibitors, especially weak or reversible inhibitors, due to the protection of the immobilization matrix. Thus, by using the enzyme in an immobilized form, most of the worries about inhibitors are eliminated. From personal experiences in designing and using almost 20 different enzyme electrodes, not a case of enzyme inhibition interfering with an assay has been noted. However, one should always be aware of this problem, especially in assaying solutions containing heavy metal ions, and especially pesticides.

III. COMPARISON OF ELECTRODES

A. Enzyme Electrodes for Substrates

Table 5 presents a comprehensive listing of the various enzyme electrodes that have been reported to March, 1974. Under each type of electrode is listed the enzyme(s) used to make the electrode, the type of immobilization used, the stability of the electrode, its response time, the approximate amount of enzyme used, the range of assay, the wash time required to reach base line, and the reference. The reader is invited to use this list as a general guide in developing electrodes.

Let us now review these various electrodes, comparing them to see which designs are best. First, it can be said that none of these electrodes are available as self-contained units that can be purchased separately. As we have mentioned, Yellowsprings Instrument Company is marketing an instrument for glucose that uses an enzyme electrode as an integral part of the instrument, and others for glucose and urea can be expected.

Many urea electrodes based on pH, cation (NH_4^+), CO_2 and NH_3 electrode probes have been described (38,48-50, 52-55). From a practical viewpoint, the best is the NH_3 electrode-based probe (38,54) because it has no interference problems from Na^+ and K^+ in solution (49,50, 52-53) With it and a chemically bound enzyme, more than 500 assays per electrode are possible at 1 to 4 minutes/assay (54). The CO_2-based sensor (55) is also quite selective, but has slower response, and the disadvantage of compensating for the CO_2 present in blood before assay. The use of a glass electrode as a probe for urea (48) is a sound idea, but it takes a long time (5 to 10 minutes) to get a stable pH change and the electrode has a limited range of concentrations within which it is useful. In the future, all urea electrodes may be based on some type of NH_3 sensor.

Undoubtedly, the best electrodes for glucose are those based on the Pt sensors. Those based on the iodide electrode are definitely not useful because they are subject to many interferences, and have slow response and a narrow linear range (34). The glass electrode-based sensor (48), thought interesting, has long response times and long wash times and is not as generally useful as the Pt-based sensor. However, Mosbach et al (48) present a new vista of electrodes by showing it is possible to measure any acid-producing enzyme reaction with good accuracy. The concept should prove useful in many systems, i.e., penicillin.

Much credit in the enzyme electrodes field is due Leland Clark, one of the pioneers in this field, who described the first "enzyme" electrode, the glucose electrode, using a Pt sensor and the soluble enzyme (41,43). He has also used the polarographic H_2O_2 monitoring system for other electrodes for use with alcohols (67), and amino acids (69).

The Williams, Doig and Korosi (44) electrode, based on measurement of the oxidation of hydroquinone formed when quinone replaces O_2 as the hydrogen acceptor, is theoretically interesting.

The Pt-based sensors, based on peroxide measurement (41,43,45,46) and on measurement of O_2 consumption (20), are the most sound and generally useful and may form the basis for future instruments, although the use of an O_2 gas electrode with immobilized glucose oxidase (42) will attract some attention by instrument builders.

The Pt electrode (60), the gaseous CO_2 electrode (55), the cation (NH_4^+) electrode (56,57-59), the I^- electrode (57), the antibiotic NH_4^+ electrode (57), and the Pt-based O_2 electrode (20) have been tried as base sensors for measuring amino acids. Electrodes of the future will probably be based on the CO_2 electrode, either regular gas type or air gap, using specific decarboxylases to build electrodes for the various amino acids. An alternative will be the Pt-based sensors, for either peroxide or O_2, with L-amino acid oxidase as a general type probe. Using either the air gap or the Pt sensors, a fast response over a wide range is possible.

The Pt electrode-based sensor for lactic acid electrode (44) has limited use and will not attract any attention.

For alcohols, the Pt-based sensors will be used, with the best results probably coming from O_2 measurement at -0.6 V vs. S.C.E. The electrode has fast response, and is more sensitive than the H_2O_2 measurement system at +0.6 V vs. S.C.E.

Penicillin electrodes based on the pH sensor will continue to attract attention, and, the excellent preliminary work of Mosbach et al (48) will probably be followed up by others.

The uric acid electrode based on O_2 measurement using a Pt electrode without a membrane is a good electrode and should be widely studied (19). Alternatively, a CO_2 sensor could be used to measure the other product of this reaction.

For amygdalin, the electrode of Mascini and Liberti (64) is the best; further studies with chemically bound enzymes should give a good final product, although measurements of amygdalin have very limited analytical use.

Other electrodes in the near future will probably use creatinase to measure liberated NH_3 with an air gap electrode or Orion gas electrode; an L-phenylalanine electrode using a lyase from sweet potato and a nitrate electrode using nitrate reductase, both of which will probably use a gaseous NH_3 detection sensor to determine the product of the reaction; and a cholesterol electrode using cholesterol oxidase and a Pt sensor. The number of new sensors that may be introduced is limited only by the imagination.

B. Substrate Electrodes for Enzymes

Only a little work has been done in this field, and that which has been done is of little practical use. As was discussed in Section I, subsection E, the problems with substrate electrodes are: 1) the substrate is consumed in the assay, while the enzyme and the electrode must be renewed periodically; 2) because the enzyme acts continuously on substrate to produce product, a kinetic assay of the reaction must be made within a short time;

(30) the substrate, if immobilized, must be tailored so it still fits into the enzyme's active site.

The most obvious solution to the problem is to use a sensor electrode over which fresh substrate is continuously passed. However, this is an expensive approach because it requires a large supply of substrate, and the resulting "electrode" is anything but simple, requiring pumps and regulators to control substrate flow. This approach was taken by Montalvo in a urea electrode for urease (67,68) and by Guilbault et al (28) and Crochet and Montalvo (69) for a cholinesterase sensing electrode.

The electrode for urease has a thin layer of urea solution trapped between the glass electrode and a dialysis membrane. Upon placing the electrode, which has urea continuously pumped about the surface, in a solution containing urease, the urea diffuses out of the membrane into solution and reacts with urease, which cannot diffuse through the membrane, to form ammonium ion. The ammonium ion diffuses back through the membrane, where it is sensed by the cation glass electrode. The electrode has a slow response time, yet a potential vs. urease concentration curve was obtained.

Guilbault, Hussein and von Storp (28) described a cholinesterase assay system using a sulfide electrode and acetylthiocholine chloride as the substrate. The thiol produced after reaction with cholinesterase was measured by the sulfide sensor, and the resulting potential was a function of the enzyme concentration:

$$\text{Acetylthiocholine Chloride} \xrightarrow{\text{ChE}} \text{Thiocholine Chloride} \quad (55)$$

The response time was about 90 seconds.

Montalvo and Crochet (69) described a cholinesterase sensing electrode prepared by coupling a pH sensing electrode to a thin dialysis membrane separating two solutions. A layer of cholinesterase to be assayed was passed over the inside layer closest to the electrode, and a layer of acetylcholine and polyethylenimine buffer (to react with the free acid produced by nonenzymic hydrolysis) was passed over the outside layer. The acetic acid, produced when the acetylcholine diffuses from the outer layer to the inner layer and reacts with cholinesterase, is sensed by the glass electrode. The potential change, ΔE/min., is a function of the cholinesterase concentration. The electrode has a response time of 1.5 to 5 minutes, and requires two pumps, one for enzyme injection, the other for a continuous substrate layer.

Although interesting, all this work has little practical value, since an enzymic assay can be performed cheaper and faster by other techniques with simpler equipment. If, however, the substrate could be immobilized over a sensor electrode and the enzyme could react with it to produce products that can be quickly sensed, a practical electrode would result. The first attempts at such a system were made by Guilbault and Gibson (70) who placed substrate, immobilized as either an insoluble salt (acetylcholine reineckate) or an ion-exchange resin (acetylcholine$^+$ or acetylthiocholine$^+$ Resin$^-$). The enzyme reacts with the substrate to produce either acid which is sensed by a pH electrode, or -SH which is sensed by a sulfide electrode, and the rate of potential change, ΔE/min., is equated to enzyme concentration.

Any future work on substrate electrodes will have to be based on "immobilized" substrate to be of any practical use, and offer advantages over existing methods. However, because the substrate membrane must be periodically replaced, this field will not blossom as widely as the enzyme electrode area.

IV. FUTURE OF ENZYME ELECTRODES

The future of enzyme electrodes is bright. Many electrodes have already been developed, and it is quite likely many more will be described in the near future. The existence of hundreds of enzyme systems and many good sensing probes makes it quite possible that electrodes can be produced for most of the inorganic and organic substances of importance in the areas of clinical, environmental, medical, and agricultural chemistry, to name but a few. The electrodes are so simple to prepare, that it is hoped that this descriptive chapter will provide stimulus to others to develop new sensors.

A major problem in this area so far has been the lack of commercially available products based on the ideas so far developed. Instead of marketing self-contained electrodes that many would want to buy and use, most companies are working toward total instruments that use the "enzyme electrode" as the central ingredient (i.e., Yellowsprings Instrument Company). It is the author's hope that at least one company will introduce a line of individual electrodes. In the meantime, the concepts are so simple, that the reader is encouraged to try making his own.

ACKNOWLEDGMENT

The financial assistance of the National Science Foundation (Grant No. GP-31518) and the National Institutes of Health (Grant No. 1 RO1 GM17268-01) in support of the experimental work of this author in the area of enzyme and substrate electrodes is gratefully acknowledged.

REFERENCES

(1). M. Cremer, Z. Biol, 47, 562 (1906).
(2). F. Haber and Z. Klemasiewicz, Z. Physik. Chem. 67, 385 (1909).
(3). G. Karremas and G. Eisenman, Bull. Math. Biophys. 24, 413 (1962).
(4). O. K. Stephanova, M. M. Shultz, E. A. Matarova and B. P. Nicolsky, Vestn. Lenings. Univ. 4, 93 (1963).
(5). E. Pungor, Anal. Chem. 39, (13), 28A (1967).
(6). G. A. Rechnitz, Chem. Eng. News, June 12, 1967, p. 146.
(7). R. A. Durst, ed., "Ion-Selective Electrodes," Special Publication, 314, National Bureau of Standards, Washington, D. C., 1969.
(8). G. J. Moody and J. D. R. Thomas, Talanta 19, 623 (1972).

(9). G. J. Moody, R. B. Oke and J. D. R. Thomas, Analyst 95, 910 (1970).
(10). G. H. Griffiths, G. J. Moody and J. D. R. Thomas, Analyst 97, 420 (1972).
(11). J. E. W. Davies, G. J. Moody and J. D. R. Thomas, Analyst 97, 87 (1972).
(12). J. Pick, K. Toth, E.Pungor, M. Vasak and W. Simon, Anal. Chim. Acta 64, 477 (1973).
(13). R. M. Garrels in "Glass Electrodes for Hydrogen and Other Cations," G. Eisenmann, ed., Marcel Dekker, Inc., New York, 1967, p. 344.
(14). R. N. Khuri, "Ion-Selective Electrodes," R. A. Durst, ed., op. cit., p. 287.
(15). D. Ammann, E. Pratsch and W. Simon, Anal. Letters 5, 843 (1972).
(16). L. Pioda, M. Wachter, R. Dohner and W. Simon, Helv. Chim. Acta 50, 1373 (1967).
(17). G. G. Guilbault and G. Nagy, Anal. Chem. 45, 417 (1973).
(18). E. Eyal and G. A. Rechnitz, Anal. Chem. 43, 1090 (1971).
(19). G. G. Guilbault and M. Nanjo, Enzyme Electrode for Uric Acid, Anal. Chim. Acta, in press.
(20). G. G. Guilbault and M. Nanjo, Enzyme Electrodes for Glucose, Amino Acids and Alcohols, Anal. Chim. Acta, in press.
(21). J. Ruzicka and E. H. Hansen, A New Potentiometric Gas Sensor - The Air Gap Electrode, Anal. Chim. Acta, Feb., 1974.
(22). E. W. White, F. McCapra and G. F. Field, J. Am. Chem. Soc. 85, 337 (1963).
(23). S. P. Colowick and N.O. Kaplan, eds., Methods of Enzymology, p. 107, Academic Press, New York (1957).
(24). G. G. Guilbault, Enzymic Methods of Analysis, Pergamon Press, New York and London, 1971.
(25). G. D. Christian, Adv. Biomed. Eng. Med. Phys. 4, 95 (1971).
(26). A. H. Kadish and D. A. Hall, Clin. Chem. 9, 869 (1965).
(27). Y. Makino and K. Koono, Rinsho Byori 15, 391 (1967) (Japan).
(28). W. R. Hussein, L. H. vonStorp and G. G. Guilbault, Anal. Chim. Acta 61, 89 (1972).
(29). R. A. Llenado and G. A. Rechnitz, Anal. Chem. 44, 1366 (1972).
(30). L. H. vonStorp and G. G. Guilbault, Anal. Chim. Acta 62, 425 (1972).
(31). R. A. Llenado and G. A. Rechnitz, Anal. Chem. 45, 826 (1973).
(32). R. A. Llenado and G. A. Rechnitz, Anal. Chem. 44, 468 (1972).

(33). R. A. Llenado and G. A. Rechnitz, Anal. Chem. 45, 2165 (1973).
(34). G. Nagy, L. H. vonStorp and G. G. Guilbault, Anal. Chim. Acta 66, 443 (1973).
(35). W. F. Gutknecht and G. G. Guilbault, Environmental Letters 2, 51 (1971).
(36). G. G. Guilbault, R. K. Smith and J. G. Montalvo, Anal. Chem. 41, 600 (1969).
(37). J. Montalvo, Anal. Chim. Acta 65, 189 (1973).
(38). T. Anfalt, A. Granelli and D. Jagner, Anal. Letters 6, 969 (1973).
(39). E. Hansen and J. Ruzicka, Anal. Chem., in press.
(40). G. Baum, Anal. Biochem. 39, 65 (1971).
(41). L. C. Clark and C. Lyons, Ann. N. Y. Acad. Sci. 102, 29 (1962).
(42). S. J. Updike and G. P. Hicks, Nature 214, 986 (1967).
(43). L. C. Clark, U. S. Patent 3,539,455 (1970).
(44). D. L. Williams, A. R. Doig and A. Korosi, Anal. Chem. 42, 118 (1970).
(45). G. G. Guilbault and G. J. Lubrano, Anal. Chim. Acta 60, 254 (1972).
(46). G. G. Guilbault and G. J. Lubrano, Anal. Chim. Acta 64, 439 (1973).
(47). G. G. Guilbault, G. J. Lubrano and D. Gray, Anal. Chem. 45, 2255 (1973).
(48). H. Nilsson, A. Akerlund and K. Mosbach, Biochim. Biophys. Acta 320, 529 (1973).
(49). G. G. Guilbault and J. G. Montalvo, J. Am. Chem. Soc. 91, 2164 (1969).
(50). G. G. Guilbault and J. G. Montalvo, J. Am. Chem. Soc. 92, 2533 (1970).
(51). G. G. Guilbault and E. Hrabankova, Anal. Chim. Acta 52, 287 (1970).
(52). G. G. Guilbault and G. Nagy, Anal. Chem. 45, 417 (1973).
(53). G. G. Guilbault, G. Nagy, and S. S. Kuan, Anal. Chim. Acta 67, 195 (1973).
(54). G. G. Guilbault and M. Tarp, Anal. Chim. Acta, in press.
(55). G. G. Guilbault and F. Shu, Anal. Chem. 44, 2161 (1972).
(56). G. G. Guilbault and E. Hrabankova, Anal. Letters 3, 53 (1970).
(57). G. G. Guilbault and G. Nagy, Anal. Letters 6, 301 (1973).
(58). G. G. Guilbault and E. Hrabankova, Anal. Chim. Acta 56, 285 (1971).
(59). G. G. Guilbault and F. R. Shu, Anal. Chim. Acta 56, 333 (1971).
(60). G. G. Guilbault and G. J. Lubrano, Anal. Chim. Acta, Feb., 1974.

(61). G. G. Guilbault and G. J. Lubrano, Anal. Chim. Acta, Feb., 1974.
(62). G. A. Rechnitz and R. Llenado, Anal. Chem. 43, 283 (1971).
(63). Ibid., p. 1457.
(64). M. Mascini and A. Liberti, Anal. Chim. Acta 68, 177 (1974).
(65). G. J. Papariello, A. K. Mukherji and C. M. Shearer, Anal. Chem. 45, 790 (1973).
(66). W. J. Blaedel, T. R. Kissel and R. C. Boguslaski, Anal. Chem. 44, 2030 (1972).
(67). J. G. Montalvo, Anal. Biochem. 38, 359 (1970).
(68). J. G. Montalvo, Anal. Chem. 42, 2093 (1969).
(69). K. L. Crochet and J. G. Montalvo, Anal. Chim. Acta 66, 259 (1973).
(70). K. Gibson and G. G. Guilbault, Anal. Chim. Acta, in press.
(71). P. V. Sundaram, A. Tweedale, and K. J. Laidler, Can. J. Chem. 48, 1498 (1970).
(72). G. J. H. Melrose, Rev. Pure Appl. Chem. 21, 83 (1971).
(73). E. Katchalski, "Struct. - Funct. Relat. Proteolytic Enzymes," (ed: P. Desnuelle), Academic Press, New York, 1970, p. 198.
(74). K. Mosbach, Sci. Amer. 224, 26 (1971).
(75). J. Gryszkiewicz, Folia Biol. 19, 119 (1971).
(76). A. N. Emery, and C. A. Kent, Birmingham Univ. Chem. Eng. 21, 71 (1970).
(77). L. Goldstein, "Fermentation Advances," (ed: D. Perlman), Academic Press, New York, 1969, p. 391.
(78). L. Goldstein and E. Katchalski, Fresenius Z. Anal. Chem. 243, 375 (1968).
(79). A. H. Sehon, Symp. Ser. Immunobiol. Stand. 4, 51 (1967).
(80). I. H. Silman and E. Katchalski, E. Ann. Rev. Biochem. 35, 873 (1966).
(81). G. Manecke, Naturwissenschaften 51, 25 (1964).
(82). I. Chibata and T. Tosa, Kagaku to Seibutsu 7, 147 (1969).
(83). T. M. S. Chang, Science 146, 524 (1964).
(84). T. M. S. Chang, Ph. D. Thesis, McGill University, Montreal, 1965.
(85). T. M. S. Chang, F. C. MacIntosh, and S. G. Mason, Canad. J. Physiol. Pharmacol. 44, 115 (1966).
(86). T. M. S. Chang, Science J., 62 (July, 1967).
(87). T. M. S. Chang, Science Tools 16, 35 (1969).
(88). T. M. S. Chang, Biochem. Biophys. Res. Comm. 44, 1531 (1971).
(89). P. R. Rony, Biotechnol. Bioeng. 13, 431 (1971).
(90). J. P. Hummel and B. S. Anderson, Arch. Biochem. Biophys. 112, 443 (1965).

(91). G. Lindau and R. Rhodius, Z. Physik. Chem. A172, 321 (1935).
(92). E. S. Vorobeva and O. M. Poltorak, Vestn. Mosk. Univ., Khim. 21, 17 (1966).
(93). M. G. Goldfeld, E. S. Vorobeva and O. M. Poltorak, Zh. Fiz. Khim. 40, 2594 (1966).
(94). E. F. Gale and H. M. R. Epps, Biochem. J. 38, 232 (1944).
(95). A. Y. Nikolayev, Biokhimiya 27, 843 (1962).
(96). T. Tosa, T. Mori, N. Fuse and I. Chibata, Biotechnol. Bioeng. 9, 603 (1967).
(97). G. Hamoir, Experimentia 2, 257 (1946).
(98). E. F. Jansen and A. C.Olson, Arch. Biochem. Biophys. 129, 221 (1969).
(99). R. Haynes and K. A. Walsh, Biochem. Biophys. Res. Comm. 36, 235 (1969).
(100). A. F. S. A. Habeeb, Arch. Biochem. Biophys. 119, 264 (1967).
(101). F. M. Richards, Annu. Rev. Biochem. 32, 268 (1963).
(102). I. H. Silman, M. Albu-Weissenberg and E. Katchalski, Biopolymers 4, 441 (1966).
(103). R. Goldman, H. I. Silman, S. R. Caplan, O. Kedem, and E. Katchalski, Science 150, 758 (1965).
(104). F. J. Wold, Biol. Chem. 236, 106 (1961).
(105). G. Manecke and G. Gunzel, Naturwissenschaften 54, 647 (1967).
(106). H. Zahn and H. Meienhofer, Makromol. Chem. 26, 126 and 153 (1958).
(107). D. J. Herzig, A. W. Rees and R. A. Day, Biopolymers 2, 349 (1964).
(108). H. F. Schick and S. J. Singer, J. Biol. Chem. 236 2447 (1961).
(109). G. Kay and E. M. Crook, Nature 216, 514 (1967).
(110). B. P. Surinov and S. E. Manoylov, Biokhimiya 31, 387 (1966).
(111). P. Bernfeld and J. Wan, Science 142, 678 (1963).
(112). P. vanDuijn, E. Pascoe, M. vanderPloeg, J. Histochem. and Cytochem. 15, 631 (1967).
(113). W. N. Arnold, Arch. Biochem. Biophys. 113, 451 (1966).
(114). G. P. Hicks and S. J. Updike, Anal. Chem. 38, 726 (1966).
(115). G. R. Penzer and G. K. Radda, Nature 213, 251 (1967).
(116). G. G. Guilbault and J. G. Montalvo, J. Amer. Chem., Soc. 92, 2533 (1970).
(117). K. Mosbach, Acta Chem. Scand. 24, 2084 (1970).
(118). E. K. Bauman, L. H. Goodson, G. G. Guilbault and D. N. Kramer Anal. Chem. 37, 1378 (1965).
(119). G. G. Guilbault and D. N. Kramer, Anal. Chem. 37, 1675 (1965).
(120). F. L. Aldrich, V. R. Usdin and B. M. Vasta, U.S. Army Report DA-18-108-405 CML-828 (1963).

(121). S. N. Pennington, H. D. Brown, A. B. Patel and C.O. Knowles, Biochim. Biophys. Acta 167, 479 (1968).
(122). H. D. Brown, A. B. Patel and S. K. Chattopadhyay, J. Biomed. Mater. Res. 2, 231 (1968).
(123). H. H. Weetall, Nature 223, 959 (1969).
(124). H. H. Weetall, Science 166, 615 (1969).
(125). J. K. Inman and H. M. Dintzis, Biochem. 8, 4074 (1969).
(126). S. A. Barker, P. J. Somers, R. Epton and J. V. McLaren, Carbohyd. Res. 14, 287 (1970).
(127). R. P. Patel, D. V. Lopiekes, S. R. Brown and S. Price, Biopolymers 5, 577 (1967).
(128). B. F. Erlander, M. F. Isambert and A. M. Michelson, Biochem. Biophys. Res. Comm. 40, 70 (1970).
(129). A. B. Patel, S. N. Pennington and H. D. Brown, Biochim. Biophys. Acta 178, 26 (1969).
(130). W. E. Hornby, H. Filippuson and A. McDonald, FEBS Letters 9, 8 (1970).
(131). W. E. Hornby and H. Filippuson, Biochim. Biophys. Acta 220, 343 (1970).
(132). D. J. Inman and W. E. Hornby, Biochem. J. 129, 255 (1972).
(133). K. P.Wheller, B. A. Edwards, R. Whittam, Biochim. Biophys. Acta 191, 187 (1969).
(134). G. R. Craven and V. Gupta, Proc. Nat. Acad. Sci. U. S. 67, 1329 (1970).
(135). K. Mosbach and B. Mattiasson, Acta Chem. Scand. 24, 2093 (1970).
(136). D. Gabel, P. Vretbald, R. Axen and J. Porath, Biochim. Biophys. Acta 214, 561 (1970).
(137). R. Axen and J. Porath, Nature 210, 367 (1966).
(138). L. Goldstein, Methods Enzym. 19, 935 (1970).
(139). M. L. Green and G. Crutchfield, Biochem. J. 115, 183 (1969).
(140). D. Gabel and B. Hofsten, Eur. J. Biochem. 15, 410 (1970).
(141). G. Kay and M. D. Lilly, Biochim. Biophys. Acta 198, 276 (1970).
(142). B. L. Vallee and J. F. Riordam. Ann. Rev. Biochem. 38, 733 (1969).
(143). J. Sir Ram, M. Bier and P. H. Maurer, Advan. Enzymol. 24, 105 (1962).
(144). H. Fraenkel-Conrat, "The Enzymes," (eds: P. D. Boyer, H. Lardy and K. Myrback), Academic Press, New York, 1959, Vol. 1, p. 589.
(145). G.Lubrano, "Amperometric Enzyme Electrodes," Ch. 4, p. 34, Thesis submitted to Louisiana State University in New Orleans for Ph.D. Thesis (1973).
(146). D. Fraser and H. G. Higgins, Nature 172, 459 (1953).
(147). M. Tabachnick, and H. Sabotka, J. Biol. Chem. 235, 1051 (1960).

(148). L. Goldstein, M. Pecht, S. Blumberg, D. Atlas and D. Levin, Biochem. 9, 2322 (1970).
(149). J. Salak and Z. Vodrazka, Collect. Czeck. Chem. Commun. 32, 3451 (1967).
(150). H. G. Higgins and D. Fraser, Aust. J. Sci. Res. Series A 5, 737 (1952).
(151). H. G. Higgins and K. J. Harrington, Arch. Biochem. Biophys. 85, 409 (1959).
(152). H. Ozawa, J. Biochem. (Japan) 62, 419 (1967).
(153). R. Axen and J. Porath, Acta Chem. Scand. 18, 2193 (1964).
(154). R. P. Parel and S. Price, Biopolymers 5, 583 (1967).
(155). K. L. Carraway and R. B. Triplett, Biochim. Biophys. Acta 160, 272 (1970).
(156). K. L. Carraway and R. B. Triplett, Biochim. Biophys. Acta 200, 564 (1970).
(157). H. D. Brown, A. B. Patel, S. Chattopadhyay and S. N. Pennington, Enzymol. 35, 215 (1968).
(158). Y. Levin, M. Pecht, L. Goldstein and E. Katchalski, Biochem. 3, 1905 (1964).
(159). A. Schejter, A. Bar-Eli, Arch. Biochem. Biophys. 136, 325 (1970).
(160). F. A. Quiocho and F. M. Richards, Biochem. 5, 4062 (1966).
(161). K. Ogata, M. Ottesen and I. Svendsen, Biochim. Biophys. Acta 159, 403 (1968).
(162). F. M. Richards, and J. R. Knowles, J. Mol. Biol. 37, 231 (1968).
(163). A. F. S. A. Habeeb and R. Hiramoto, Arch. Biochem. Biophys. 126, 16 (1968).
(164). R. Axen, J. Porath and S. Ernbach, Nature 214, 1302 (1967).
(165). A. Bar-Eli and E. Katchalski, J. Biol. Chem. 238, 1690 (1963).
(166). E. Katchalski, "Solid Phase Proteins: Their Preparation, Properties and Applications," (ed: A. S. Hoffman), Battelle Seattle Research Center, 1971, p. 1.
(167). L. Clark, Biotechnol. Bioeng. Symp. No. 3, 377 (1972).
(168). J. G. Montalvo and G. G. Guilbault, Anal. Chem. 41, 1897 (1969).
(169). L. Clark, Proc. Internat. Union Physiological Sciences, Vol. 9, 1971.
(170). G. G. Guilbault, R. McQueen and S. Sadar, Anal. Chim. Acta 45, 1 (1969).
(171). A. Johansson, J. Lundberg, B. Mattiasson and K. Mosbach, Biochim. Biophys. Acta 304, 217 (1973).

Chapter 8

AFFINITY CHROMATOGRAPHY

George Baum, Ph.D., and Stanley J. Wrobel, Ph.D.

Corning Glass Works
Corning, New York

I. INTRODUCTION

Within the last few years, a powerful new separation technique known as affinity chromatography or biospecific adsorption has become available to biochemists. The technique is based on the observation that many biological events involve the formation of reversible complexes between two or more interacting species. Enzymes are well known for binding substrates, inhibitors, cofactors and allosteric effectors. Hormones form stable complexes with specific binding proteins and with cellular receptors. Genes interact with nucleic acids and repressors. Antibodies form complexes with appropriate antigens. Lectins bind to specific surface antigens on red blood cells and to certain carbohydrates. These interactions can be exploited to purify a wide variety of biological materials.

Immunologists have employed a related process in purifying specific antibodies from a heterogeneous globulin fraction by passing the crude sample through a particulate column containing adsorbed or covalently coupled antigen. The corresponding antibody which is selectively retarded, is subsequently removed by further elution.

Affinity chromatography is thus a separation process whereby the material to be separated is selectively retarded by complex formation to an insolubilized binding ligand. High selectivity is achieved by using binding ligand materials with specific or highly selective affinity for the desired substance.

The reaction between two interacting molecules can be expressed as A + B = AB. In affinity chromatography, either A or B can be attached either to an insoluble support such as a polysaccharide, or to an inorganic material such as controlled-pore glass. If, for example, the purification of enzyme A is desired, B may be the specific inhibitor that is employed as the ligand and is attached in some way to an insoluble support. A solution containing A is offered to the insolubilized ligand. The strength of the association between the enzyme and the ligand will influence the manner by which the enzyme is removed. Impurities not bound to the ligand or carrier are easily eluted from the column. The retarded or retained enzyme is then removed by changing the conditions of the medium such that the interactions are less favored, or by introducing a competitive soluble ligand that has a greater affinity for the enzyme than the bound inhibitor.

This review discusses various classes of ligands that have been used and the elution techniques by which desired species are removed. Our primary concern will be the range of experimental techniques employed rather than a comprehensive review of reported separations. Separations that depend upon ion exchange processes and immunoadsorption will not be covered. This review has two major sections. The first half deals with critical factors in affinity column design, and the latter half deals with specific applications and an evaluation of reported experimental results.

A number of reviews on affinity chromatography have already appeared. Each has a special emphasis that can supplement the reader's acquaintance with developments in this area. P. Cuatrecasas, who has been the leading practitioner, has published several reviews dealing mainly with experimental results, applications and derivatization of agarose (1,2). A limited, but more rigorous discussion of the chemical processes of support activation and ligand design is given by Guilford (3). Weetall's review (4) deals primarily with support activation and extensively with preparing controlled-pore glass derivatives. Although physical principles have been treated by Cuatrecasas, a more critical discussion is presented by O'Carra (5). Although many successful separations have been reported and numerous reviews have appeared, this field is far from mature. No rigorous physical treatment of the separation process has been presented. Many debate the roles of spacer and support on separation efficiency. Quantitative studies and carefully designed experiments to measure the contribution of physical and chemical processes to eluate retention and elution are just beginning to appear.

II. GENERAL CONSIDERATIONS

The satisfactory demonstration of a biospecific affinity separation must satisfy several criteria. Exposing the ligand-support to a solution containing the molecule to be purified must remove from solution that biological activity associated with the molecule. If it is an enzyme purification, then enzyme activity should decrease in the solution presented to the derivatized support; if it is a binding protein, specific ligand binding capability in the solution should be reduced.

Activity loss within the solution must occur only when the solution is exposed to a support with a ligand specific for the molecule and its biological activity. Thus, appropriate controls are necessary to satisfy this criterion. Controls with nonspecific ligands and controls with chemically modified supports minus the ligand must be used.

The loss of activity must be due to the formation of a reversible specific complex between the molecule of interest and the immobilized ligand, and the molecule to be purified should be recoverable free of other molecules preferably by conditions that selectively cause complex dissociation. An important point here is that the activity loss in the solution results from the formation of a complex between the solute of interest and the ligand bound to the support. Activity loss can only be an apparent loss resulting, for example, from the release of ligand from the support and from resultant complex formation in solution.

It should be apparent by now that affinity chromatographic purifications may require extensive chemical manipulations to prepare specific or selective adsorbents for each purification and to prepare adsorbents to serve as controls. The satisfactory demonstration of a biospecific or bioselective interaction between a derivatized support and the molecule to be purified must be followed by determining the conditions that will reverse the complex formation to allow recovery in the eluting solution of the adsorbed molecule with full biological activity.

It would be misleading to give the impression that affinity chromatographic purifications require nothing more than a routine chemical immobilization of a ligand, adsorption of the complementary binding component, and elution by some general change in the composition of the solution in contact with the support. Each purification should be approached as one with its own characteristics and possible pecularities. Many complications in the general scheme for purification by affinity chromatography may occur. The solute to be purified may bind so tightly to the immobilized ligand that the conditions for recovery may result in denaturation. Impurities may bind to the support or to the derivatized portions and may then be eluted together with the substance to be purified. The ligand may dissociate from the support and interfere with the purification. Indeed this list of potential pitfalls is not easily concluded. The following discussions should point out some of the generalities and pitfalls that can be encountered.

Once the conditions (support, ligand, adsorption and eluting conditions) for a particular purification are determined, however, the advantage of affinity chromatography becomes quite evident. It can eliminate the many steps associated with conventional methods requiring considerably more time and result in lower product yields due to material loss during multi-step operations.

III. THE SUPPORT

The immobilization of one of the interacting components is generally accomplished by attachment to an insoluble polymeric substance or to an inorganic substance, such as glass. It is necessary that any substance considered as a support possess certain characteristics necessary for attaching ligand and for good chromatographic separations.

The support must possess reactive surface groups that may be used for reaction with the ligand, or else the surface of the support must be easily modified to introduce reactive groups. For general applicability, the support must be stable to a wide range of reaction and chromatographic conditions, and it must retain a favorable structure following the reactions used for ligand immobilization. After immobilization reactions it must retain surface properties that allow proper interaction with the chromatographic medium.

The support must possess a surface concentration of reactive groups necessary for effective concentrations of the immobilized component and, once derivatized, the support should exhibit little or no nonspecific affinity for soluble molecules. If porous, the support should be highly permeable to the macromolecules of interest, and it should retain its permeability following derivatization. Finally, it must be rigid enough to withstand chromatographic flow rates and repeated reuse.

Supports that have received the most attention for affinity chromatographic purposes are agarose, polyacrylamide and glass.

A. Polyacrylamide

Polyacrylamide is prepared by copolymerizing acrylamide and N,N'-methylenebisacrylamide to provide a polymer with a hydrocarbon framework carrying carboxamide side chains that serve as binding sites for other molecules. It is available as spherical beads if various

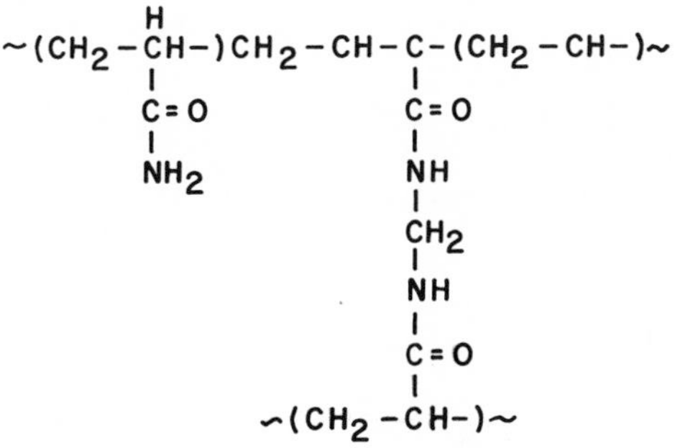

```
          H
~(CH2 -CH-)CH2 -CH-C-(CH2 -CH-)~
          I            I
         C=O          C=O
          I            I
         NH2          NH
                       I
                      CH2
                       I
                      NH
                       I
                      C=O
                       I
              ~(CH2 -CH-)~
```

(1)

sizes and porosities, is stable between pH 1 and pH 10, tolerates concentrated salt, urea, and guanidine solutions, and is not subject to microbial attack. Inman and Dintzis (6) developed procedures for modifying polyacrylamide for covalent bonding. The carboxamide-NH_2 can be displaced by other amines, such as ethylenediamine or hydrazine. Displacement of hydrazine generates the acyl hydrazide, which can be converted to an acyl azide with nitrous acid, acylated with a p-nitrobenzoyl group to provide, after reduction, an aryl amine derivative, or treated with succinic anhydride to generate a terminal carboxylic acid functional group. Polyacrylamide's principal advantage is that it enables a much higher degree of ligand substitution because of its large number of carboxamide groups. However, should the acyl azide be prepared, some shrinkage and decrease in porosity of the gel can be expected (7).

Procedures for preparing only two polyacrylamide derivatives are given: acyl hydrazide (Eq. 2), and acyl azide (Eq. 3). Procedures for preparing other acyl amine derivatives, such as those with ethylenediamine, are

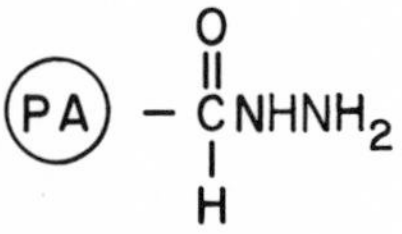

$$\text{(PA)}-\overset{\overset{\displaystyle O}{\|}}{\underset{\underset{\displaystyle H}{|}}{C}}NHNH_2 \qquad \text{ACYL HYDRAZIDE} \tag{2}$$

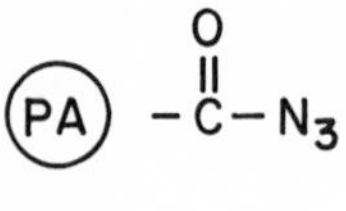

$$\text{(PA)}-\overset{\overset{\displaystyle O}{\|}}{C}-N_3 \qquad \text{ACYL AZIDE} \tag{3}$$

similar to that given for hydrazine. The acyl azide procedure is a general one for bonding primary amine-bearing molecules to polyacrylamide. Some experimental procedures for further reactions involving the acyl hydrazide can be adapted from procedures described elsewhere in this chapter or book.

1. Preparing the Hydrazide Derivative of Polyacrylamide.

Water-swollen polyacrylamide beands in a volume of water equal to 1.3 times the volume of the hydrated gel are stirred in a closed flask in a fume hood at 45° to 50°C (constant temperature bath) with 3-6 M hydrazine hydrate. The extent of reaction desired determines the reaction time; a linear relationship exists between the degree of polyacrylamide substitution and reaction time at least over a period of eight hours. As a guide, 6 M hydrazine hydrate and a reaction temperature of 47°C produces a polyacrylamide derivative with about three millimoles of acylhydrazine per gram of polymer; 3 M hydrazine hydrate at 50°C, produces a polymer with about two millimoles of hydrazide per gram of polyacrylamide.

The hydrazide derivative is washed in a fume hood with 0.1 M sodium chloride on a Buchner funnel or by batch washing with decantation of the saline washes. The washing is continued until the washes are free of hydrazine as determined by a negative color test with 2,4,6-trinitrobenzene-sulfonate. The washed gel is stored in an aqueous medium, in the presence of a preservative, such as sodium azide.

It must be noted that because polyacrylamide gel particles adhere to glass surfaces, siliconized glassware or polyethylene labware should be used whenever possible to facilitate transferring and mixing operations. In addition, when conducting the trinitrobenzenesulfonate color test, add a few drops of a freshly prepared 3% solution of trinitrobenzenesulfonate and 1 ml of saturated sodium tetraborate to 5 ml of sample solution.

2. Coupling of Primary Amines via Acyl Azides

The washed hydrazide polymer (50 ml packed volume) is suspended in 100 ml of 0.3 N hydrochloric acid. The ice-chilled suspension is treated with 10 ml of 1 M sodium nitrite with stirring for two to five minutes. At this point, the gel may be washed rapidly with 0.3 N HCl, 0.1 M sulfamic acid, and cold water, or it can be used directly for coupling amine. In the latter case, the amine must be added with an alkaline buffer and/or the pH of the reaction medium must be raised (e.g., with sodium hydroxide solutions) to a value favorable for reaction involving the unprotonated amine.

After the reaction has proceeded for one to two hours at 4°C, the gel is washed with distilled water and is suspended in 2 M ammonium chloride - 1 M ammonium hydroxide at pH 8.8 for five hours to convert unreacted hydrazide and azide to the carboxamide form.

3. Coupling of a Protein (bovine serum albumin) to Polyacrylamide via an Acyl Azide

A 100-mg portion of a polyacrylamide hydrazide derivative (1.3 mmole of hydrazide/g) that was washed with 0.3 N hydrochloric acid was suspended in 15 ml of the same acid and treated with 1 ml of 1 M sodium nitrite for 20 minutes. The gel was collected on a Buchner funnel and washed quickly with 0.3 N HCl, 0.1 M sulfamic acid, and cold water. The gel was immediately suspended in 12 ml of a 0.5% solution of bovine serum albumin, labeled with ^{131}I, in 0.1 M sodium tetraborate at 4°C. The mixture was stirred in an ice bath for one hour. To convert any remaining acyl azide groups back to the unreactive amide, 2 ml of 1 M ammonium hydroxide and 1 M ammonium chloride were added and the mixture was stirred for another two hours. The gel was washed with 0.2 M sodium chloride and with water. The amount of bound protein, determined by radioactivity count, was 67 mg/g of the untreated polyacrylamide.

B. Agarose and Other Polysaccharides

Agarose has been the most frequently used support to date. It is a linear polysaccharide composed of (1 → 3) - linked β-D-galactopyranose and (1 → 4) - linked 3,6-anhydro-α-L-galactopyranose (Eq. 4). The commercially available beaded forms of this polysaccharide are stable

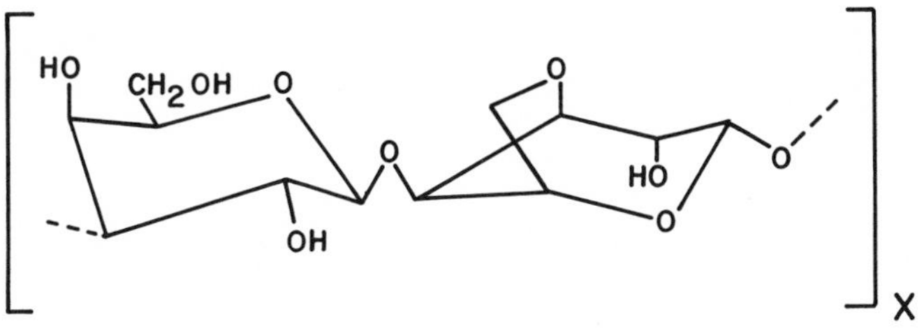

(4)

between pH 4 and 9 and have very open structures that allow access of very large molecules.

Agarose and other polysaccharide supports such as cross-linked dextran and cellulose can be readily activated for binding molecules by treatment with cyanogen bromide. Although other reagents can be used for attaching molecules to polysaccharides, the reaction with cyanogen bromide has gained wide popularity and is convenient and quick. However, since caution is required because cyanogen bromide is toxic, those who would prefer not to handle this reagent can obtain cyanogen bromide-activated agarose from Pharmacia Fine Chemicals. The mechanism for activating polysaccharides with cyanogen bromide has been studied and the reader is referred to other references (3,8-11) for discussions of this reaction. The reaction pathway that has been proposed is shown in Fig. 1.

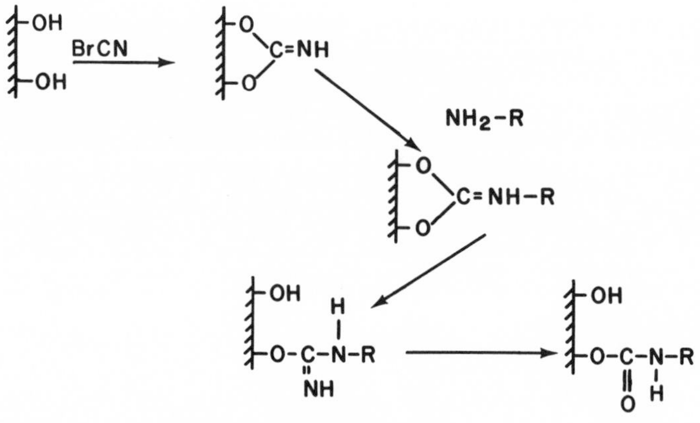

Fig. 1

Polysaccharides are generally activated with cyanogen bromide in a medium maintained at the desired alkaline pH by constantly adding sodium hydroxide solution. The extent of activation and, consequently, of ligand coupling is controlled by the pH of the medium (12).

For example, from pH 9.6 to 11.0, the weight (mg) of bound chymotrypsin per gram of dry chymotrypsin-Sephadex conjugate varied from 35 to 90 (while the activity ratio relative to free enzyme varied from 35% to 15%, respectively). Recently, Porath et al described a modified procedure that utilizes a buffered solution to maintain the desired pH conditions for activation (13). This procedure was introduced to eliminate the need for vigorous stirring to prevent pH gradients within the reaction mixture. The degree of activation can be controlled by the concentration of cyanogen bromide added to the reaction mixture. As an indication of the extent of activation, the amount of glycylleucine bonded to the agarose under different conditions was determined. Under one set of conditions, a highly activated gel coupled 250 μ mole per gram of dry conjugate; a moderately activated gel, 140; and a weakly activated gel, 105.

The extent of ligand attachment to the activated polysaccharide may also be regulated by the amount of free ligand that is offered to the activated support and by the pH of the coupling reaction.

The cyanogen bromide-activated polysaccharides will react with unprotonated primary and secondary amines. For most efficient coupling, the pH of the reaction medium should be higher than the pK_a of the amine-bearing ligand; a pH greater than 10 should be avoided because the activated polymer's stability decreases at these higher pH values. For aromatic amines, the reaction pH should be 8 to 9; for amino acids, 9.5 to 10.0; for primary aliphatic amines, 10 (1). It should be apparent, therefore, that by lowering the pH of the reaction medium below the pK_a of the amine, the degree of coupling can be reduced because the amount of unprotonated amine is diminished.

1. Cyanogen Bromide Activation of Agarose

The following operation should be conducted wholly within a well-ventilated hood. A volume of agarose that has been washed well with water is mixed with an equal volume of water, and the stirred suspension is treated with finely ground cyanogen bromide (50 to 300 mg per ml of packed gel). The pH is immediately raised to 11, and is maintained there by constantly adding a sodium hydroxide solution (2-8 _M_ depending on the scale of the reaction). The temperature is maintained at about 20°C by adding pieces of ice. After 8 to 12 minutes, reaction completion is indicated by the complete dissolution of the cyanogen bromide and the cessation of a rapid fall in pH. The gel is collected on a sintered glass funnel (coarse) and is washed with cold water and with a chilled buffer solution that will be used in the coupling reaction.

It should be noted that the cyanogen bromide may also be added as an aqueous solution. In addition, the extent of activation of the polysaccharide can be controlled by the pH of the reaction medium. The lower the pH, the lower the extent of activation.

It is also important that the activation be conducted for the short times prescribed because prolonged alkaline conditions can convert the activated polysaccharides into inactive carbamylated polymers.

2. Activation of Agarose (4%) by Modified Cyanogen Bromide Procedure

a. Highly activated. The gel is washed with distilled water, and the interstitial water is removed on a Buchner funnel. Ten grams of the gel is suspended in 10 ml of a cold (5 to 10°C) potassium phosphate buffer (3.33 moles K_3PO_4 + 1.67 moles of the K_2HPO_4 per liter of solution), and the suspension is diluted with water to bring the total volume to 20 ml. An aqueous cyanogen bromide solution (4 ml of a solution of 100 mg of cyanogen bromide per ml) is added in small portions over a period of two minutes. After the reaction mixture is gently stirred at 5 to 10°C for an additional eight minutes, the gel is collected and washed with cold water on a sintered glass funnel. The whole operation should be performed in a well-ventilated hood.

b. Moderately activated. The same procedure is used, but the volume of the phosphate buffer is 2 ml, and the volume of the cyanogen bromide solution is 0.8 ml.

c. Weakly activated. The procedure for highly activated agarose is used, but the volume of phosphate buffer is 0.48 ml, and the volume of the cyanogen bromide solution is 0.2 ml.

3. Chemical Coupling of Amine-Bearing Molecules to Cyanogen Bromide-Activated Agarose

The activated agarose is washed with the cold buffer used for the coupling reaction. A cold solution of ligand in buffer, equal in volume to that of the packed gel, is added to the activated gel and maintained with gentle stirring at $4^{o}C$ for about 16 hours. In some cases, the reaction may be complete within two to four hours, but the longer reaction time helps insure maximum coupling. The derivatized agarose is then washed on a sintered glass funnel or in a column with appropriate buffers until ligand is no longer detected in the washes. The gel is stored in buffer in the cold with a preservative such as sodium azide (at least 0.02%).

To insure complete removal of activated groups after the coupled reaction, it may be desirable to further treat the support with reagents such as ethanolamine or 1-amino-propane-2,3-diol.

4. Coupling of Concanavalin A to Cyanogen Bromide Activated Agarose

Drained Sepharose 4B (25 g) was washed with four volumes of distilled water and then suspended in 20 ml of water. A solution of cyanogen bromide (4 g in 25 ml of water) was added to the stirred mixture, and the pH was raised to 11 and maintained there by manually adding 4 M sodium hydroxide. After about 10 to 12 minutes, the gel was collected on a sintered glass funnel (coarse) and washed under vacuum with 100 ml of cold water and 100 ml of 0.1 M cold sodium bicarbonate.

The activated agarose was added to a 40-ml solution of Con A that was prepared by dissolving 250 mg of Con A in a mixture of 20 ml of 1 M sodium chloride and 20 ml of 1 M sodium bicarbonate. The resulting suspension was gently stirred at 4°C overnight. The gel was then collected on a coarse sintered glass funnel and washed with two liters of 0.1 M sodium bicarbonate. The wash solution was collected in fractions and assayed for protein.

C. Glass

Besides the organic polymeric supports such as polyacrylamide, agarose, cross-linked dextran, and cellulose, inorganic supports, such as glass, are available for immobilizing ligands to be used for biospecific separations. Commercially available porous glass is beginning to receive greater attention as a support material. This controlled-pore glass is available in different mesh sizes and in different pore diameters.

Controlled-pore glass support is resistant to microbial attack and to changes in acidity and solvent. Its rigid structure is a desirable quality for column operation under pressure, for repeated use, and for changes in the nature of the chromatographic medium. The porous nature of the glass particles offers a large surface area and, depending upon the pore diameter, open channels that allow access of even very large macromolecules.

The glass surface is activated by treating the glass with a silane that bears a reactive functional group, such as an amine. A very useful silane for this purpose is γ-aminopropyltriethoxysilane. Ligands can be chemically coupled to this activated support by bifunctional reagents, such as glutaraldehyde, or by coupling reagents that promote amide formation. Amine-derivatized glass may be further modified to provide other reactive groups. Experimental procedures for preparing various derivatized controlled-pore glass supports are provided in another chapter of this book.

IV. MODIFIED SUPPORTS AND SPACERS

A general and versatile derivative of any support is one bearing an amine group. Activation of the inorganic supports involved converting the surface silanol groups to amine groups by reaction with an amine-bearing silane. This modified support is now suited for coupling the ligand and is capable of being further modified by a host of chemical reactions. Polysaccharide supports, once activated with cyanogen bromide, may be converted to amine derivatives by reaction with such bifunctional chemicals as α, ω-diamino-alkanes. Finally, the polyacrylamide supports can be manipulated in the same manner. Reaction of the polymer's carboxamide groups with hydrazine or α, ω-diaminoalkanes will provide the derivatized gel with an amine group.

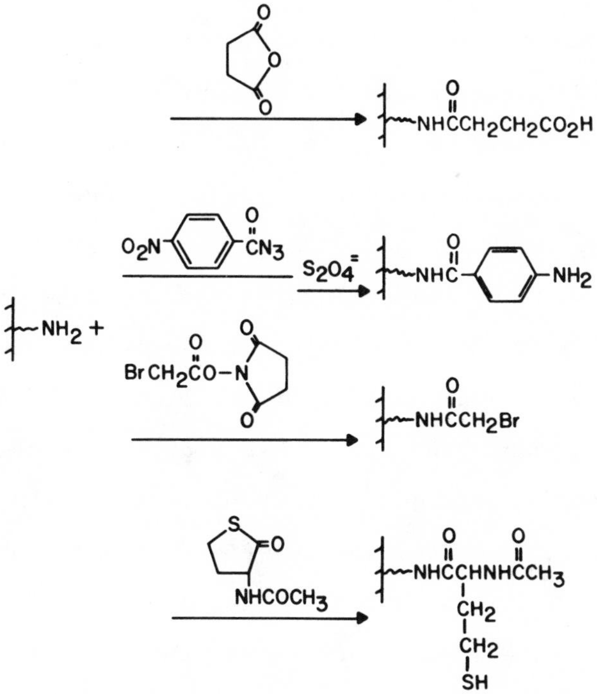

$$\text{}\!\sim\!NH_2 + \text{succinic anhydride} \longrightarrow \text{}\!\sim\!NH\overset{O}{\overset{\|}{C}}CH_2CH_2CO_2H$$

$$\text{}\!\sim\!NH_2 + O_2N\text{-}C_6H_4\text{-}\overset{O}{\overset{\|}{C}}N_3 \xrightarrow{S_2O_4^{=}} \text{}\!\sim\!NH\overset{O}{\overset{\|}{C}}\text{-}C_6H_4\text{-}NH_2$$

$$\text{}\!\sim\!NH_2 + BrCH_2\overset{O}{\overset{\|}{C}}O\text{-}N(\text{succinimide}) \longrightarrow \text{}\!\sim\!NH\overset{O}{\overset{\|}{C}}CH_2Br$$

$$\text{}\!\sim\!NH_2 + \text{N-acetylhomocysteine thiolactone (}NHCOCH_3\text{)} \longrightarrow \text{}\!\sim\!NH\overset{O}{\overset{\|}{C}}\underset{CH_2CH_2SH}{CH}NH\overset{O}{\overset{\|}{C}}CH_3$$

Fig. 2

Cuatrecasas (1,7) has demonstrated the conversion of the amine-supports to a number of other functional groups as shown in Fig. 2. Reaction with succinic anhydride introduces a carboxyl function; with a carboxyl-activated *para*-nitro benzoic acid, an aryl amine, after dithionite reduction; with N-acetylhomocysteinethiolactone, a thiol function; and with O-bromoacetyl-N-hydroxysuccinimide, a bromoacetyl group.

The need for such chemical modification of the support arises from the necessity of introducing on the surface of the support groups that can react with available reactive functional groups on the ligand. At times, the ligand must be attached to the support at a distance from the surface. The chemical reactions discussed earlier can be used to introduce "spacer" molecules between the support surface and the ligand. Such spacers may be necessary to provide more efficient interaction between the immobilized ligand and the solute molecules due to the increased steric availability of the ligand to the solute. This may be because the ligand has a greater degree of movement in interacting with a complementary binding site, because the surface does not interfere with the ability of the binding site "cavity" of the solute to fully interact with the ligand, or because the ligands are essentially separated by a greater distance from each other when attached to long, flexible chains, which reduces the chances of masking of some ligands by solute molecules bound to adjacent ligands.

The need for a spacer is found in a study that examined the binding of anthranilate 5-phosphoribosylpyrophosphate phosphoribosyltransferase to a Sepharose-anthranilate derivative (14). A derivatized support prepared by bonding anthranilic acid through its amino-group (Eq. 5) to cyanogen bromide-activated Sepharose 2B failed to bind a large quantity of the enzyme; only

$$\text{Support}-NH-C_6H_4-CO_2H \quad (5)$$

260 of 2800 units of enzyme activity was adsorbed and recovered from this support. Insertion of a spacer consisting of hexamethylenediamine and succinic acid (Eq. 6) enhanced the ability of the support to specifically bind

$$\text{Support}-NH(CH_2)_6NH\overset{O}{\overset{\|}{C}}CH_2CH_2\overset{O}{\overset{\|}{C}}NH-C_6H_4-CO_2H \quad (6)$$

the enzyme. Under the same conditions, approximately 2200 units of activity out of a total of 4000 units applied to the support were recovered in a specific manner.

The nature of spacers used in preparing biospecific adsorbents has varied to a very limited extent. Such bifunctional amines as hexamethylenediamine and di(3-

aminopropyl)amine have been used frequently, and have been further modified and extended by reaction with succinic anhydride (Fig. 2). In general, spacers have been composed of hydrophobic residues with a possible hydrophilic residue, such as an amide function, or with other functional groups that have ion exchange capabilities. Because hydrophobic and ionic groups are present in the spacers, these spacers may contribute to nonspecific interactions between the support and solute moledules and cannot be considered inert. In fact, hydrophobic supports have recently received attention for purifying proteins that may have hydrophobic pockets capable of interacting with hydrocarbon chains attached to the surface of the support. This topic is discussed further in a later section.

The selection of spacer seems to be based either upon precedent or preparative simplicity. Each purification, however, must be approached as a separate dissimilar purification and if spacer is necessary, spacer length should be determined for maximum separation efficiency. After a spacer of specific length has been shown to improve a separation, increasing spacer length further may lead to greater nonspecific adsorption, or conceivably to a decrease in column efficiency, because coiling or folding of flexible linear spacers may decrease the availability of ligand to the solute molecules. Hydrophilic spacers may be more desirable if they do not carry ionic groups. Derivatives of glycidylether have been employed (15) (Eq. 7).

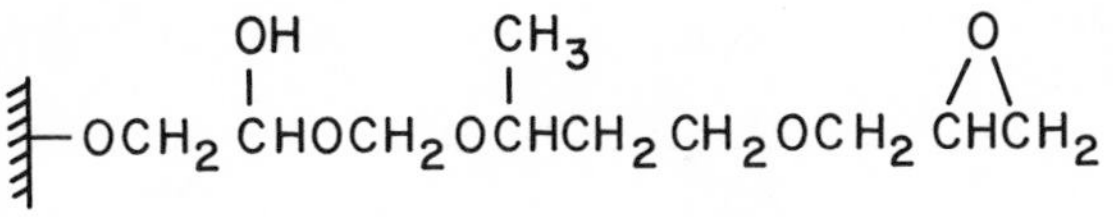

$$\text{Support}-OCH_2\overset{\overset{OH}{|}}{C}HOCH_2O\overset{\overset{CH_3}{|}}{C}HCH_2CH_2OCH_2\overset{O}{\widehat{CHCH_2}} \tag{7}$$

Rigid spacers such as benzidine (Eq. 8) have also been used (16).

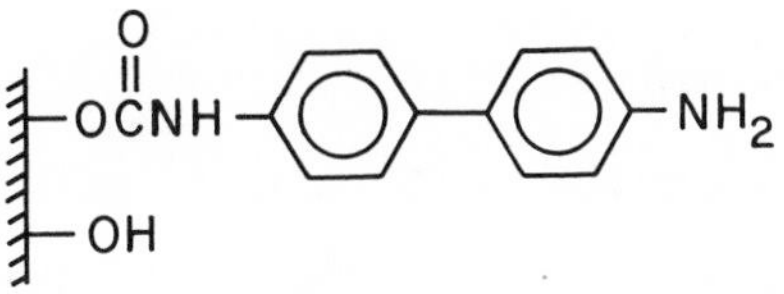

$$\text{Support}\begin{cases}-O\overset{\overset{O}{\|}}{C}NH-C_6H_4-C_6H_4-NH_2 \\ -OH\end{cases} \tag{8}$$

In most cases, the spacer is attached first to the activated support. If the spacer is composed of more than one molecular species, it is added to the support in a stepwise fashion. Finally, the ligand is attached to the prepared support. As the number of reactions for attaching ligand increases, there is a concomitant increase in the likelihood of leaving undesirable hydrophobic or ionic groups or both on the carrier surface to interact in a nonspecific manner with the solute molecules.

This may result because of incomplete reactions at each step in the synthetic scheme. Should the use of a spacer be necessary, it may be more favorable to first modify the ligand by attaching the spacer to it and then attach the ligand-spacer to the activated support. This may be more demanding, but it allows for proper purification and characterization of intermediates, and reduces the chance of leaving residual ionic groups on the support surface.

It must be emphasized that proper controls are required to determine to what extent nonspecific interactions to the spacer are occurring. This means that one control should consist of a support with spacer only. It has been shown that attaching a supposedly noninteracting ligand onto a spacer can cause retention of the molecule to be purified apparently due to nonspecific interactions of a hydrophobic nature (5,17).

It was mentioned earlier that the bond between ligand and support must be stable under chromatographic conditions. Loss of ligand during chromatography can result in complexation between the ligand and solute of interest, a result that may lead to a misleading interpretation of the experiment and possibly may result in a totally ineffective affinity separation. The immobilization procedures for small ligands that have been discussed thus far result in monovalent attachment of the ligand or spacer to the support. The instability of such monovalent linkages has been demonstrated by some investigators (18) and has at least been considered by others as a possible serious limitation for affinity chromatography.

One suggested approach to overcoming this difficulty consists of using a polyvalent spacer between the support and the ligand. It is believed that higher stability could be expected of such derivatives due to multipoint attachment of the polyfunctional spacer. Spacers of this variety that have received some experimental attention are polylysine (19,20), poly(lysyl-alanine) (19,20, 21), albumin (20,21), and denatured albumin (21).

Additional benefits might be expected from using polyvalent spacers. A higher degree of ligand substitution may be realized since the polyvalent spacer has a number of reactive groups that may be available for multipoint attachment to the support, and for bonding several ligand molecules. The ligand is at the same time separated from the support surfaces by large distances and may find itself in more favorable environments for interaction with its complementary binding component in solution.

V. LIGANDS

The uniqueness of affinity chromatography is the presence in a support column of a molecular species (ligand) which possesses a specific affinity for the solute to be purified. The ligand is usually chosen from a biological source which has a natural pairing tendency with the desired product. The ligand is usually a relatively simple compound and the physiochemical basis for pairing is at least rationalized. The specificity of the ligand requires that a specific column be prepared for each solute. The introduction mentioned a number of biologically paired species. Each member may be used as a ligand for isolating its natural partner. The degree of solute retention depends on the affinity of the solute for the immobilized ligand. O'Carra (5) has proposed the relationship

$$R = \frac{[L]}{K}, \tag{9}$$

where R is the retention in units of bed volume, [L] is the molar ligand concentration, and K is the dissociation constant. Deviations from this relationship are attributed by O'Carra to adsorption effects.

A. Enzyme Ligands

A particularly striking characteristic of enzymes is their high substrate specificity. Since an early event in the enzyme-substrate reaction is that of binding to form an activated complex, the substrate itself may be considered an appropriate affinity ligand. Substrate affinity ligands are particularly useful for isolating enzymes which have not been well characterized, when inhibitors are unknown or difficult to prepare, or when a new enzyme is sought. It has also been proposed that the product of an enzymatic reaction has enough affinity for the enzyme to be used as a ligand (22). Examples of substrate ligands are given in Section A of Applications.

There are obviously some difficulties with this approach. The reaction that binds the ligand to the carrier may modify the substrate so the complex does not form. The binding may impose new steric restrictions on the enzyme-substrate reaction. The greatest disadvantage is that one must expect some conversion of the ligand, even at low temperature, and multiple use of the column is usually not feasible. Under favorable conditions, column regeneration may be possible.

Much of our present understanding of enzyme action mechanisms and the character of the active site may be attributed to chemical studies of enzyme interaction with specific inhibitors. The competitive inhibitor may be chemically tailored to irreversibly (or nearly so) bind to the active site of the enzyme and block complex formation with the corresponding substrate. The high affinity of these inhibitors for certain enzymes suggests their use as affinity ligands. There are two major concerns in using this approach. The enzyme may bind so tightly to the immobilized ligand that it may not be dissociated without loss of activity; or the inhibitor may be eluted from the column and deactivate the released enzyme. The use of inhibitor ligands to purify acetylcholinesterase and β-galactosidase is illustrated in Section B of Applications.

To overcome the disadvantage of having to prepare a specific column for each solute, investigators have turned to preparing general ligands, which may be considered to be group specific (23,24). Bound enzyme cofactor can be used to isolate dehydrogenases. Special elution methods may be used to release specific binding components from the affinity support. The preparation of a variety of cofactor ligands is described in Section F of Applications.

B. Hydrophobic Ligands

The spacer between support and ligand cannot safely be considered an innocuous, inert element. Even if the capacity for forming hydrogen bonds or electrostatic or charge-transfer sites is reduced or eliminated, the spacer may still have high affinity for hydrophobic regions on proteins. Although these regions are often located within proteins and contribute to the maintenance of tertiary structure, nonpolar side chains - ala, val, leu, phe - may reside in accessible regions of the protein and be attracted to the nonpolar hydrocarbon portions of the spacer. The hydrophobic character of the spacer allows its use as a ligand for isolating proteins having particular affinity for hydrophobic sites (5,17).

Hydrophobic chromatography has been used to purify glycogen phosphorylase (25), glycogen synthetase (26), aspartate carbamoyltransferase (28) and α-chymotrypsin (27). Sodium chloride gradient elution was used to remove the retained protein. Actually, proteins could be expected to salt out onto the hydrophobic support as the ionic strength is increased. There is a reference to unpublished work by Porath in the lowering of the ionic strength to desorb protein from a hydrophobic column (28). The use of more effective eluent solutions - alcohols, amines and surfactants - is discussed by Hjerten (29). Purifying glycogen phosphorylase on hydrophobic ligands is discussed in Section C of Applications.

The use of hydrophobic supports as structural probes can be expected to receive considerable attention.

C. Covalently Reactive Ligands

Ligands that actually react with the solute to form covalent bonds may also be used to effect separation. In covalent chromatography, the ligand selectively reacts with an activated site on the protein to form a covalent bond. Cleavage of this bond must then occur in the elution step to release the bound solute. A sulfhydryl-containing ligand will react with active sulfhydryl groups of a protein, such as papain. Examples of such separation schemes are presented in Section D of Applications.

D. Ligands for Binding Proteins

A particularly rewarding application for affinity chromatography is the isolation of binding proteins. This class of materials includes plasma and cellular binding proteins. These biological materials have relatively well-defined binding partners that can serve as affinity ligands. It may be necessary to dissociate the binding protein from its partner before attempting the isolation on an affinity column. Dissolution of the support or leaching of the ligand are general hazards in isolating these materials. The presence of binding protein in the eluate is usually detected by binding to its tagged partner. As a consequence of the minute amounts of material used in such studies, any solubilized ligand would mask the presence of binding protein. In attempting to isolate steroid binding protein with a deoxycorticosterone (DOC) agarose column, Ludens and co-workers (30) found that cytosol promotes the release of the DOC into the eluate and prevents the detection of steroid binding capacity in the eluted fractions.

It must also be positively determined that the retained protein is specifically bound to the binding partner and not by some type of nonspecific binding. Otherwise, the adsorbed protein released during elution may be incorrectly considered the binding protein. The isolation of lectins, vitamin B_{12} intrinsic factor, and hormone receptors is described in Section E of Applications.

VI. BIO-AFFINITY ADSORPTION AND ELUTION

When an affinity adsorbent has been prepared for isolating a single component from a crude mixture, the sample is applied to the affinity column under those conditions of pH, ionic strength, and temperature that favor efficient adsorption of the solute to be purified. Attention must also be given to the flow rate so that the proper interaction between the immobilized ligand and solute molecules may take place. In some cases, where the sample is highly contaminated or the association constant is low, it may be desirable to allow the applied sample to remain in contact with the packed affinity support for a period of time before beginning column flow and elution. Those solute materials that have no affinity for the support are then washed off. Continued elution with the same solution should theoretically result in recovery of the bio-specifically adsorbed material in the effluent. Although this procedure may be time-consuming and may result in a large dilution of the adsorbed material in the effluent, some successful purifications have been reported.

In most cases, recovery of the adsorbed material is accomplished by changing the pH, ionic strength, or buffer composition of the eluent. Such compositional changes of the eluent are intended to affect the conformation of the adsorbed macromolecules in such a way as to alter the binding site and consequently reduce affinity for the immobilized ligand. Similar results can be obtained by incorporating structure-breaking agents, such as urea and guanidine hydrochloride, into the eluting solution. Examples of these elution methods will be found throughout the Applications section.

It is important that the conditions used to disrupt the immobilized complex are such that the recovered material retains its full biological activity when normal solution conditions are restored. In many cases, readjusting the solution composition, adjusting solution pH to neutrality, and removing agents, such as urea and guanidine hydrochloride, will result in restoring the biological activity of the material under consideration.

If the affinity support has general binding capabilities, it may adsorb from solution several different specific binding solute species. These different adsorbed materials may be separated and recovered as individual classes of compounds by stepwise or gradient elution. Gradients of increasing ionic strength have been used to elute different adsorbed species based upon their affinity. For the immobilized ligand, the stronger the association, the greater the ionic strength required for dissociation. One interesting reported example of gradient elution involves separating isoenzymes of lactate dehydrogenase that had been adsorbed onto N-(6-aminohexyl)

-AMP bound to Sepharose and eluted with a concave gradient of NADH (31).

Another occasionally used approach consists of cleaving the bond between ligand and support, thus releasing the liberated complex into solution. Azo linkages may be broken with reducing agents (7), and ester and amide bonds may be cleaved by appropriate alkaline or acidic conditions or by treatment with efficient nucleophiles. Recovery of the material to be purified then requires the dissociation of the complex and the separation of the two binding components.

The results obtained by changing eluent pH or ionic strength or by adding urea or guanidine hydrochloride must be interpreted cautiously until it can be demonstrated unequivocably that the adsorption and elution process was specific. Under such elution conditions, solute molecules nonspecifically adsorbed to the derivatized support may be eluted together with specifically bound material. To insure recovery of specifically adsorbed material, competitive inhibitors or soluble ligand should be added to the eluent. For example, the complex formed between an immobilized concanavalin A and a soluble glycoprotein may be dissociated with glycoprotein recovered in the effluent by elution with a solution of methyl-α-D-glucopyranoside (32,33) which competes with the glycoprotein for carbohydrate binding sites on the immobilized concanavalin A. In many cases where the binding component of the complex is being adsorbed onto an immobilized ligand, elution with the soluble ligand or with a competitive ligand requires the additional step of dissociating the complex in the effluent. However, such recovery of the adsorbed material would demonstrate convincingly that specific adsorption and elution have been involved in the purification procedure.

VII. APPLICATIONS

The chemical methods that have been used to immobilize a variety of classes of chemical compounds are quite limited. We will present some examples of specific ligands coupling which represent the type of coupling reactions that have appeared to be successful and valuable for a number of investigators. Where the ligands have some general interest, the preparative method, or at least an outline of the procedure, will be given.

A. Substrate Affinity Ligands

1. Isolation of Hydrolases

A number of metabolic disorders have been shown to result from inherited genetic defects. In the inborn lysomal diseases such as Tay-Sachs or Fabrys, accumulations of glycosphingolipids occur (34). The accumulations result from low turnover in tissue and are due to defects or a deficiency of the appropriate catabolic enzymes. Investigations into isolating these enzymes are greatly facilitated by affinity chromatography using the stored glycolipid or a closely related analog as the substrate affinity ligand.

In Fabry's disease, the major accumulated neutral glycolipid is galactosyl α(1→4)-galactosyl β(1→4) glucosyl β(1→1) ceramide (Eq. 10). Mapes and Sweeley (35) were unable to prepare the appropriate trihexoside in sufficient quantity to study its suitability as an affinity ligand. The substituted disaccharide p-amino phenylmelibioside (Eq. 11) was used instead.

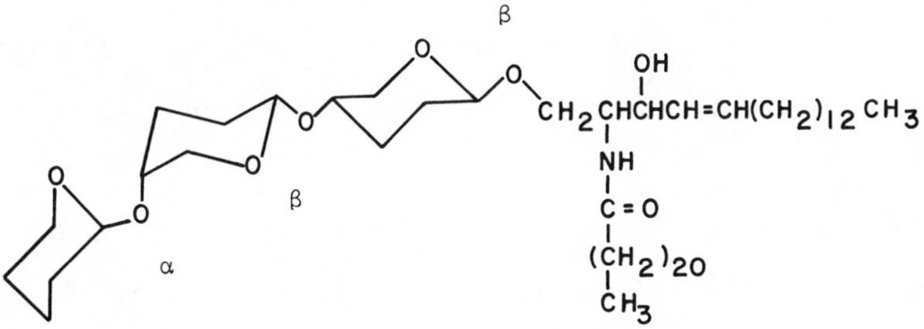

β OH
$CH_2CHCHCH{=}CH(CH_2)_{12}CH_3$
NH
C=O
$(CH_2)_{20}$
CH_3
α β

(10)

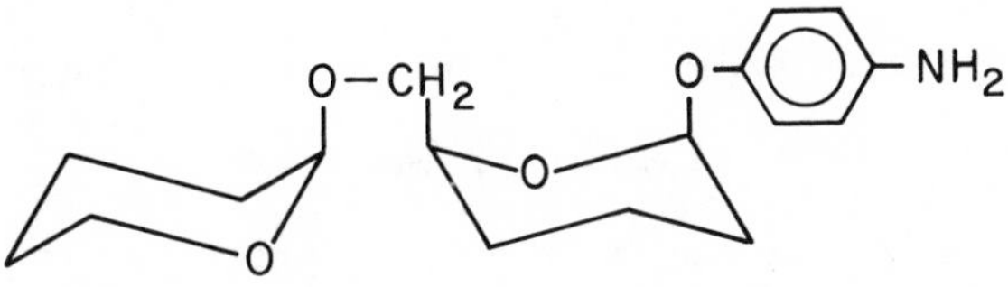

$O-CH_2$ O NH$_2$

(11)

Samples of α-galactosidase activity were applied to a pH equilibrated column (0.6 cm dia. x 6.5 cm high) at 4°C. The sample was allowed to equilibrate in the column for 30 minutes. Nonadsorbed proteins were removed with the equilibrating buffer. The bound protein was released with about 100 ml of 0.001 M 2-(N-morpholino)ethane sulfonic acid buffer containing 0.1% Triton X-100. Multiple forms (A and B) of two principal α-galactosidases were eluted. Form B is missing in the Fabry. Phosphatidylcholine in both the equilibration buffer and elution buffer was beneficial for certain of the enzyme forms. Although these authors have contributed much to the examination of substrate specificity of some of this enzyme's multiple forms, further studies are required to determine the origin of these enzymes, binding constants and other characteristics.

In Tay-Sachs disease, the enzyme β-N-acetylhexosaminidase is missing and massive accuumulation of the trihexose ceramide (Eq. 12) occurs (36).

```
gal  - gal  - glu - ceramide
 |      |      |
NHAc   NHAc neuraminic acid.          (12)
```

The enzyme was isolated from normal tissue by Dawson, Propper and Dorfman via affinity chromatography (37) using glycopeptide derived from bovine nasal septum as the substrate affinity ligand. The ligand is believed to be

```
gal - (glu-gal)n-glu-(gal)2-xyl-ser-peptide.
 |       |    |    |                                  (13)
NHac    UA  NHAc  UA
```

Unfortunately, the authors give few experimental details. The ligand was coupled to Sepharose 4B via the CNBr method. The column was eluted with 0.1% Triton X-100, pH 4.4 to remove the retained proteins.

A related enzyme, β-N-acetylglucosaminidase was purified by Junowicz and Paris by the use of p-aminophenyl-β-N-acetylglucosaminide as the affinity ligand coupled to succinyl-diaminodipropylamino-Sepharose 4B (38) (Eq. 14). However, a satisfactory separation from β-glucuronidase was not achieved. The authors used a continuous NaCl gradient elution to remove the retained enzymes. Elution by mono- and disaccharides was not successful. No mention is made of column reuse, although the authors do report extensive binding studies with enzyme mixtures.

$$\text{Ⓢ}\sim NH(CH_2)_3NH(CH_2)_3NHCO(CH_2)_2CO_2H. \qquad (14)$$

a. Preparation of p-aminophenylmelibioside (35). Melibiose was converted to the octaacetate using acetic anhydride in dry pyridine at 80°C. The melibiose octaacetate was brominated with 48% HBr in acetic acid at room temperature. p-nitrophenol was coupled to acetobromomellibiose in acetone containing 1.0 M sodium carbonate. The remaining acetate groups were removed with 0.5% sodium methylate in methanol, and the product was reduced to p-aminophenylmelibioside with sodium hydrosulfite at pH 7.4.

b. Preparation of p-aminophenylmelibioside-Sepharose (35). p-aminophenylmellibioside was coupled to succinoylamino alkyl agarose. The support had 8 μ moles of carboxyl group content per ml of packed bed. A portion (10-ml packed bed) of Affinose 202 was washed with 20 bed volumes of 0.1 M sodium chloride (pH 6.0). p-aminophenylmelibioside (1.6 g) was dissolved in 150 ml of 40% dimethylformamide and was added to a suspension of the washed resin in 50 ml of 40% dimethylformamide. After adjusting to pH 5.0 with 0.1 N HCl, a solution of 1-ethyl-3(3-dimethylaminopropyl)-carbodiimide (1 g) in 8 ml of water was added over a 10-minute period. The reaction was allowed to proceed at room temperature for 24 hours with occasional shaking. The product of the coupling reaction was packed into a column and washed successively with 40% dimethylformamide, water, and 0.001 M MES buffer (pH 5.4), until ultraviolet-absorbing materials and carbohydrates, as measured by the anthrone reaction (39), were absent from the eluate.

2. Isolation of an Amidase

An interesting example of a substrate affinity ligand used to isolate a new enzyme is provided by

LeGoffic, Labia and Andrillon (40). Bacterial resistance to penicillin is believed to result from production of β-lactamase, which cleaves one amide bond to penicillin to yield inactive penicilloic acid (Eq. 15). The ligand ampicillin (Eq. 16) was bound to the polysaccharide carrier Indubiose via glutaraldehyde. The supernate from sonicated E. coli K_{12} cells was placed on the column equilibrated with 0.01 N NaCl. Elution of the desired enzyme was achieved with 1.0 M NaCl, a purification factor of about 500 X is claimed. The column was presumably operated at room temperature and no mention of column reuse is made.

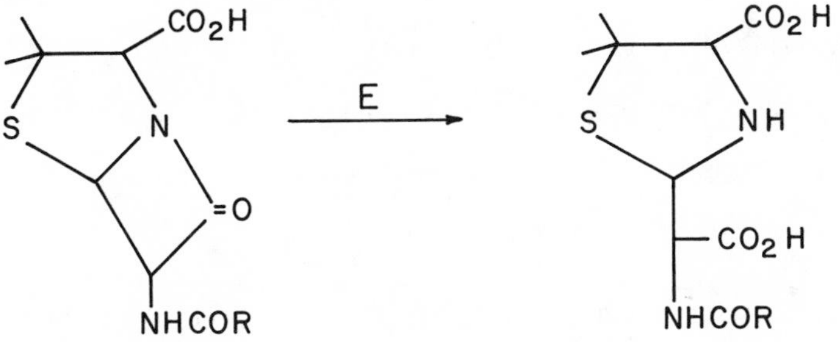

(15)

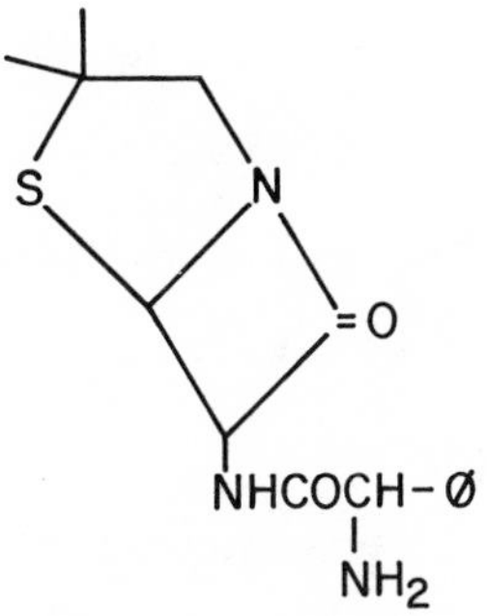

(16)

a. Preparation of Ampicillin-glutaraldehyde-Indubiose. Indubiose ACA-3/4 (2 g dried adsorbent) in 50 ml distilled water is treated three times at 20°C with 10 ml of a 25% aqueous solution of glutaraldehyde, the pH being kept at 11 by continually adding 1 M NaOH. The adsorbent is extensively washed with water (4 liters) to remove excess glutaraldehyde and then resuspended in 100 ml of distilled water. This mixture is gently rocked with 0.4g ampicillin trihydrate at 4°C for 48 hours, with the pH kept at 7. It is then washed well with water (8 liters).

3. Isolation of a Deaminase

Another example of using a substrate ligand to isolate the enzyme(s) involved in a metabolic pathway is given by Chibata's group at Tanabe (41). Aspartic acid is coupled to Sepharose 6B via the spacer of hexamethylenediamine-succinic acid (Eq. 17).

$$\sim NH(CH_2)_6NHCO(CH_2)_2CONHCH(CO_2H)\ CH_2CO_2H. \qquad (17)$$

The column adsorbs the enzymes aspartase, asparaginase and aspartate β-decarboxylase. Fumarase was not adsorbed onto the column. Independent experiments using a linear NaCl gradient elution from 0 to 0.1 N NaCl at pH 7.0 indicated the following enzyme affinities for the aspartic acid ligand:

aspartase > aspartate β-decarboxylase > asparaginase.

The presence of 0.1 M L-aspartic acid in the elution buffer permitted release of the enzyme at a lower ionic strength and increased considerably the recovery values of enzyme activity. Asparaginase from *P. vulgaris* and *E. coli* exhibited very similar elution characteristics even though the enzymes are not immunologically equivalent.

a. Preparation of N-(ω-aminohexyl)-L-aspartic acid-Sepharose. L(+)-monochlorosuccinic acid was prepared by the dropwise addition of 6.7 g D(-) aspartic acid and 20 g NaCl in 50 ml 1 M HCl to 50 ml of 4% NaNO at 0 to 5°C with stirring. After stirring for 20 hours, 50 ml of concentrated HCl was added, and the reaction product was extracted 3X with diethylether. Monochlorosuccinic acid (2.6 g) was isolated in crystalline form by evaporation of the ether.

Monochlorosuccinic acid (2.6 g) was dissolved in 50 ml of absolute alcohol. Hexamethylenediamine (7.5 g) in 50 ml ethanol was slowly added and the mixture was refluxed with stirring for 20 hours. Then 100 ml of water was added and an oily product was obtained by concentration under vacuum. The product was passed through an Amberlite IRA 401 (OH^- type) ion-exchange column. The column was washed with water and developed with 14% NH_4OH. The eluate was evaporated to dryness and recrystallized from 200 ml methanol and 10 ml of water to yield 0.5 g of N-(ω-aminohexyl)-L-aspartic acid.

ω-aminohexyl-L-aspartic acid was coupled to CNBr-activated Sepharose 6B.

4. Isolation of a t-RNA Synthetase (42)

The enzyme leucyl-transfer ribonucleic acid synthetase (LRS) was successfully purified by using the substrate ligand leucine-transfer ribonucleic acid bound directly to CNBr-activated agarose beads. The amount of nucleotide bound to the matrix was determined by digesting the derivatized gel with T_2 ribonuclease and measuring the amount of nucleotide released (A_{260}). The coupling efficiency was about 30%. The enzyme was eluted with a linear gradient of 0 to 0.5 M ammonium chloride (Fig. 3). The column also exhibited retention of isoleucyl-tRNA synthetase and phenylalanyl-tRNA synthetase. However, the adsorption of leucyl-tRNA synthetase was not affected by the presence of the other enzymes. It thus appears that the ligand binds the leucyl-tRNA synthetase specifically. However, the column after activation (but not before) exhibits nonspecific adsorption toward the synthetases. Ion-exchange sites may have remained after incomplete reaction of the activated agarose.

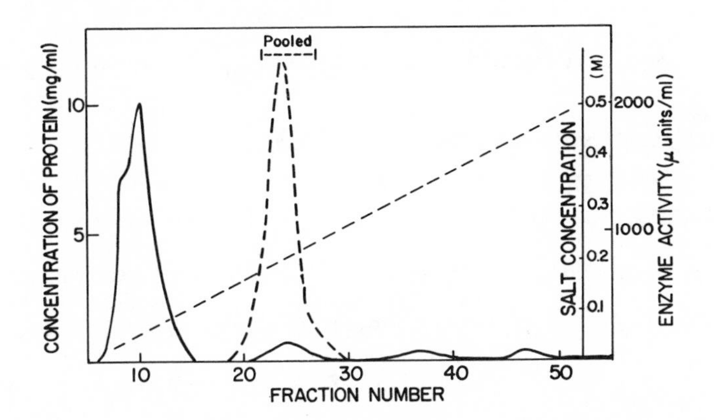

Fig. 3. Separation of leucyl-tRNA synthetase on Sepharose; $tRNA^{Leu}$ matrix. A partially purified preparation of LRS (20 munits) was separated $tRNA^{Leu}$ matrix. Column size, 1.0x25cm; Flow rate, 15 ml/hr. Elution was carried out with a linear concentration gradient of NH_4Cl in 0.05M phosphate buffer (pH 7.0), 0.01M 2-mercaptoethanol, and 10% v/v glycerol. ——, amount of protein, ----, LRS activity; — — —, concentration of NH_4Cl.

B. Inhibitor Ligands

1. Inhibitor Ligands for Acetylcholinesterase (AChE)

Phenyltrimethylammonium ion is a good competitive inhibitor for acetylcholinesterase (43). Three different laboratories have used a related inhibitor ligand for purifying acetylcholinesterase from the electric tissue of Electrophorus electricus. Comparing their results reveals the importance of spacer length, spacer composition and elution solution. The inhibitor (Eq. 18) was coupled to

$$NH_2-C_6H_4-\overset{+}{N}(CH_3)_3 \qquad (18)$$

Sepharose via the spacer (Eq. 19) derived from 3,3'-diaminodipropylamine and succinic acid. Although K_i for

$$-NH(CH_2)_3NH(CH_2)_3NHCO(CH_2)_2CO_2H. \qquad (19)$$

the ligand is ~ 10^{-6} M, the spacer-ligand combination did not sufficiently retard AChE in 0.1 N NaCl to effect a significant purification. With a dimer of (Eq. 19) as the spacer, the enzyme is firmly retained (44). Figure 4 shows the retention of acetylcholinesterase with several spacers and inhibitor ligands. Elution was achieved using the competitive inhibitor Tensilon ($K_i = 10^{-6}$ M). High recovery of the added enzyme activity was obtained. It is interesting to note that erythrocyte acetylcholinesterase is so firmly bound to the column that at half the added eel enzyme activity only 76% could be reocvered by elution with 10 column volumes of 10^{-2} M Tensilon in 0.04 M $MgCl_2$ - 0.1 N NaCl, pH 7.8. Very high specific activity AChE was isolated. With 6-aminocaproic acid as a spacer (Eq. 20) the inhibitor ligand

$$\sim NH(CH_2)_6CO_2H \qquad (20)$$

(Eq. 18) afforded a column having high affinity for acetylcholinesterase (45). This spacer is much shorter than that described by Eq. 19. The enzyme was successfully eluted with 1 N NaCl; however, the enzyme was about 40% denatured. Doubling the spacer arm gave improved, but not dramatically different results (46). The origin of the difference between these two sets of experiments are not entirely clear. In the latter experiments, salt extracts of fresh tissue were used as the crude enzyme source. In the former reports, acetylcholinesterase was purchased from commercial sources. Acetylcholinesterase may be obtained in several molecular forms having different sedimentation coefficients depending on the method of isolation. These forms differ in their abilities to aggregate in low ionic strength solutions. The difficulties observed with spacers 19, 20, and ligand 18 were circumvented by using a more potent acetylcholinesterase inhibitor. A derivative of N-methyl acridine (Eq. 21) having a K_i of 0.3×10^{-6} was coupled to CNBr-activated

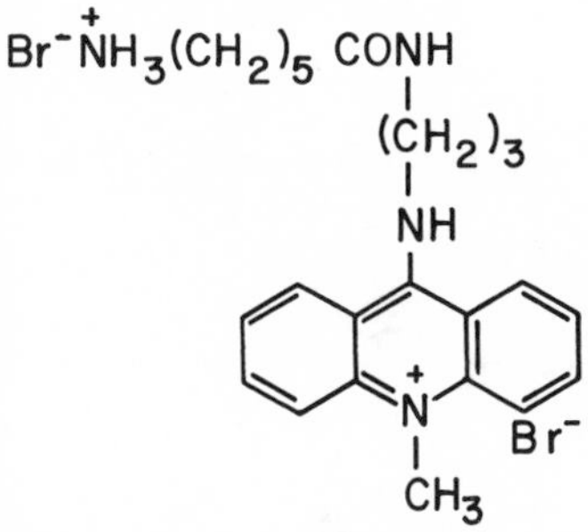

$Br^- \overset{+}{N}H_3(CH_2)_5CONH$

$(CH_2)_3$

NH

$\overset{+}{N}$ Br^-

CH_3

(21)

Sepharose via the 6-aminocaproyl spacer (47). More than 50% of the applied activity units were recovered upon elution with the inhibitor decamethonium bromide in 1 M NaCl in phosphate buffer. By this procedure, two molecular forms having sedimentation coefficients of 14S and 18S, respectively, were isolated.

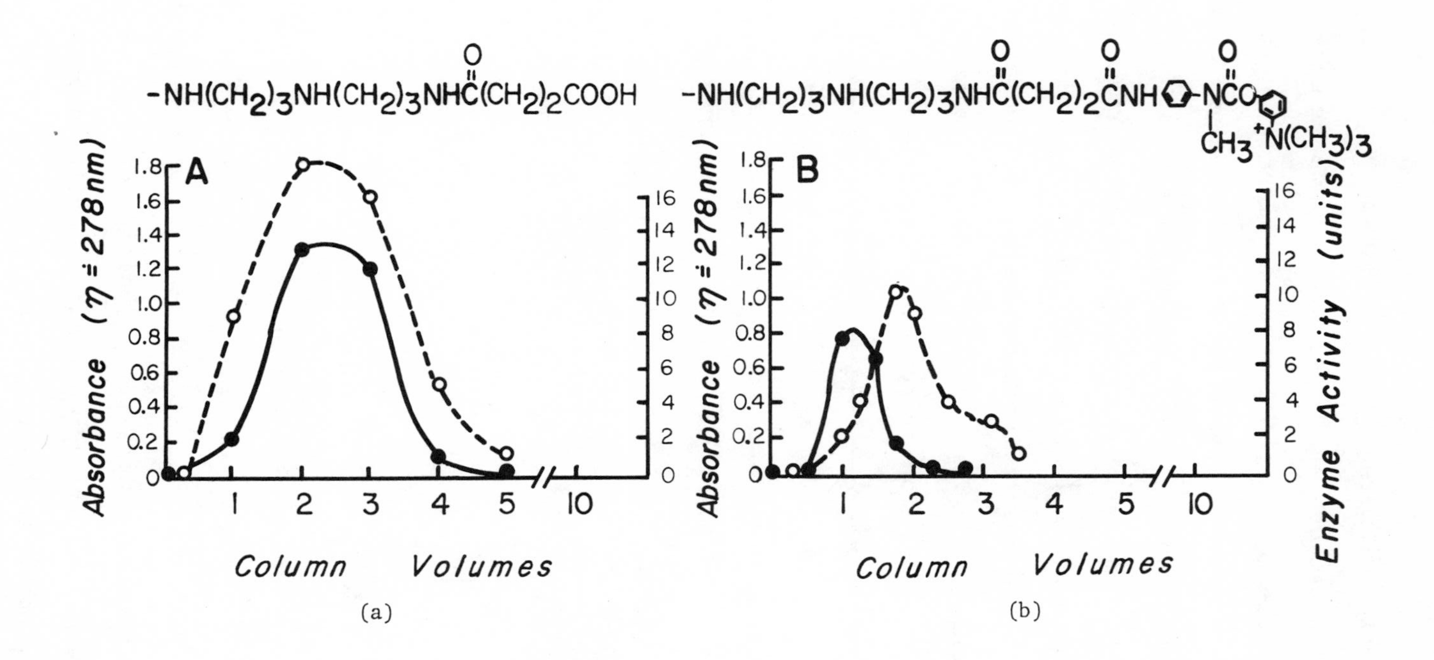
-NH(CH2)3NH(CH2)3NHC(CH2)2COOH
A
Absorbance (η = 278nm)
Column Volumes
-NH(CH2)3NH(CH2)3NHC(CH2)2CNH-NCO
CH3 +N(CH3)3
B
Absorbance (η = 278nm)
Enzyme Activity (units)
Column Volumes

(a)

(b)

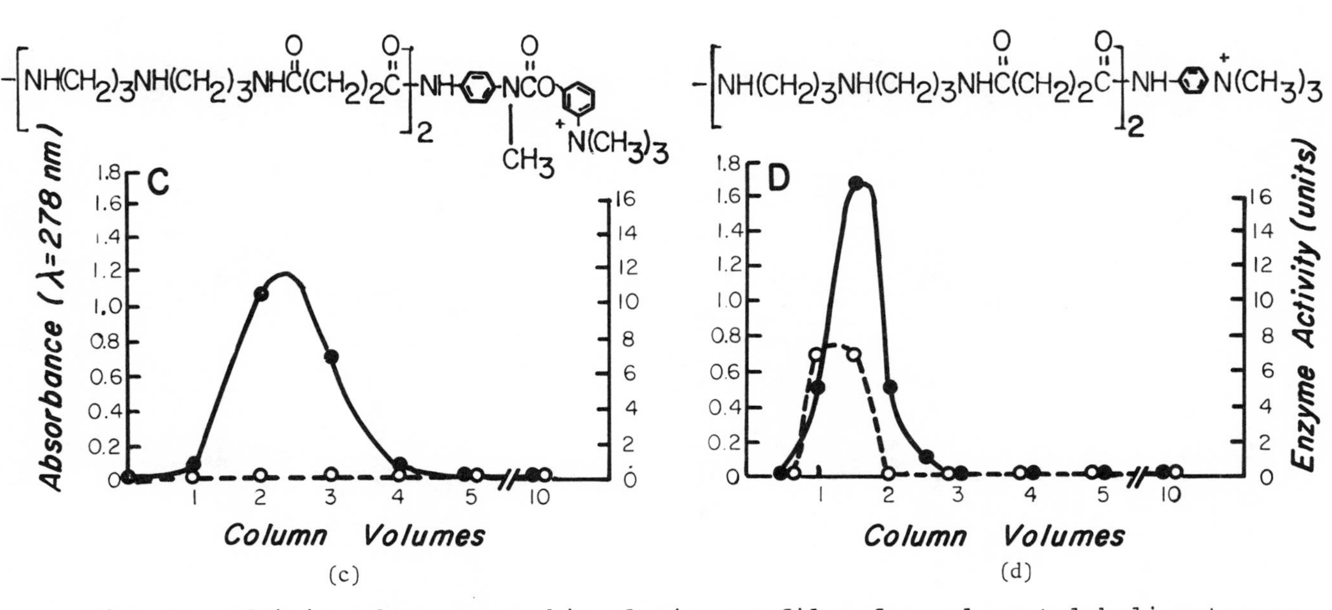

Fig. 4. Affinity chromatographic elution profiles for eel acetylcholinesterase. Column dimensions: 0.9 cm X 1-7 cm, packed with derivatives of 2% agarose; flow rate = 3 ml/hr; temperature, $4^{o}C$; solvent: 0.1 M NaCl-0.04 M $MgCl_2$ adjusted to pH = 7.8 with $NaHCO_3$; enzyme activity, (O-O); absorbance, (●-●). Inhibitor ligands together with "side arm" extensions are shown for each figure. Bovine serum albumin was added to each enzyme sample as carrier protein. A, 3.2 mg of albumin (40 enzyme units) applied, column volume 0.5 ml; B, same as A, column volume 3 ml; C, 4.5 mg of albumin (110 units), column volume 1.5 ml; D, 7.5 mg of albumin (2000 units), column volume 4 ml.

Dialysis against 0.1 M NaCl yielded the 11S molecular form. By using the potent acridinium inhibitor ligand, it was possible to apply the crude enzyme to the column at high ionic strength (1.0 M NaCl) and still obtain a high degree of retention.

2. Inhibitor Ligand for β-Galactosidase

Although it is desirable that the ligand be a strong inhibitor for the desired enzyme, favorable results have been obtained with relatively weak inhibitors. The p-aminophenyl-β-D-thiogalactopyranoside (Eq. 22), a competitive inhibitor of β-galactosidase with a K_i of only

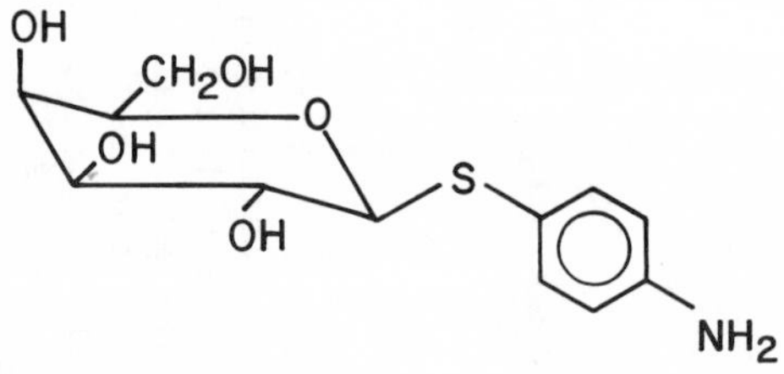

OH, CH_2OH, O, OH, S, OH, NH_2

(22)

5×10^{-6} M was successfully used by Steers, Cuatrecasas and Pollard (48). Their studies also demonstrated, at least for this one system, the importance of proper spacer length. When a short spacer of about 10 Å was used, no significant retention of enzyme was observed. With a spacer of about 21 Å (Eq. 19), the authors report very high retention of enzymic activity. The enzyme could be only poorly eluted with substrate. Optimal elution was achieved with alkaline buffers. The crude enzyme from E. coli is placed on the column in 0.1 M NaCl, 0.05 N TRIS · HCl, pH 7.5, and eluted with 0.1 M sodium borate at pH 10.05. The borate in some way facilitates release of the bound enzyme. We (49) and other investigators (50) observed with 0.1 M borate buffer that pH 8.0 is a satisfactory elutant for the enzyme. However, the enzyme is not eluted with TRIS buffer up to pH 9.5.

Quite different results were obtained using porous glass as a support (49). The length of the spacer seemed to have no effect on the binding strength of the ligand. No significant differences were observed with either a short malonic acid spacer (Eq. 23) or a long azelaic spacer (Eq. 24).

$$HO_2CCH_2CO_2H \qquad (23)$$

$$HO_2C(CH_2)_7CO_2H \qquad (24)$$

When the inhibitor or derivative was replaced with the neutral ligand aniline, the column still selectively retained β-galactosidase. These results suggested that the spacer having a significant hydrophobic segment possesses an affinity for the enzyme. More compelling evidence that the spacer rather than the ligand was actually the primary attractant was given by O'Carra (5). When a glucoside was coupled to the end of the spacer employed by Steers et al., β-galactosidase is still retained on the column. The improvement noted when the spacer length was extended may be attributed to increasing the hydrophobic character and binding capacity of the spacer. Controlled-pore glass was also used as a matrix by Woychik and Wondolowski (52). The same inhibitor (*p*-aminothiogalactopyranoside) was coupled to the arylamine glass derivative via the diazonium salt.

The thiogalactopyranoside is an expensive inhibitor for β-galactosidase. Recently, Kanfer, Petrovich and Mumford reported very successful results using D-galactose-α-lactone as the ligand and benzidine as the spacer to Sepharose (Eq. 25) (51). Unreacted sites on the gel were

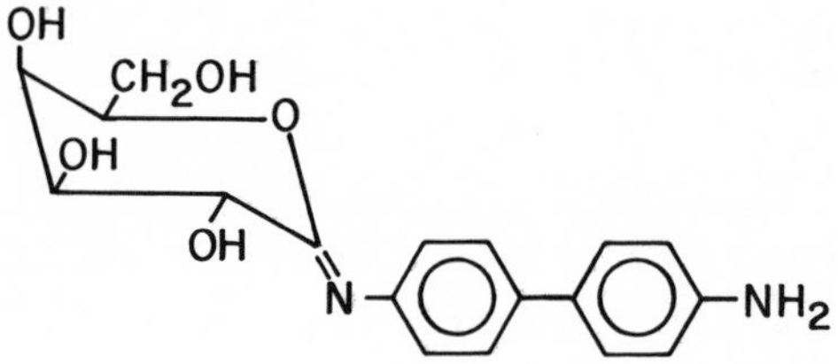

(25)

quenched with acetic anhydride. Crude β-galactosidase from jack-bean meal was put on the column in 0.01 *M* phosphate buffer, pH 7.0, and eluted with 0.1 *M* borate buffer, pH 10.0. The lactones are inexpensive and are readily available. It is anticipated that these inhibitors will be very important for isolating or purifying many carbohydrate hydrolases. Both α-and β-galactosidases were purified with the galactonate column.

Table 1 gives the results of purifying β-galactosidase by several investigators. Although an exact comparison cannot be made due to variations in assay conditions, the galactolactone is certainly a very satisfactory ligand.

Table 1

Affinity Chromatography of β-galactosidase

Enzyme Source	Affinity Ligand	Activity Before chromat.	μ moles min-mg After chromat.	Ref.
E. coli	PAPTG	-	320[a]	48
E. coli	PAPTG	45	450[b]	50
Jack Bean	galactolactone	12.6	600[c]	51
A. niger	PAPTG	60	120[d]	52
A. niger	PAPTG	102	240[e]	49

PAPTG = *p*-aminophenyl-β-D-thiogalactopyranoside.
All assays utilized o-nitrophenylgalactopyranoside as substrate. a) 28°C, 0.23 *mM* substrate, pH 7.0, b) 37°C 0.23 *mM* substrate, pH 7.0, c) 37°C, 10 *mM* substrate (*p* isomer), pH 3.5, d) 37°C, 5 *mM* substrate, pH 4.0 (value corrected to conditions of e), e) 60°C, 8.3 *mM* substrate, pH 4.0.

a. Preparation of Galactonolactone-Benzidine-Sepharose. A suspension of 250 ml of cyanogen bromide-activated agarose beads and 817 mg benzidine in 200 ml dioxane was stirred overnight at room temperature. The benzidine-gel derivative was then washed on a fritted glass funnel with 2.5 liters of dioxane followed by 1.4 liters of water.

The derivatized gel (60 ml) was suspended in 100 ml water, pH adjusted to 4.0, and D-galactono-α-lactone (5.85 g) and 1-ethyl-3(3-dimethyl-aminopropyl) carbodiimide-HCl (500 mg) are added with stirring. After adjusting pH to 5.0, the mixture is stirred overnight. After repeating this procedure twice, the gel is finally filtered and washed with one liter of water. The gel is suspended in 100 ml water and sonicated with 2 ml of acetic anhydride. The gel is then filtered and washed thoroughly with water.

C. Hydrophobic Ligands

In an earlier discussion, the contribution of the spacer to the retardation of β-galactosidase was noted. In studies of the role of hydrophobic bonding in the retention of proteins, the spacer is the affinity ligand. By varying the length of a lipophilic spacer, one obtains a probe the size of the available hydrophobic pocket on a given protein. An exact interpretation is not now available. For example, in the isolation of glycogen phosphorylase (25) on a series of alkane-substituted agaroses (Eq. 26), the degree of retention increases from none at $n = 1$, to slight retardation at $n = 3$, to reversible binding at $n = 4$, to very tight binding at $n = 6$.

$$\sim NH(CH_2)_nH \qquad (26)$$

Ligands having alkane chains shorter than n = 4 may not be sufficiently hydrophobic to promote retention, since in the other works cited, a C_4 chain is the minimum ligand length for any retardation. An interesting separation of glycogen synthetase and glycogen phosphorylase from muscle extract can be accomplished on ω-aminoalkyl-agarose (Eq. 27). The extract is passed through a column of ω-amino-butyl-agarose where glycogen synthetase but not glycogen phosphorylase is retained. The excluded protein is then passed through a column of ω-amino-hexyl-agarose where glycogen phosphorylase is retained. Release of the retained enzymes was accomplished by a linear NaCl gradient elution.

$$\sim NH(CH_2)_nNH_2 \qquad (27)$$

D. Covalently Reactive Ligands

Relatively few examples of covalent chromatography have been published. Organophosphate inhibitors of acetylcholinesterase are known to react at a specif serine residue with phosphorylation of the hydroxyl group. Ashani and Wilson (53) coupled to p-nitro-phenylphosphate derivative to Sepharose via an extended arm (Eq. 28). The product is capable of phosphorylating the serine residue at the

$$\sim N(CH_2)_2OP(=O)(CH_3)O\phi NO_2 \quad (28)$$

active site of acetylcholinesterase. The enzyme is released from the column by elution with oximes such as 2(hydroximinomethyl)-1-methylpyridinium iodide (Eq. 29) or 1,1'-trimethylene bis (4-hydroximinomethylpyridinium) dibromide (Eq. 30).

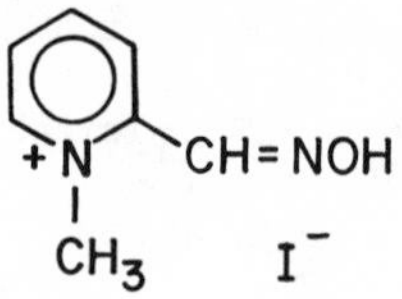

$$\text{2-(CH=NOH)-1-}CH_3\text{-pyridinium } I^- \quad (29)$$

$$[\text{4-(CH=NOH)-pyridinium-1-}CH\text{—}]_2 CH_2 \quad 2\,Br^- \quad (30)$$

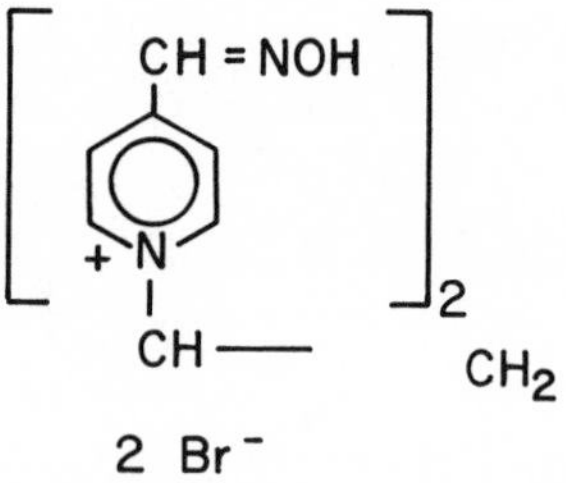

These oximes are strong nucleophilic reagents which will regenerate inactive phosphorylated enzyme. The column is active against esterolytic enzymes which have a serine residue at the active site. Ashani and Wilson have also trapped α-chymotrypsin with this column.

2,2'-dipyridyl disulfide (2-Py-S-S-2-Py) at pH of 3.5 to 4.0 reacts at least 10^2 faster with the activated thiol of papain than with simple thiols. Brocklehurst et al (54) prepared the conjugate Sepharose-(glutathione-2-pyridyl disulfide (Eq. 31) for the covalent chromatography of papain. The enzyme is put on the column at pH 4.0 at room temperature. Elution with the same buffer is continued until no protein appears in the eluent. The covalently bound papain is removed at pH 8.0 and the simple thiolate ion reacts 10 to 20 times faster with disulfide than papain.

$$\sim N\text{-}CH(CO_2H)\text{-}(CH_2)_2CONH\text{-}CH(CONHCH_2CO_2H)\text{-}CH_2\text{-}S\text{-}S\text{-}Py \qquad (31)$$

The column is regenerated by reduction with dithiothreitol and is converted to the initial ligand by treatment with 2-Py-S-S-Py-2. The column could possibly be used to purify other enzymes, such as urease, which possess an activated sulfhydryl group.

1. Preparation of the Sepharose-glutathione-2-pyridyl Disulfide Conjugate

Sepharose (50 g) is activated with cyanogen bromide in the usual manner. The gel is washed rapidly with 0.1 M $NaHCO_3$ at 4°C on a sintered-glass funnel and transferred to a 200-ml stoppered flask containing glutathione (1 g)

dissolved in 20 ml of 0.1 M $NaHCO_3$ adjusted to pH 8.5. The volume was brought to 90 ml and shaken for 20 hours at 22°C. The gel is then washed with 0.1 M $NaHCO_3$ on a sintered-glass funnel.

The Sepharose-glutathione gel is then washed with 0.1 M TRIS-HCl buffer, pH 8.0, containing 0.3 M NaCl and 1 mM EDTA. The gel is then suspended in the same buffer containing 20 mM dithiothreitol and vacuum filtered. The reduction process is repeated twice. The reduced gel is washed in turn with 1 M NaCl, 0.5 M NaHCO , 0.1 M TRIS-HCl, pH 8.0. Each of these solutions contained 0.3 M NaCl and 1 mM EDTA. The reduced gel is allowed to react with 1.5 mM 2-Py-S-S-2-Py in 0.1 M TRIS-HCl buffer pH 8.0, containing 0.3 M NaCl and 1 mM EDTA, with the gentle stirring for 30 minutes. The gel was then washed as above until no disulfide could be detected by measuring extinction at 281 nm.

Organomercurial derivatives have also been used for isolating sulfhydryl proteins (55). *p*-chloromercurobenzoic acid can be coupled to aminoalkyl agarose via the carbodiimide reaction (Eq. 32). The column was used to fractionate thyroglobulin preparations. *p*-aminophenyl mercuric acetate can be coupled directly to CNBr-activated agarose to yield a similar product which was used to purify papain.

$$\sim N(CH_2)_3NHCO\phi HgCl \qquad (32)$$

Penicillin kills bacteria by preventing the cross-linking of nascent cell wall by a transpeptidase. It is believed that the membrane contains a penicillin-binding protein which undergoes penicilloylation. 6-Aminopenicillanic acid (Eq. 33) was bound to succinylated-diamino-dipropylamino-Sepharose (56). A semiquantitative assay for the bound ligand was performed using the hydroxamate assay for penicillins. Several mg of penicillin derivative was bound per ml of gel.

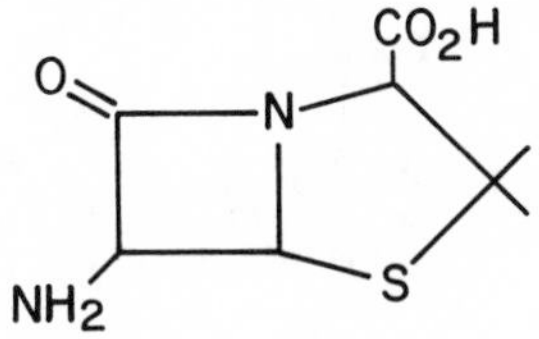

(33)

The membrane extract containing binding protein was incubated with a slurry of the affinity gel with gentle agitation for 30 minutes at 25°C. The mixture was then loaded into a squat column and rapidly washed with a buffer solution of 1.0 M NaCl, 0.05 M KPO_4, pH 7.0, and 0.1% Nonidet P-40. The covalently bound proteins were eluted with a buffer solution of 0.8 M hydroxylamine, 0.5 M NaCl, 0.05 M KPO_4, pH 7.0, 1% Nonidet P-40 and 1 mM 2-mercaptoethanol. The eluent contained five protein components separable by polyacrylamide gel electrophoresis (See Fig. 5). All five components bind ^{14}C-penicillin G.

To date, however, only one component was found to exhibit enzymatic activity, a D-alanine carboxypeptidase.

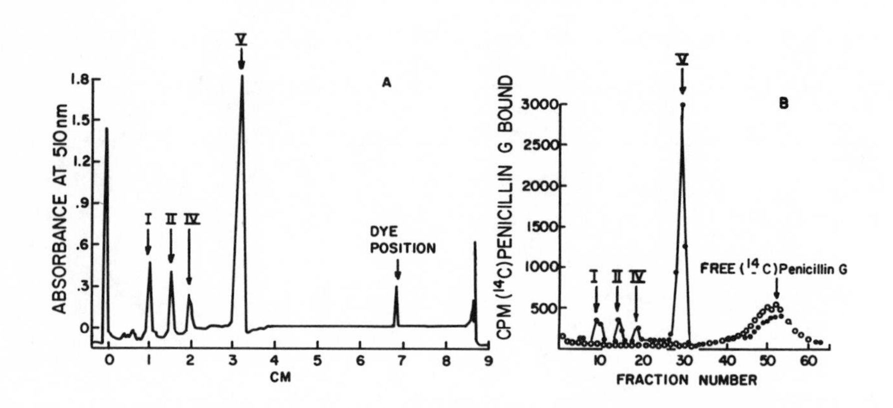

Fig. 5. Binding of (^{14}C) penicillin G to the penicillin-binding components isolated by affinity chromatography. (A) Densitometry of an SDS gel of isolated binding components stained with Coomassie brillant blue. (B) SDS gel of the isolated binding components to which (^{14}C) penicillin G had been bound.

E. Isolation of Binding Proteins

1. Lectins

Inhibitor ligands may also be used to isolate nonenzymatic substances. Extracts of *Bandeiraea simplicifolia* have exhibited anti-blood group B agglutinating activity. The agglutinating activity, however, was inhibited by galactose and galactose-containing oligosaccharides.

Hayes and Goldstein (57) succeeded in isolating the α-D-galactosyl-binding lectin by affinity chromatography on a column containing a melibiose derivative as the inhibitor ligand (Fig. 6).

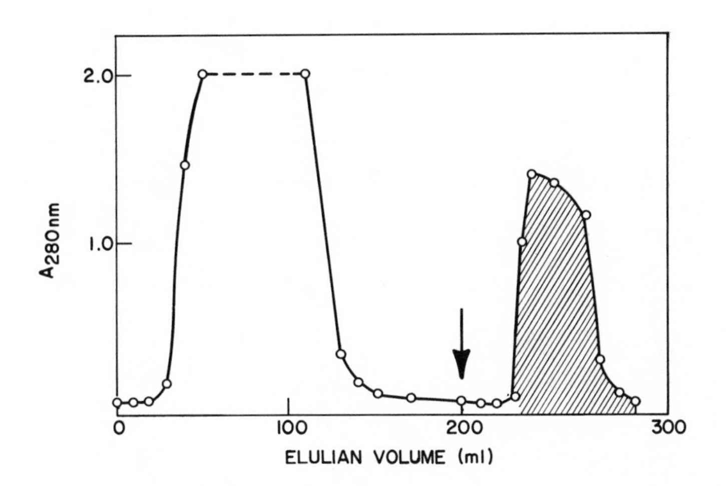

Fig. 6. Elution profile of B. simplicifolia extract. An aliquot (32 ml, 230 mg) of the ammonium sulfate fraction was layered onto thecolumn. The column was washed and subsequently eluted with 0.5 M D-galactose in the equilibrating buffer, indicated by the arrow, resulted in displacement of the lectin.

Melibiose was oxidized to melibionic acid (Eq. 34) and the corresponding potassium salt was bound to aminoethyl-Bio-Gel P-300 with carbodiimide. Sepharose derivatives were unsatisfactory for this separation. Neither p-aminophenyl α-D-galactopyranoside coupled to cyanogen bromide-activated Sepharose nor melibionic acid coupled to aminoethyl-Sepharose derivatives exhibited sufficient retention.

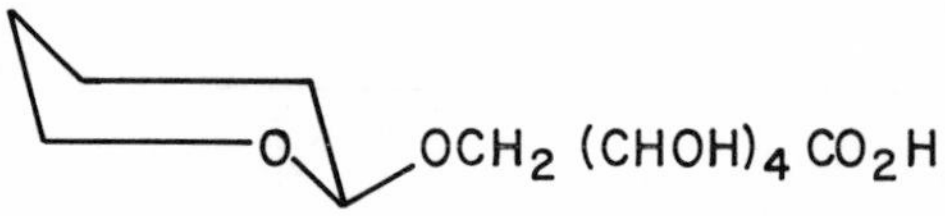

$$\text{(Sepharose–O–)}\ OCH_2(CHOH)_4CO_2H \tag{34}$$

Rodd and Wilkinson (58) attempted to use a Sepharosylglycyltyrosyl N-(p-aminophenyl) oxamic acid (Eq. 35) for the isolation of sialidase *Clostridium perfringes*; however, they encountered serious retention of a hemagglutinin along with several enzymatic proteins.

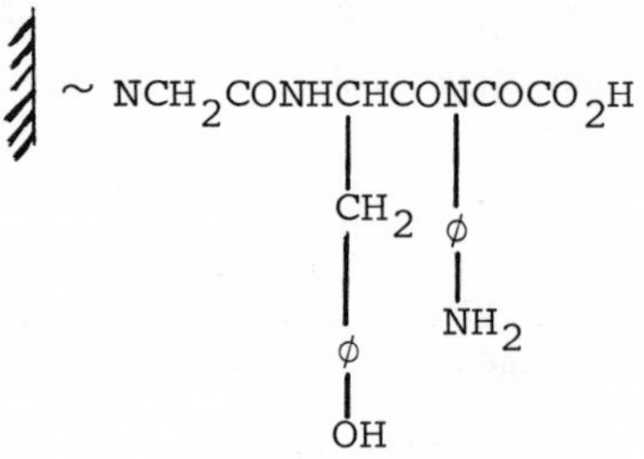

$$\sim NCH_2CONHCH(CH_2\phi OH)CON(\phi NH_2)COCO_2H \tag{35}$$

Using a noncarbohydrate support, such as polyacrylamide, is particularly advantageous for isolating a carbohydrate-binding protein. A second advantage is that investigators may assay directly the degree of saccharide ligand substitution. The lectin was easily eluted from the Bio-Gel column by adding 0.5 M D-galactose in the equilibrating buffer (0.01 M phosphate, pH 7.0, 0.15 M NaCl, 0.1 mM $CaCl_2$).

<u>a. Coupling of Melibionate to Polyacrylamide.</u> Potassium melibionate (7.5 g) was added to a suspension of 50 ml aminoethyl-Bio-Gel P-300 in 50 ml water. The solution was adjusted to pH 4.7 with 1 <u>M</u> HCl. 1-Ethyl-3 (3-dimethylaminopropyl) carbodiimide HCl (3.5 g) dissolved in water (2 ml) was added with stirring and the pH was maintained at 4.7 by adding 0.5 <u>M</u> HCl over 30 minutes. The beads were filtered, washed with water until the filtrate was free of carbohydrate, and then washed with equilibrating buffer.

Although the previously cited investigators (<u>57,58</u>) found that Sepharose derivatives were unsatisfactory for isolating lectins, Lis, Sharon, and co-workers (<u>59</u>) obtained very satisfactory purification of wheat germ agglutinin and soybean agglutinin on Sepharose derivatives. Soybean agglutinin was isolated on Sepharose-ε-aminocaproyl-β-D-galactopyranosyl-amine (Eq. 36). The

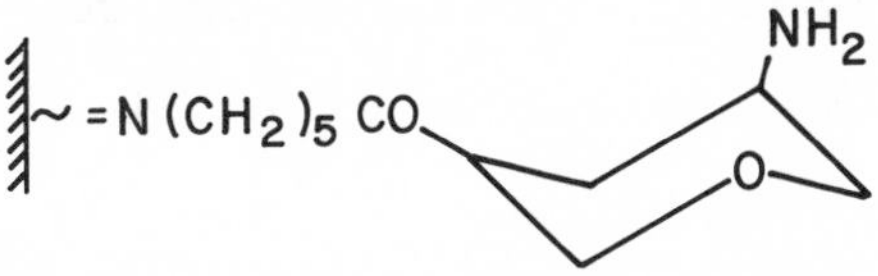

$$NH_2$$

$$\sim=N(CH_2)_5CO \quad O$$

(36)

corresponding N-acetyl-β-D-glucosamine derivative was used to purify wheat germ agglutinin. The agglutinin is extracted from defatted wheat germ, and precipitated with ammonium sulfate. The dialysis sample is placed on the affinity column at 4°C in saline solution (concentration not given). The adsorbed agglutinin is not removed with saline or N-acetyl-D-glucosamine. Elution was achieved with 0.1 <u>M</u> acetic acid.

A much simpler affinity column for wheat germ agglutinin was employed by Block and Burger (<u>60</u>). Chitin, a natural occurring polymer of N-acetyl-glucosamine (Eq. 37),

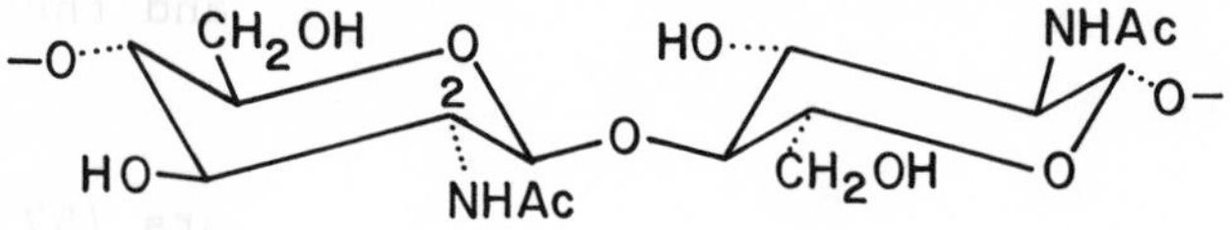

(37)

serves as both support and ligand. Elution of retained wheat germ agglutinin posed some difficulty. The column was developed with 0.01 <u>M</u>-TRIS-NCl, pH 8.5, containing 1 <u>M</u> NaCl, which removes nonspecifically bound protein. The NaCl must then be removed from the column. As with Sepharose, the agglutinin is not eluted with either salt or N-acetyl-D-glucosamine. However, removal of the retained protein can be achieved with 0.05 <u>N</u> HCl or di- and trisaccharide hydrolysates of chitin.

2. Vitamin B_{12}

Binding protein for Vitamin B_{12} is present in extremely small amounts in human plasma. Only about 8 mg of transcobalamin I is present in 100 liters of plasma. Human intrinsic factor, a glycoprotein which also binds vitamin B_{12}, is present in gastric juice in slightly greater amounts. Vitamin B_{12} contains no primary amino groups or other reactive functional groups for conveniently attaching the vitamin to an affinity support. It was, therefore, necessary to modify the ligand to permit attachment. Mild acid hydrolysis of Vitamin B_{12} cleaves one of the three propionamide side chains. The monocarboxylic acid was then coupled to 3,3'-diaminodipropylamine-substituted Sepharose with a water-soluble carbodiimide (61). The affinity column thus prepared was successfully used to isolate transcobalamin I and II, intrinsic factor from human and hog gastric mucosa, and binding protein from human granulocytes. Elution was achieved with 7.5 M guanidine HCl containing 0.1 M potassium phosphate, pH 7.5. The separation achieved is shown in Fig. 7.

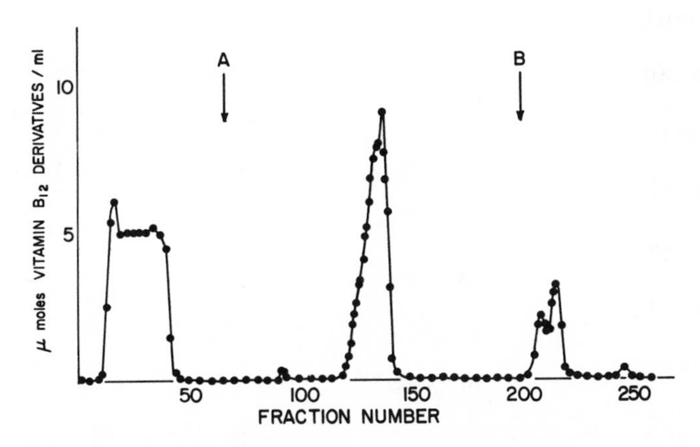

Fig. 7. QAE-Sephadex chromatography of the acid hydrolysis products of vitamin B after elution from AG1-X8. The column was eluted with 0.4 M pyridine. A designates the start of the gradient between 0.4 M pyridine and 0.4 M pyridine containing 0.16 M acetic acid. B designates the start of elution with 0.4 M pyridine containing 0.32 M acetic acid.

3. Hormone Receptors

Affinity chromatography is becoming a valuable technique for isolating cellular fragments that bind selectively to hormones. Insulin-binding proteins have been isolated from liver homogenates (62). Several derivatives gave satisfactory results. However, the preferred derivative involved binding insulin by the N-terminal residue of B chain (B1 phe) to the 3,3'-diaminodipropylamine-succinic acid ligand.

The binding protein resides in the membrane of the liver cells. The surface of adipose fat cells also has a high concentration of insulin-binding protein. An immobilized insulin agarose column was used to retain intact adipocytes and their ghosts (64).

A binding protein for estradiol resides within the cytosol of the uterus. Isolation of the protein has posed major problems because of the minute quantities present in biological tissues. An affinity column for estradiol receptor has been prepared using 17-β-estradiol-17-hemisuccinate (Eq. 38) as the ligand and 3,3'-diaminodipropylamine as the spacer (63). When estradiol was bound through the phenolic hydroxyl of ring A (Eq. 38), no receptor protein retention was observed. Thorough column washing is absolutely essential to remove nonspecifically adsorbed proteins from the column. In order to ensure thorough removal of unreacted estradiol from the column, uterine supernate was passed through a washed column, the eluate was extracted with ether and a competitive binding experiment was conducted with ^{3}H-estradiol of the ether extract and the receptor fraction. Elution of the receptor was accomplished by a combination of (a) estradiol-containing buffer, (b) dilution of the gel, and (c) increasing the temperature to shift the equilibrium of the estradiol-receptor complex.

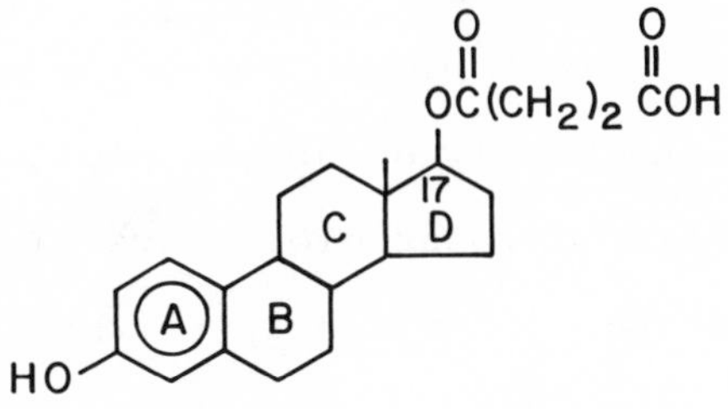

(38)

The gel is removed from the column and suspended in a two-fold bed volume of 10 *mM* TRIS-HCl (pH 7.5), 0.2 *M* KCl, 1 *mM* EDTA and 0.5 mM dithiothreitol. The suspension is incubated with ^{3}H-estradiol at 30°C for 15 minutes. The suspension is then filtered and washed with buffer. The filtrate is cooled to 4°C and incubated for an additional hour, and the free ^{3}H-estradiol is removed by extraction at 4°C for five minutes with dextran-coated charcoal.

The estradiol-receptor complex moves rapidly from the cytoplasm of the target cell into the nucleus where binding sites for the complex are believed to exist. Puca and co-workers (65) bound various nuclear fractions to CNBr-activated Sepharose, and ^{3}H 17-β-estradiol-receptor complex was passed through the column. Specific binding sites for the hormone-receptor complex were found in the low-salt insoluble material extracted from the nuclei with 2 M NaCl.

F. Cofactor Ligands

These ligands are employed for group-specific separations. By using highly selective elution procedures, further purification is possible. Several investigators have coupled NAD (Eq. 39) directly to the reactive spacer

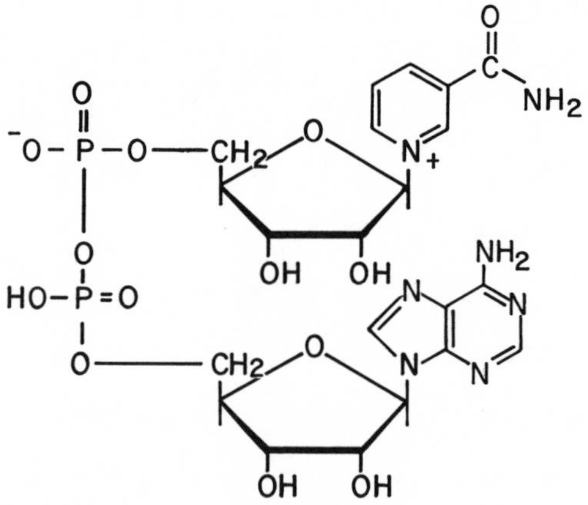

(39)

such as CNBr-activated cellulose (66), which is made possible through the 6-amino position of the purine

(Eq. 40). The diazonium salt of the arylamine derivative of CPG (67) couples to NAD possibly at position 8 of the purine. These methods did not permit attachment of various spacers. A reasonable approach (68) has been to couple ε-amino caproic acid to CNBr-agarose and to couple NAD by the carbodiimide procedure to give agarose N^6- (6-aminohexyl-NAD). The reported procedure (68) employs dicyclohexylcarbodiimide, which yields the water-insoluble dicyclohexylurea. This urea derivative is quite difficult to remove from the reaction mixture. Use of a water-soluble carbodiimide such as 1-ethyl-3(3-dimethylaminopropyl)-carbodiimide hydrochloride is suggested.

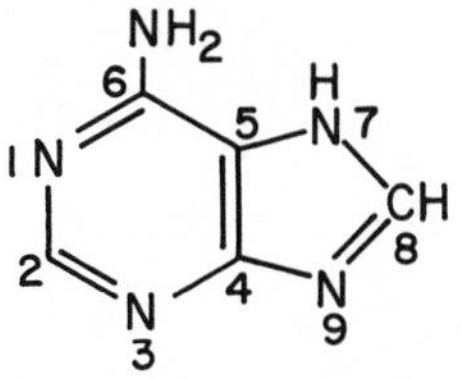

(40)

A second procedure leading to the same derivative is also given. In this method, nicotinamide mononucleotide is coupled to N^6-(6-aminohexyl)-5-AMP (69). The ligand-spacer combination can then be coupled to a matrix.

A third procedure (70) yields 8-(6-aminohexyl)amino-AMP (Eq. 41). This method appears quite satisfactory and procedures for converting the AMP derivative to ADP and ATP are available (70).

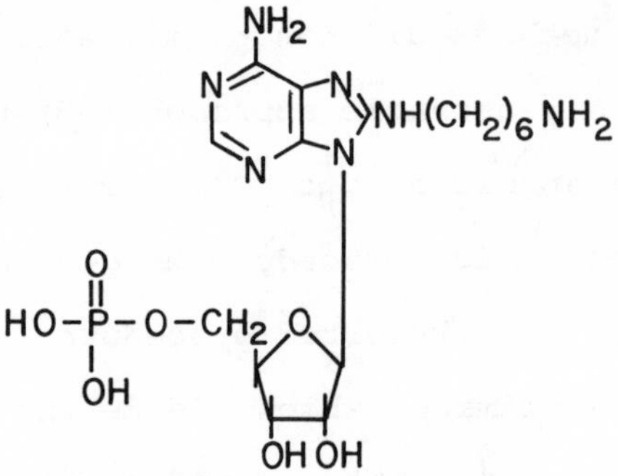

(41)

1. Preparation of Agarose N^6-(6-aminohexyl-NAD^+) (68)

To 200 ml of activated Sepharose 4B, representing 8 g dry weight was added 20 g of ε-aminocaproic acid dissolved in 200 ml of 0.1 M $NaHCO_3$, and the final pH was adjusted to 8.5 using 4 M NaOH. After stirring for 14 hours at room temperature, the gel was extensively washed with 0.1 M $NaHCO_3$, 0.01 M HCl, 0.5 M NaCl, and water. The gel was then filtered off and 80 g of the wet gel -- corresponding to 3.9 g dry weight -- was washed on filter with 80% (v/v) of aqueous pyridine. After filtration, the gel was transferred to a bottle. This was followed by successive addition of 0.8 g of NAD^+ dissolved in 24 ml of water and 40 g of DCC (dicyclohexylcarbodiimide) dissolved in 96 ml of pyridine. The suspension was gently shaken on a rotary shaker for 10 days at room temperature.

To remove formed dicyclohexylurea, the gel was filtered off and washed successively with water, ethanol, n-butanol, ethanol, water, and 80% (v/v) of aqueous pyridine. The gel was then resuspended in the first filtrate and an additional 10 g of DCC were added. After another week, the washing procedure was repeated. In addition, the gel was washed 10 minutes each in 0.1 M $NaHCO_3$ and 0.001 M HCl followed by extensive washings with 0.5 M NaCl and water. All washings were carried out at room temperature. The gel was stored in water at 4°C.

The total amount of gel-bound nucleotide material was determined spectrophotometrically. To this end, 0.25 g of gel was hydrolyzed in 5 ml 0.5 M HCl at 100°C for 15 minutes. After dilution with water to a total volume of 10 ml, the amount of nucleotide material was calculated from the absorption obtained at λ = 260 nm using hydrolysates of gel together with known amounts of NAD^+ as standards.

2. Preparation of N^6(6-aminohexyl)-NAD^+ (69)

A mixture of N^6-(6-aminohexyl)-5-AMP (45 mg), trifluoroaceticanhydride (0.5 ml) and nicotinamide mononucleotide (45 mg) was allowed to react overnight in a sealed tube at room temperature. The anhydride was then removed under reduced pressure and the viscous residue was washed extensively with ether and dried over KOH/P_2O_5

under vacuum. The product was further purified by passage through a Dowex-1-X8 column (eluted with water) followed by paper electrophoresis at pH 9.0. The yield of purified product was 20-25%.

3. Preparation of 8-(6-aminohexyl)amino-AMP (70)

Disodium AMP (0.9 g, 2 mmol) was dissolved in 1 M sodium acetate buffer, pH 4.0, (80 ml) and bromine-water (0.15 ml of bromine, 3 mmol, dissolved in 15 ml of water with vigorous shaking) was added. After thorough mixing, in a stoppered flask, the solution was stored in the dark at room temperature for 18 hours. The pale-yellow color was discharged from the solution by adding sodium metabisulphite (0.1 g). The solvent was removed by rotary evaporation, and the residue was concentrated several times from ethanol to give a white solid. This was dissolved in water (100 ml) and applied to a column (20 cm x 1.5 cm) of Dowex 1 (X8; formate cycle; 200-400 mesh), which was washed with water (300 ml) and then developed with 0.1 M formic acid. Two major u.v.-absorbing peaks were eluted. The first, which contained some 8 to 9% of the applied 260 nm-absorbing material, had a u.v. absorption spectrum identical to that of AMP. The second peak ($\lambda_{max.}^{H^+}$ 262.5 nm) was eluted between 1 and 2 liters and usually contained between 80 and 90% of the material applied (as determined by extinction (260 nm) measurements). After adding activated charcoal (1 g) to the pooled fractions from the second peak and stirring for several minutes, the charcoal was separated by filtration and the filtrate was monitored for u.v.-absorbing material. The process was continued until no appreciable material absorbing at 260 nm was found in the filtrate. Four grams of charcoal was used. The product was eluted from the charcoal by washing several times with aqueous 50% (v/v) ethanol (250 ml) containing 2% (v/v) concentrated aqueous NH_3. The combined washings (1.5 to 2 liters) containing more than 90% of the 260 nm-absorbing material eluted from the column were concentrated to dryness. The product was obtained in about 75% overall yield based on u.v. absorption.

VIII. CONCLUSIONS

This review has attempted to present a critical discussion of the practice of affinity chromatography and the selection of experimental methods and results. Affinity chromatography has progressed rapidly during the past few years and has received wide recognition because of the attractiveness of the bio-specific nature of the separation and the apparent simplicity of the procedure. We have discussed potential complications inherent in the use of affinity chromatography today. It has become apparent though, from the accomplishments in the field to date, that generalizations may not exist, and supports, coupling procedures, and spacer structure must be optimized for each individual separation.

The first stage in affinity chromatography development has ended. Many successful affinity chromatographic separations have been realized without fully understanding the reasons for the success. The field has been dominated by experimentalists. Limited progress has been made in developing a rigorous physical model to describe the molecular events occurring between solute and the derivatized support. The rapid advance of this separation procedure has now reached a stage where more analytical examinations of these molecular processes are being made. With a better understanding of these interactions between the derivatized supports and solute, it is hoped that better support designs will lead to better separations.

Although discussions in this chapter have dealt mostly with purifying soluble materials, affinity chromatography has also been used to purify particulate material. The purification of hormone tissue receptors associated with the plasma membrane using immobilized hormones has been reported (64), and immobilized concanavalin A has been utilized in fractionating vesiculated plasma membranes (71). In addition, affinity supports have been used in cell separations (72-75). If this application of specific affinity separations can be further developed, it could immensely benefit research in cellular chemistry and biology. These examples demonstrate the feasibility of specific solid-solid phase reactions for separations and purifications.

There can be no doubt that there will be continued interest in this technique for isolating and purifying biological materials. However, the techniques of affinity chromatography can most certainly be applied to other kinds of studies, such as defining structural requirements necessary for binding interactions (14,76,77), identification of metabolic enzymes acting on a particular substrate, and blood detoxification. A recent very interesting report dealing with the latter application of affinity chromatography describes the use of immobilized albumin and an extracorporeal shunt for removing bilirubin from the blood (78,79). The principles of affinity chromatography are also applicable in automating analytical procedures based upon competitive binding principles.

Finally, more extensive work is required to demonstrate the value of affinity chromatography for scaled-up isolation and purification of biological materials. A limited number of papers dealing with such scale-up operations have appeared (50), but, certainly, much remains to be done to demonstrate the industrial value of this potentially potent isolation and purification procedure.

REFERENCES

(1). P. Cuatrecasas and C. B. Anfinsen, "Methods in Enzymology", ed. S.P. Colowick and N. O. Kaplan, Academic Press, New York 1971 Vol. 22, p. 345.

(2). P. Cuatrecasas, "Biochemical Aspects of Reactions on Solid Supports", ed. G. R. Stark, Academic Press, New York, 1971, p. 79.

(3). H. Guilford, Chemical Society Review 2 249 (1973).
(4). H. H. Weetall, "Separation and Purification Methods" ed. E. S. Perry, C. J. vanOss and E. Grushka, Marcel Dekker, Inc., New York 1974, p. 199.
(5). P. O'Carra, "Industrial Aspects of Biochemistry", ed. B. Spencer, Federation of European Biochemical Societies, 1974, p. 107.
(6). J. K. Inman and H. N. Dintzis, Biochemistry, 8, 4074 (1969).
(7). P. Cuatrecasas, J. Biol. Chem., 245, 3059 (1970).
(8). G. J. Bartling, H. D. Brown, L. J. Forrester, M. T. Koes, A. N. Mather, and R. O. Stasiw, Biotechnol. Bioeng. 14 1039 (1972).
(9). R. Axen and S. Ernback, Eur. J. Biochem., 18, 351 (1971).
(10). R. Axen and P. Vretblad, Acta Chem. Scand. 25, 2711 (1971).
(11). L. Ahrgren, L. Kagedal, and S. Akerstrom, Acta Chem. Scand. 26, 285 (1972).
(12). R. Axen, Per-Ake Myrin, and J. C. Janson, Biopolymers 9, 401 (1970).
(13). J. Porath, K. Aspberg, H. Drevin, and R. Axen, J. Cromatogr. 86 53 (1973).
(14). S. L. Marcus and E. Balbinder, Anal. Biochem. 48 448 (1972).
(15). M. Einarson, Thesis Uppsala University, 1972.
(16). J. N. Kanfer, G. Petrovich, and R. A. Mumford, Anal. Biochem., 55, 301 (1973).
(17). Personal observations.
(18). G. I. Tesser, H. U. Fisch and R. Schwyzer, FEBS Letters 23, 56 (1972).
(19). M. Wilchek, FEBS Letters, 33, 70 (1973).
(20). P. Cuatrecasas, I. Parikh, and M. D. Hollenberg, Biochemistry 12, 4253 (1973).
(21). V. Sica, E. Nola, I. Parikh, G. A. Puca and P. Cuatrecasas, Nature NB 244, 36 (1973).
(22). P. Brodelius and K. Mosbach, Acta Chem. Scand. 27, 2634 (1973).
(23). C. R. Lowe and P. D. G. Dean, FEBS Letters, 14, 313 (1971).
(24). K. Mosbach, H. Guilford, R. Ohlsson and M. Scott, Biochem. J. 127, 625 (1972).
(25). Z. Er-el, Y. Zaidenzaig and S. Shaltiel, Biochem. Biophys. Res. Commun. 49, 383 (1972).
(26). S. Shaltiel and Z. Er-el, Proc. Nat. Acad. Sci. U.S. 70, 778 (1973).
(27). B. H. J. Hofstee, Anal. Biochem. 52, 430 (1973).
(28). R. J. Yon, Biochem. J. 126, 765 (1972).
(29). S. Hjerten, J. Chromatogr. 87, 325 (1973).
(30). J. H. Ludens, J. R. DeVries and D. D. Fanestil, J. Biol. Chem., 247, 7533 (1972).

(31). P. Brodelius and K. Mosbach, FEBS Letters, 35, 223 (1973).
(32). D. Allan, J. Auger, and M. J. Crumpton, Nature New Biology 236, 23 (1972).
(33). M. L. Dufau, T. Tsuruhara, and K. J. Catt, Biochim. Biophys. Acta, 278, 281 (1972).
(34). J. B. Standburg, J. B. Wyngaarden and D. S. Fredrickson, "The Metabolic Basis of Inherited Disease", McGraw-Hill, New York, 1972.
(35). C. A. Mapes and C. C. Sweeley, J. Biol. Chem., 248 2461 (1973).
(36). S. Okada and J. S. O'Brien, Science 165, 698 (1969).
(37). G. Dawson, R. L. Propper, and A. Dorfman, Biochem. Biophys. Res. Comm. 54, 1102 (1973).
(38). E. Junowicz and J. E. Paris, Biochim. Biophys. Acta 321, 234 (1973).
(39). J. E. Hodge and B. T. Hofreiter, "Methods in Carbohydrate Chemistry", eds. R. L. Whistler and M. L. Wolfram, Academic Press, New York, 1963, Vol. 1, 380.
(40). F. LeGoffic, R. Labia, and J. Andrillon, Biochim. Biophys. Acta, 315, 439 (1973).
(41). T. Tosa, T. Sato, R. Sano, K. Yamamoto, Y. Matuo, and I. Chibata, Biochim. Biophys. Acta, 334, 1 (1974).
(42). H. Hayashi, J. Biochem. 74, 203 (1973).
(43). I. B. Wilson and M. V. Wondolowski, Biochim. Biophys. Acta 289, 347 (1972).
(44). J. D. Berman and M. Young, Proc. Nat. Acad. Sci. U. S., 68, 395 (1971).
(45). N. Kalderon, I. Silman, S. Blumberg, and Y. Dudai, Biochim. Biophys. Acta 207, 560 (1970).
(46). T. L. Rosenberry, H. W. Chang, and Y. T. Chen, J. Biol. Chem. 247, 1555 (1972).
(47). Y. Dudai, I. Silman, M. Shinitzky and S. Blumberg, Proc. Nat. Acad. Sci. U. S. 69, 2400 (1972).
(48). E. J. W. Steers, P. Cuatrecasas and H. B. Pollard, J. Biol. Chem., 246, 196 (1971).
(49). Unpublished results.
(50). P. J. Robinson, P. Dunnill and M. D. Lilly, Biochim. Biophys. Acta 285, 28 (1972).
(51). J. N. Kanfer, G. Petrovich and R. A. Mumford, Anal. Biochem. 55, 301 (1973).
(52). J. H. Woychik and M. V. Wondolowski, Biochim. Biophys. Acta 289, 347 (1972).
(53). Y. Ashani and I. B. Wilson, Biochim. Biophys. Acta 276, 317 (1972).
(54). K. Brocklehurst, J. Carlson, M. P. J. Kierstan and E. M. Crook, Biochem. J. 33, 573 (1973).
(55). L. A. E. Sluyterman and J. Wijdenes, Biochim. Biophys. Acta, 200, 593 (1970).
(56). P. M. Blumberg and Strominger, Proc. Nat. Acad. Sci. U. S., 69, 3751 (1972).

(57). C. E. Hayes and I. J. Goldstein, J. Biol. Chem., 249, 1904 (1974).
(58). J. I. Rood and R. G. Wilkinson, Biochim. Biophys. Acta 334, 168 (1974).
(59). R. Lotan, A. E. S. Gussin, H. Lis and N. Sharon, Biochim. Biophys. Res. Comm. 52, 656 (1973).
(60). R. Bloch and M. M. Burger,Biochem. Biophys. Res. Comm. 58, 13 (1974).
(61). R. H. Allen and Majerus, J. Biol. Chem. 247, 7695 (1972).
(62). P. Cuatrecasas, Proc. Nat. Acad. Sci. U. S., 69, 1277 (1972).
(63). V. Sica, E. Nola, I. Parikh, G. A. Puca, and P. Cuatrecasas, Nature New Biol., 244, 36 (1973).
(64). D. D. Soderman, J. Germershausen, H. M. Katzen, Proc. Nat. Acad. Sci. U. S. 70, 792 (1973).
(65). G. A. Puca, V. Sica and E. Nola, Proc. Nat. Acad. Sci. U. S., 71, 979 (1974).
(66). C. R. Lowe and P. D. G. Dean, FEBS Letters 14, 313 (1971).
(67). M. K. Weibel, H. H. Weetall and H. J. Bright, Biochem. Biophys. Res. Comm. 44, 347 (1971).
(68). P. O. Larsson and K. Mosbach, Biotechnol. Bioeng. 13, 393 (1971).
(69). K. Mosbach, H. Guilford, R. Oklsson and M. Scott, Biochem. J., 127, 625 (1972).
(70). I. P. Trayer, H. R. Trayer, D. A. P. Small and R. C. Bottomley, Biochem. J. 139, 609 (1974).
(71). A. Zachowski and A. Paraf, Biochem. Biophys. Res. Comm. 57, 787 (1974).
(72). P. Truffa-Bachi and L. Wofsy, Proc. Nat. Acad. Sci. U. S. 66, 685 (1970).
(73). G. M. Edelman, U. Rutishauser and C. F. Millette, Proc. Nat. Acad. Sci. U. S. 68, 2153 (1971).
(74). C. Henry, J. Kimura, and L. Wofsy, Proc. Nat. Acad. Sci. U. S. 69, 34 (1972).
(75). U. Rutishauser, C. F. Millette, and G. M. Edelman, Proc. Nat. Acad. Sci. U. S., 69, 1596 (1972).
(76). I. Sjoholm and I. Ljungstedt, J. Biol. Chem. 248, 8434 (1973).
(77). K. Kasai and S. Ishii, J. Biochem. 74, 631 (1973).
(78). P. H. Plotz, P. D. Berk, B. F. Scharschmidt, J. K. Gordon, and J. Vergalla, J. Clin. Invest. 53, 778 (1974)
(79). B. F. Scharschmidt, P. H. Plotz, P. D. Berk, J. G. Waggoner and J. Vergalla, J. Clin. Invest. 53, 786 (1974).

Chapter 9

SOLID PHASE IMMUNOASSAY

William F. Line and Milton J. Becker

Metrix Division
Armour Pharmaceutical
Chicago, Illinois 60616

I. INTRODUCTION

One practical application of immobilized biochemicals technology is the use of immunoadsorbents in highly sensitive immunoassays which employ labelled antigens or antibodies for detection. Such tests are called labelled immunoassays (LIA). Currently, the most popular LIA procedures are radioimmunoassays (RIA), which employ isotopically labelled immunogens. Other labelling methods have also been described. Most LIA procedures were designed for use in clinical laboratories to detect low circulating levels (i.e. pg-ng range) of such diverse materials as hormones, drugs, vitamins and infectious agents. This chapter deals with the application of immunoadsorbents to this area.

The term immunoadsorbent, in the context of this chapter, implies an antigen or antibody which has been rendered water-insoluble. As with enzymes, immobilization may be achieved by any of four methods: adsorption; entrapment; protein crosslinking; or covalent attachment.

In adsorption to a surface, e.g. plastic tubes, ion exchange resins, etc. (1-5), the immobilized immunochemical is retained on the surface by electrostatic forces in the case of ion exchangers, or by a combination of Van der

Waals forces and hydrophobic bonds in the case of non-polar surfaces.

The entrappment technique (6) consists of forming a polymer from a solution containing the material to be insolubilized. The immunochemical then becomes entrapped in the intersices of the polymer as it is formed. Polyacrylamide gels are the most common entrapping medium.

Proteins are crosslinked with bifunctional reagents such as glutaraldehyde, ethyl chloroformate, or toluene 2, 4 diisocyanate (7,8).

Covalent attachment is done with synthetic or natural polymers (9-29).

Some of the earliest work in immobilized biochemicals technology concerned the preparation of immunoadsorbents. In 1951, Campbell et al (9) described the covalent attachment of ovalbumin to the diazotized p-aminobenzyl ether of cellulose. Isliker (10) coupled antigens such as human serum albumin and influenza virus to tanned red cell ghosts and to the acid chloride of a divinyl benzene cross-linked methacrylate (IRC-50). Subsequently, immunoadsorbents have been synthesized by covalent attachment to such carriers as: diazotized polyaminostyrene and polystyryl mercuric acetate (11); the diazotized benzidene amide of carboxymethyl cellulose (13,14); the bromoacetyl ester of cellulose (12); various derivatives of cross-linked dextrans (Sephadex) (15,16,17,19) gels; and the imidocarbonates of agarose gels (19, 21-26), and organofunctional controlled pore glass beads (26-28).

Immunoadsorbents have been formed by adsorption to plastic tubes (1-4) and latex particles (5) as well as by entrappment in polyacrylamide gels (6). Most of these earlier immunoadsorbents were used in isolating the corresponding immunogen by affinity chromatography. This material has been thoroughly reviewed elsewhere (30,31).

More recently, immunoadsorbents have been applied in LIA procedures. The majority of these solid-phase immunoassays employ insoluble antibodies to detecting antigen in competitive binding RIA (1,2,8,15,16,20,21, 25,26). Other solid-phase immunoassays have been described which do not involve radiation (4,7,29), while still others do not rely on competitive binding (3,4,21, 32). Table 1 lists some of the solid phase LIA protocols which have been described.

Table 1

Solid-Phase Immunoassays

Attached Immunogen	Carrier	Attachment	Carrier Functional Group	Assay For	Detection Method	Method of Assay	Ref.
Human -globulin (HGG) Human Albumin (HSA)	cellulose	covalent	diazotized aryl amine	anti-HGG anti-HSA	RIA	sandwich	32
Antibodies to 1. chorionic gonadotropin (HCG) 2. lutenizing hormone (LH)	dextran-Sephadex	covalent	isothio-cyanate	HCG LH	RIA	competitive binding	15
anti-HCG	dextran	covalent	isothio-cyanate imidocar-bonate	HCG	RIA	competitive binding	16
antibodies to 1. growth hormone (HGH) 2. insulin 3. HCG	dextran	covalent	isothio-cyanate	HGH insulin HCG	RIA	competitive binding	20

Table 1 (continued)

Attached Immunogen	Carrier	Attachment	Carrier Functional Group	Assay For	Detection Method	Method of Assay	Ref.
insulin				insulin antibody		competitive binding	21
anti-HCG	dextran	covalent	imidocarbonate	HCG	RIA	sandwich	
antiserum to 1. HSA	agarose	covalent	imidocarbonate	HSA	enzyme linked	competitive binding	7
2. HCG	protein		cross-linked	HCG			
HCG	protein	cross-linked	-----	anti-HCG	enzyme linked	competitive binding	7
antibody to streptolysin				streptolysin			
antiserum to insulin	dextran	covalent	isothiocyanate	insulin	enzyme linked	competitive binding	29
HSA dinitophenol	plastic tube	adsorbed	-----	anti-HSA anti-DNP	enzyme linked	sandwich	4
antisera to:							
1. placental lactogen (HPL)	plastic	adsorbed	-----	HPL			

Table 1 (continued)

Attached Immunogen	Carrier	Attachment	Carrier Functional Group	Assay For	Detection Method	Method of Assay	Ref.
2. HGH	tube			HGH	RIA	competitive binding	1,2
3. Insulin				insulin			
antiserum to HBAg-hepatitis associated antigen	plastic tube	adsorbed	-----	HBAg	RIA	sandwich	3
antibodies to thyroglobulin	protein agarose	covalent	cross-linked imidocarbonate	thyro globulin	RIA	competitive binding	8
antiserum to triiodothyronine (T_3)	agarose	covalent	imidocarbonate	T_3	RIA	competitive binding	25
Antiserum to digoxin	agarose glass	covalent	imidocarbonate isothiocyanate	digoxin	RIA	competitive binding	26

II. LIA TECHNIQUES

A. Competitive Binding Analysis

The principles of competitive binding analysis, are outlined below.

$$Ab + Ag^* \longrightarrow Ab—Ag^*,\ B_o/T$$

$$Ag + Ag^* + Ab \longrightarrow Ab—Ag—Ag^* + Ag^*,\ B/T$$

Reaction 1 & 2, Competitive Binding Analysis. Ag* = labeled antigen; Ag = antigen; Ab = antibody; B_o/T = fraction available label bound in the absence of Ag; and B/T = fraction label bound in the presence of Ag. $B/T < B_o/T$.

When labeled antigen (Ag*) is incubated with a limiting quantity of antibody (Ab), a labeled antigen-antibody (Ag*Ab) complex is formed. When unlabeled antigen is introduced (e.g. as an unknown in a patient's body fluids), it competes with Ag* for the limited number of Ab binding sites. Thus, the quantity of Ag* that is displaced is proportional to the quantity of Ag present in the unknown. To assure that Ab is limiting, the antiserum concentration is adjusted so that 40 to 60% of the label is bound in the absence of unlabeled antigen.

Competitive binding analysis was first applied to immunoassays by Yalow and Berson (33) in 1960, who described an insulin RIA employing ^{125}I-labeled insulin as the marker antigen. Since then, many competitive binding RIA procedures have been described for circulating hormone levels (1,2,20,25,34-36), drugs (26,37-40), and such infectious agents as hepatitis-associated antigen HBAg (3,41-43).

Other methods of antigen (or antibody) labeling may be substituted for radiolabeling according to the sensitivity demanded by the system. One technique that reportedly approaches the sensitivity of RIA is the enzyme-linked immunoassay (ELISA) (4,7,29). Here, the antigen or antibody to be determined is conjugated to an enzyme which has a high turnover and for which a simple determination is available. The quantity of antibody-bound antigen labeled in this manner may then be determined by assaying either Ag-Ab complex or the unbound fraction for conjugated enzyme activity. Alkaline phosphatase and horseradish peroxidase are among the enzymes which have been employed for labeling (4,44).

It is obvious that the process can be reversed so that labeled antibody may be used to determine circulating antibody levels in either LIA system.

From the foregoing, it is apparent that the mode of detection in competitive binding immunoassays is the release of labeled antigen from the complex with increasing concentration of unlabeled antigen. That is, the amount of Ag* complexed decreases with increasing Ag. In RIA, three conventions are commonly used to express the quantity of label that is antibody bound:

1. % B/F is the fraction of antibody-bound label (B) over that which remains free x 100. Thus, if half the isotope is bound, % B/F = 100; if one quarter is bound, % B/F = 33.3; and if 3/4 is bound, % B/F = 300.

2. % B/T is simply the fraction of the total isotope complexed x 100. Thus, if half the label is bound, % B/T = 50; if a quarter is bound % B/T = 25; and if 3/4 is bound, % B/T = 75.

3. % B/B_o is the ratio of the label that is antibody-bound at a given Ag concentration to that bound when the antibody is incubated with Ag* alone x 100. That is, % $B/B_o = \frac{B/T}{B_o/T} \times 100$.

In the author's opinion, the second convention is preferred, because it directly defines the operational variable (percent of total added isotope that is antibody-bound).

In enzyme-linked systems, the quantity of antibody-bound label may be expressed as the fraction or percent of conjugated enzymatic activity. More often, it is expressed in terms of the experimental variable used to measure the conjugated enzymatic activity (e.g., $E_{1\ cm}^{410\ nm}$

to hydrolyze p-nitrophenyl phosphate when the conjugated enzyme is phosphatase).

Displacement curves in competitive binding immunoassays generally yield linear functions when plotted on a semilogarithmic scale (e.g., % B/T vs log Ag) or by plotting the percent antibody-bound label vs 1/(Ag) (Sipp's plot) (37).

B. Sandwich Technique

The sandwich or direct technique is the LIA analog of the double antibody method in which the second antibody is labeled. It may be employed in systems of polyvalent antigens (i.e., antigens capable of binding antibody at greater than a 1:1 ratio). A bivalent antigen could bind two molecules of antibody/molecule antigen ($AgAb_2$), trivalent $AgAb_3$, etc. Generally, these antigens are macromolecules (e.g. proteins, polynucleotides, etc.). The sandiwch methods described have employed either immobilized antigen or antibody. The method is illustrated in Reaction 3:

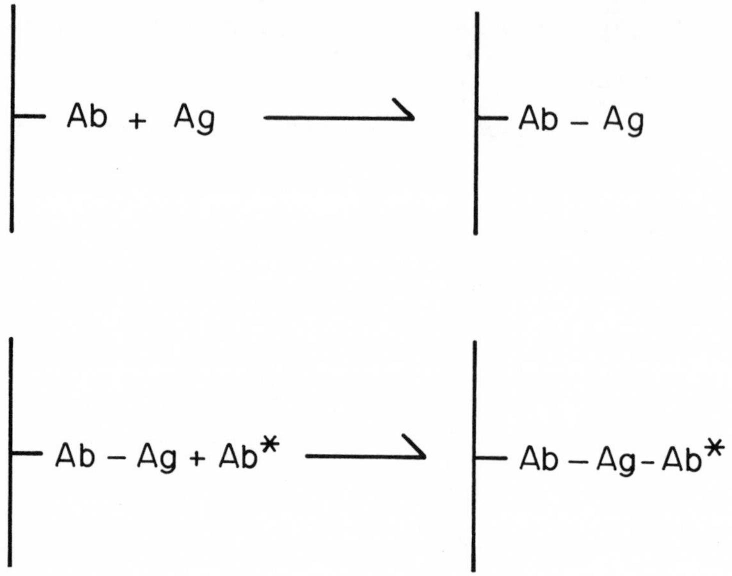

$$\vdash Ab + Ag \longrightarrow \vdash Ab - Ag$$

$$\vdash Ab - Ag + Ab^{*} \longrightarrow \vdash Ab - Ag - Ab^{*}$$

Reaction 3. Direct or Sandwich Method. Ab = antibody attached to an insoluble carrier; Ag = the antigen to be determined; and Ab* = labeled antibody.

As shown here, a solid-phase antibody is incubated with the antigen (generally introduced as an unknown quantity in the patient's fluids) to form a solid-phase complex. Following removal of the incubation medium, the complex is incubated with labeled antibody (Ab*). The quantity of Ab* bound is proportional to the quantity of Ag present in the sample. Thus, the polyvalent antigen is sandwiched between the immobilized antibody and the labeled antibody. Using immobilized and labeled antigen, the same procedure may be employed to determine antibody.

The sandwich method was first applied to LIA by Gurvich (32), employing antigens covalently attached to cellulose ethers. Later, it was used to detect hepatitis-associated antigen (Australian antigen) (HVAg) (3) using HBAg antiserum adsorbed to plastic blood collection tubes.

In the authors' opinion, the sandwich method offers two distinct advantages over competitive binding analysis. First, the quantity of antibody-bound label increases with increasing Ag concentration which is in contrast to competitive binding techniques where increasing Ag results in an increasing displacement of label. Second, when employed for antigen detection, labeling in confined to the IgG subfraction of the antiserum γ-globulin. This is in sharp contrast to competitive binding methods in which it is necessary to device methods for labeling many diverse antigen molecules.

III. CARRIERS

A. Advantages of Solid-Phase Immunoassay

A necessary step in most competitive binding immunoassays is the separation of antibody bound from excess free antigen. Many of the assays described have been for determining low-molecular-weight antigens. In these systems, separation is generally affected by treating the incubation mixture with adsorbents (e.g. charcoal, dextran-coated charcoal, ion exchange resins, etc.) capable of binding free antigen but not antigen that is complexed with the macromolecular antiserum γ-globulin (33,37-39).

When the antigen is macromolecular, other separation methods are employed. These include second antibody techniques (34), salting out the γ-globulin fraction of the antiserum along with bound antigen with $(NH_4)_2SO_4$ or Na_2SO_4 (45), and gel filtration of the incubate (46). Using immobilized antisera makes secondary separations unnecessary, since antigen binding and separation of bound from free antigen occur simultaneously. As a corollary, careful carrier selection provides a common mode of performing LIA procedures. That is, specific antisera attached to the same carrier could be used to detect small antigens (e.g. steroids, small peptides, etc.) and macromolecular antigens (e.g. proteins, nucleic acids and whole microorganisms).

Finally, LIA procedures are capable of measuring the primary binding of antigen and antibody in contrast to quantitative visual tests (e.g., precipitin, agar-gel diffusion, etc.) which require higher orders of binding to form visible precipitates. Therefore, the former should permit more rapid determination. On the contrary, many LIA procedures, either homogenious or solid-phase, that are reported in the literature require from 16 to 18 hours (3,15,16,20) to as long as several days (34). Such extended performance periods have hindered the acceptance of these tests as routine diagnostic procedures in the clinical laboratory. Some of the more recent homogeneous (36-38) and solid-phase LIA (7,25,26) protocols may be performed in more reasonable periods of time.

One of the reasons for using such prolonged incubation periods may be to conserve difficult-to-raise antisera. Immunoadsorbents formed by covalent methods may be regenerated by brief treatment at pH 2 or 3 as outlined in affinity chromatography publications (22-24). This permits the investigator to recover immunoadsorbents formed from antisera that are expensive, hard to obtain, or both, and thus enables the use of antibody excess to perform determinations within reasonable periods of time.

It is not surprising that considerable industrial research performed in this area has resulted in the publication of several patents (7,20,21,27,28) and the appearance of several solid-phase RIA kits, a few of which are listed in Table 4.

B. Carrier Requirements

Labeled immunoassays have been developed for a wide range of antigens (drugs, peptide and steroid hormones, bacterial and viral particles, etc.). To insure the widest applicability, chosen support systems should be equally applicable to all antigens. In addition, the support system should be able to be applied in chromatographic purifications based on biological affinity (i.e., affinity chromatography). Carriers for the preparation of immunoadsorbents should have several properties.

Large internal pore diameter, i.e., support material that is macroporous or macroreticular, is important. A large internal pore diameter is necessary to permit the immobilization of relatively large quantities of macromolecular antibody molecules (M.W. about 150,000), and to permit a reasonable diffusion of macromolecular antigens throughout the carrier pores resulting in maximum antigen immobilized antibody contact.

The support material should have low background adsorption. The carrier itself should not nonspecifically bind significant quantities of the antigen to be determined. This requirement eliminates highly ionic supports such as various maleic anhydride copolymers that have been described for enzyme attachment (47,48), as well as high or moderate-capacity ion exchangers. It would also eliminate highly aromatic polymers which tend to form hydrophobic bonds (particularly with aromatic antigens like thyroxin and triiodothyronine).

Since water is the universal solvent for most biological systems, an obvious carrier requirement is that it be hydrophillic. Good storage and operational stability are also required. Stability to regeneration and reuse is particularly desirable for LIA immunoadsorbents synthesized from expensive or difficult to raise antisera. Stability is also a requirement for affinity supports and is why covalent attachment has been employed for this purpose (22-24).

And finally, the support must have good mechanical properties; that is, the support should have acceptable filtration and flow properties. It should also prove resistant to moderate pressure gradients. This requirement must be met for a support to be employed in affinity chromatographic purifications employing columns.

Currently, three support systems appear to fulfill most of the requirements above: Covalent attachment to the imidocarbonates of cross-linked dextran (Sephadex) or agarose gels (15,16,20,21,25,26); covalent attachment to organofunctional controlled-pore glass (26,28); covalent attachment to organofunctional polyacrylamide gels (49).

Recently, macroporous polymers of hydroxyalkyl methacrylates, which would appear to fulfill all the requirements listed for the preparation of immunoadsorbents, have been described for the covalent coupling of enzymes (50).

In addition to covalent methods, many LIA systems employ antisera adsorbed to plastic tubes as the immobilized phase. Representative methods for immobilizing antisera on three carriers -- agarose gels, controlled-pore glass, and plastic tubes -- will be presented later.

C. Methods of Coupling

1. Cyanogen Bromide Activated Agarose Gels

Polymeric alcohols and primary amines in general and

cross linked dextran (Sephadex) and agarose gels in particular yield highly reactive polymers with cyanogen halides, which are capable of attaching to proteins primarily through their pendant primary amino groups (provided for the most part by the protein's side chain lysyl residues) (22-24, 51-53). CNBr is the cyanogen halide most commonly employed for polymer activation.

Although the activation and coupling mechanisms have not been firmly established, the most recent evidence (51,52) indicates that CNBr adds across vicinal OH groups (vicinal in three-dimensional space; i.e., they may be on the same or separate polymer chains) to form both the unreactive amide (Structure 3) and the reactive imidocarbonate (Structure 2), as shown in Reaction 4. The activated gel then reacts with proteins (through their pendant primary amino groups as noted previously) to form either an N-substituted imidocarbonate (Structure 4), amid (Structure 5), or most likely the isourea (Structure 6) (52). Activation of agarose gels is generally carried out in aqueous media at pH 11 (18, 22-24), while, with cross linked dextrans, a lower pH of 9.5 to 10.0 is employed due to massive gel crosslinking at higher pH (20).

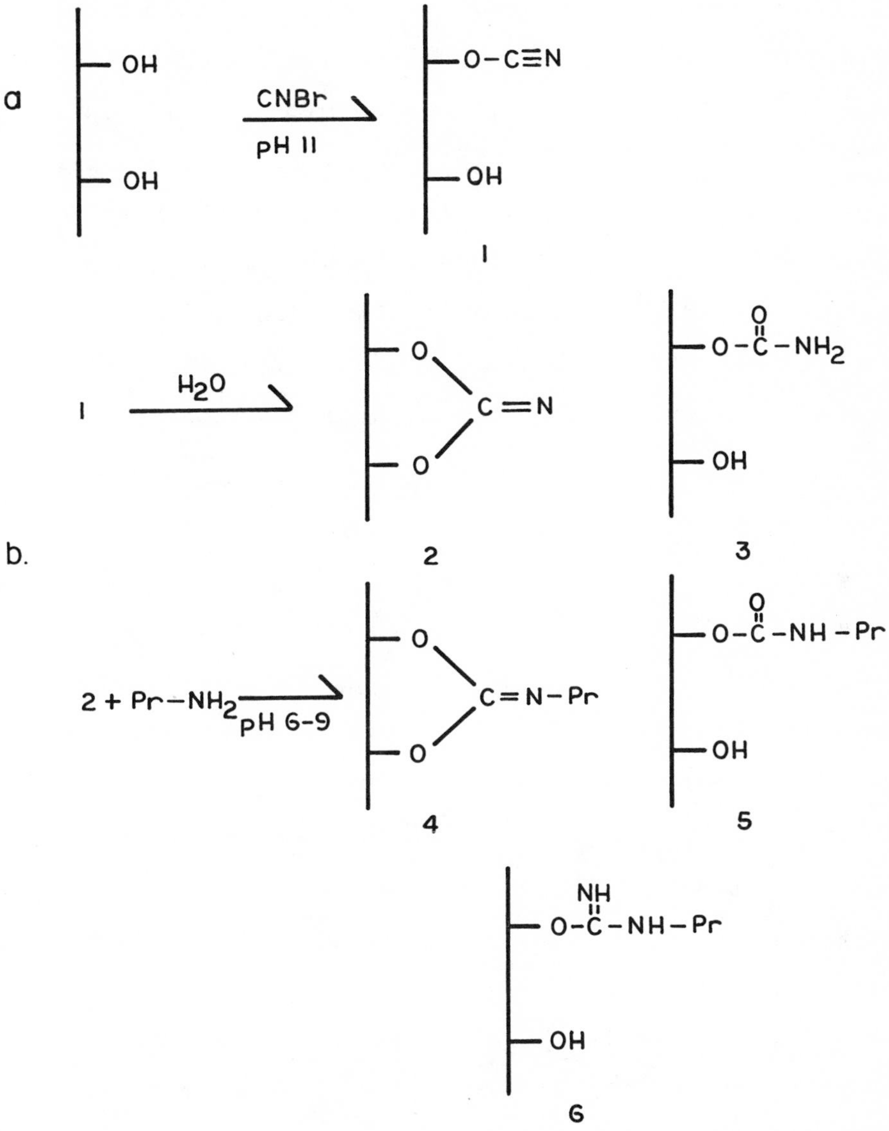

a

OH, OH $\xrightarrow[\text{pH 11}]{\text{CNBr}}$ O–C≡N, OH

1

b.

1 $\xrightarrow{H_2O}$ O, O >C=N (2); O–C(=O)–NH_2, OH (3)

2 + Pr–NH_2 $\xrightarrow[\text{pH 6-9}]{}$ O, O >C=N–Pr (4); O–C(=O)–NH–Pr, OH (5)

O–C(=NH)–NH–Pr, OH (6)

Reaction 4: Cyanogen Halide Activation of Polysaccharides. a. Activation. CNBr adds across vicinal -OH groups to produce both the reactive imidocarbonate (2), and the unreactive carbamate (3). b. Coupling with Proteins. The imidocarbonate (2) reacts with proteins through their pendant primary amines (P-NH_2) to form (4) the N-substituted imidocarbonate, (5) the N-substituted car bamate, or (6) the isourea.

In some cases, it has been proven desirable to place a spacer group or hydrocarbon leash between the matrix and the protein or ligand to be coupled (22-24). This is generally accomplished by adding a bifunctional primary amine of varying chain length to the activated gel. This also serves to introduce other functional groups into the gel which are capable of coupling protein pendant groups other than their primary amino groups. A few of these derivatives are shown in Reaction 5:

a. $2 + H_2N-(CH_2)_n-NH_2 \longrightarrow$ $-NH-(CH_2)_n-NH_2$, $-OH$

$A-NH_2$

b. $A-NH_2 +$ (succinic anhydride) $\longrightarrow A-NH-\overset{O}{\overset{\|}{C}}-(CH_2)_2-COOH$

c. $A-NH_2 + O_2N-C_6H_4-\overset{O}{\overset{\|}{C}}-N_3 \xrightarrow[2)\ Na_2S_2O_7]{} A-NH-\overset{O}{\overset{\|}{C}}-C_6H_4-NH_2$

Reaction 5 - Functional Group Derivatives of CNBr activated agarose. a) Synthesis of alkyl amino functional agarose (A-NH). Aliphatic diamines react with activated agarose under the same conditions as proteins. This reaction is employed to introduce spacer groups between the matrix and the adduct to be immobilized.

b) Synthesis of carboxyl agarose from the alkylamine by reaction with succinic anhydride.

c) Synthesis of arylamino agarose with p-nitrobenzoyl azide. The p-nitogenzoyl amide formed is used to generate the p-aminobenzoyl amide by reducing the nitro group with $Na_2S_2O_7$.

Thus, proteins may be coupled to either alkyl or aryl amino gels through their side chain carboxyls by activating the carboxyls with a water-soluble carbodiimide. Treatment of the arylamino gel with HNO_2 produces a diazonium-containing polymer which can form an azo compound with the pendant phenolic (tyr.) or imidazole (his.) protein side chains. The preparation of various organofunctional polysaccharide derivatives of imidocarbonates has been reviewed elsewhere (22-24).

Agarose gels of varying molecular weight exclusion limits may be obtained from either Pharmacia or Bio-Rad Laboratories. Table 2 lists several gels and their molecular weight operating ranges (when used in gel permeation chromatography). In addition, several reactive agarose derivatives are commercially available. A few of these are listed in Table 3.

Table 2

Commercial Macroporous Supports

Support	Vendor	Trade Name	Molecular Weight[a] Exclusion Limit	Molecular Weight[a] Operating Range
Agarose	Bio-Rad Labs. Richmond, CA.	Bio-Gel A, 0.5 m	500,000	10,000 - 500,000
"	"	Bio-Gel A, 1.5 m	1,500,000	10,000 - 1,500,000
"	"	Bio-Gel A, 5 m	5,000,000	10,000 - 5,000,000
"	"	Bio-Gel A, 15 m	15,000,000	40,000 -15,000,000
"	"	Bio-Gel A, 50 m	50,000,000	100,000 -50,000,000
"	"	Bio-Gel A, 150 m	150,000,000	1,000,000-150,000,000
"	Pharmacia Piscataway, N.J.	Sepharose 2 B	40,000,000	100,000 -40,000,000
"	"	Sepharose 4 B	20,000,000	100,000 -20,000,000
"	"	Sepharose 6 B	4,000,000	100,000 - 4,000,000

Table 2 (continued)

Support	Vendor	Trade Name	Molecular Weight[a] Exclusion Limit	Molecular Weight[a] Operating Range
Porous Glass	Bio-Rad Labs. Richmond, CA.	Bio-glas 200	30,000	3,000 - 30,000
"	"	Bio-glas 500	100,000	10,000 - 100,000
"	"	Bio-glas 1000	500,000	50,000 - 500,000
"	"	Bio-glas 1500	2,000,000	400,000 -2,000,000
"	"	Bio-glas 2500	9,000,000	800,000 -9,000,000
Porous Glass	Pierce Chemical[b] Rockford, ILL.	CPG-550 Å	350,000	
"	"	CPG-1350 Å		

a. As determined with proteins when employed in molecular permeation chromatography.

b. Distributor-produced by Corning Glass Works, Corning, New York.

Table 3

Commercial Functional Group Carriers for Covalent Protein Attachment

Support	Vendor	Trade Name	Support Functional Group
Agarose	Pharmacia Piscataway, N.J.	CNBr activated Sepharose	imidocarbonate
Agarose	Bio-Rad Labs. Richmond, CA.	Affi-Gel 10	N-hydroxysuccinimide ester
Controlled-Pore Glass	Pierce Chemical[c] Rockford, ILL.	Aminopropyl CPG	primary alkyl amine
"	"	Aminoaryl CPG	primary aryl amine
"	"	p-Nitroaryl CPG	$-NO_2$-p-nitro benzoyl amide of aminopropyl CPG
"	"	active ester CPG	N-hydroxysuccinimide ester
"	"	thiol CPG	thiol-cysteinyl amide of aminopropyl CPG
"	"	carboxyl CPG	COOH-succinyl amide of aminopropyl CPG

Table 3 (continued)

Support	Vendor	Trade Name	Support Functional Group
polyacrylamide gels	Bio-Rad Labs. Richmond, CA.	Aminoethyl Bio-Gel P	primary alkyl amine
"	"	Hydrazine Bio-Gel P	hydrazide

c. Distributor-manufactured by Corning Glass Works, Corning, New York.

In coupling antisera to agarose (25,26), materials needed include Bio-Gel A, 0.5 m, 50 to 100 mesh particle size (Bio-Rad Laboratories); and technical grade cyanogen bromide (Eastman Kodak).

Relatively small-scale attachment of crude antisera (up to one liter settled gel bed volume) to the gels may be conveniently carried out using a coarse sintered glass funnel as the reaction vessel (Fig. 1). This allows activation, washing, and coupling reactions to be performed without having to transfer the gel. The stem of the funnel is equipped with a rubber tube leading to an aspirator, and the tube is equipped with a Day pinch-clamp allowing for quick aspiration and resealing of the system. A mechanical stirrer and pH electrode (a combination electrode is particularly useful here because of the small space requirement) are mounted above the funnel. Gel activation should be carried out within a well-ventilated hood, because CNBr is toxic and quite acrid. The procedure described below has been used to prepare immunoadsorbents of triiodothyronine (21) and digoxin (22) antisera, and is of general utility.

Twenty ml of agarose (two grams dry weight by manufacturer's specification) are transferred to a 60-ml sintered glass funnel and washed by stirring several times in water to remove the buffer salts and preservative (NaN_3) in which the gel is shipped. After the final water wash (generally about 10 volumes should suffice), the gel is aspirated to a moist filter cake and overlayed with 20 ml 0.1 M Na_2CO_3, pH 11.0, and some finely crushed ice so the outside of the funnel is cool to the touch. With rapid stirring, six grams of solid CNBr are added. The pH is maintained at 11.0 ($\pm$ 0.5) by the dropwise addition of 6 N NaOH for 45 minutes during which all the CNBr dissolves and no further pH decline is observed. Crushed ice is added periodically to keep the reaction mixture cool.

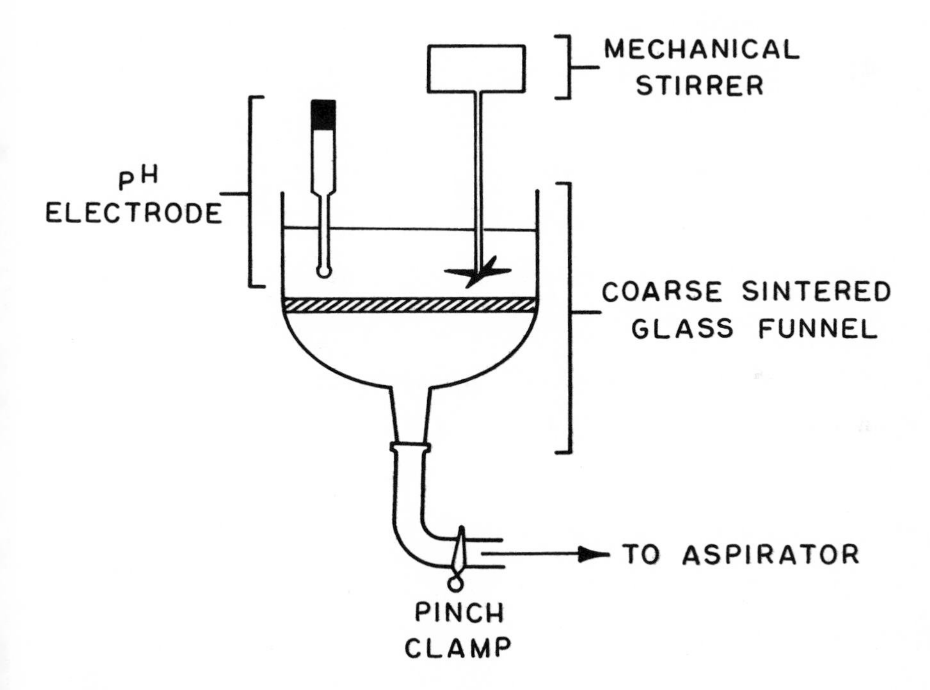

Fig. 1: Experimental design for small-scale coupling to CNBr-activated agarose gels.

The activated gel is then harvested by filtration and washed several times by stirring in cold 0.5 M NaHCO , pH 9.0. After the final buffer wash (again about 5 to 10 volumes), the gel is aspirated to a moist filter cake and overlayed with 20 ml 0.5 M $NaHCO_3$, pH 9.0, containing 0.6-3.0 ml antiserum. The protein concentration of rabbit serum is 60 to 70 mg/ml. The suspension is allowed to stir gently at room temperature overnight.

After reaction, the conjugate is harvested by aspirating to a moist filter cake. It is then washed by stirring several times in each of the following solutions to remove protein that is not covalently bound: 0.1 M phosphate buffer - 0.3 M NaCl, pH 8.0; distilled water; 10^{-3} M HCl-0.1 M NaCl; and distilled water.

The conjugate is then aspirated to a moist filter cake. It may be stored in a closed container at 4°C as a moist filter cake or in aqueous suspension till use. No satisfactory method of drying agarose-protein conjugates has yet been discovered. Attempts at lyophilization result in a decrease in the hydrated volume of the gel and activity loss of attached antibody. It is thought that both result from the collapse of the internal pore structure.

Proteins and other ligands have been coupled to CNBr activated agarose in the pH range 6 to 9.5. Following activation, the gel should be washed with 5 to 10 volumes of the buffer in which the protein to be coupled is dissolved. The stirred gel equilibrates rapidly with the coupling buffer as evidenced by the rapid drop from pH 11 to that of the coupling buffer.

Many other variations have been described for gel activation and coupling. The quantity of CNBr used for activation may vary from 50 to 300 mg per ml gel (settled volume) (19, 22-24). The quantity is frequently adjusted by adding CNBr as a concentrated aqueous solution, rather than as a solid as described earlier (19,22-24). Also as described earlier, coupling of labile proteins is frequently performed at 4°C, rather than at room temperature (22-24).

2. Organofunctional Controlled-Pore Glass

Like Sephadex and agarose gels, controlled-pore glass beads may serve as supports for molecular permeation chromatography. Glass beads of varying molecular weight operating ranges are available from Bio-Rad Laboratories and Pierce Chemical. Table 3 lists the properties of various commercial preparations.

Messing and Weetall (27,28) developed methods for covalently coupling proteins to controlled-pore glass (and other metal oxides). The glass is first rendered organofunctional by reaction with a bifunctional silane, with 3-triethoxysilyl-1-aminopropane being used most commonly. The glass is refluxed in a toluene solution of the silane to yield the aminopropyl glass as outlined in Reation 6.

a. $|-O-\underset{|}{\overset{|}{Si}}-O-(C_2H_5O)_3-Si-(CH_2)_3-NH_2 \xrightarrow[\text{REFLUX}]{\text{TOLUENE}}$

$|-O-\underset{|}{\overset{|}{Si}}-O-\underset{|}{\overset{|}{Si}}-(CH_2)_3-NH_2$

$G-NH_2$

b. $G-NH_2 \xrightarrow[\text{2) } Na_2S_2O_7]{\text{1) } p\text{-}O_2N-C_6H_4-C(=O)-Cl} G-NH-C(=O)-C_6H_4-NH_2$

c. $G-NH_2 + CSCl_2 \xrightarrow[\text{REFLUX}]{CHCl_3} G-N=C=S$

Reaction 6 - Functional derivatives of Controlled-Pore Glass. a) Synthesis of aminopropyl derivate by silylation with 3-aminopropyl triethoxysilane. b) Synthesis of the aryl amine by acylation of the aminopropyl derivative with p-nitrobenzoyl chloride. The resulting p-nitrobenzoyl amide is reduced to the p-aminobenzyl amide with $Na_2S_2O_7$. c) Synthesis of the isothiocyanate by thiophosgination of the aminopropyl derivative.

Proteins may be coupled directly to the aminopropyl glass through their pendant carboxyl groups by activating the -COOH with a water-soluble carbodimide (54). The aminopropyl glass may be used as the base with which to introduce other reactive functional groups. For example, reaction with p-nitrobenzoyl chloride yields the p-nitrobenzoyl amide (Reation 6), from which the p-aminobenzoyl amide is generated by reduction with $Na_2S_2O_7$ (27). Reaction with HNO_2 yields the diazonium salt, which forms an azo linkage through the protein's pendant tyrosyl and histidyl residues.

Antisera have been attached to the isothiocyanate derivative in high yield and with good retention of antibody activity (26,28). The isothiocyanate is generated from the alkylamino glass by reaction with thiophosgene in $CHCl_3$ as outlined in Reaction 6.

The glass isothiocyanate (ITC-glass) couples with proteins through their pendant primary amino groups with the formation of the N,N' substituted thiourea adduct. This method is described in detail later.

Several organofunctional derivatives of controlled-pore glass are available from Pierce Chemical. These are listed in Table 3.

In coupling antisera to ITC-glass (26), materials needed include aminopropyl controlled-pore glass (Aminopropyl CPG) with 550-angstrom internal pore diameter in varying particle sizes and available from Corning Glass Works and Pierce Chemical (Table 3). Line, et al (26) employed aminopropyl glass beads in one micron-sized particles.

To prepare ITC-glass, 10 grams of aminopropyl glass are overlayed with 160 ml of a 10% w/v $CHCl_3$ solution of $CSCl_2$ in a two-necked roundbottom flask. One neck is fitted with a sealed mechanical stirrer and the other with a reflux condenser. The glass is refluxed with stirring for five hours on a steambath. The ITC-glass is then harvested by filtration and washed with $CHCl_3$, acetone, and finally $CHCl_3$. It is then dried over P_2O_5 *in vacuo* and stored dessicated until use. All steps of this reaction should be performed within a well-ventilated fume hood.

To couple antisera to ITC-glass, one to three ml (equivalent to 65 to 195 mg serum protein) of the antiserum is dissolved in 20 ml 0.5 M $NaHCO_3$, pH 9.0, and chilled in an ice bath for about 30 minutes. Two grams of the ITC-glass is then added to the mechanically stirred antiserum solution. The suspension is allowed to come to room temperature and stir overnight.

Following reaction, the conjugate is harvested by centrifugation and is washed by stirring several times in 0.1 M phosphate buffer - 0.3 M NaCl, pH 8.0, and then in distilled water to remove protein which is not covalently bound. It may be stored as an aqueous suspension in a closed vessel at 4^oC until use. As with agarose conjugates, no satisfactory method of drying glass conjugates has yet been discovered.

3. Adsorption to Plastic Tubes

Antisera may be adsorbed to different types of plastic blood collection tubes (e.g. polystyrene, polyethylene, polypropylene, etc.) by incubating an appropriate dilution of the antiserum in them for varying periods of time at mildly alkaline pH. The following example illustrates the method.

Aliquots of an appropriate dilution of an antiserum against hepatitis B antigen are delivered into a 15 x 86-mm polypropylene tube. The tubes are capped and allowed to stand at room temperature for 24 hours. The tube contents are then aspirated and rinsed five times with 1 ml of 0.01 M tris-HCl-0.02% w/v NaN_3, pH 7.1. The tubes are then capped and stored at $4^{o}C$ until use.

The quantity of antiserum necessary to produce a satisfactory assay when adsorbed to plastic tubes and employed in LIA systems depends to a great extent on the titer of the antiserum (which may be determined by titration in the appropriate soluble immunoassay system). It also depends on the quantity of antiserum protein that can be adsorbed to the external surface of the tube. The latter will probably vary with the plastic employed. In the HBAg system cited earlier (3), the authors used 0.3 ml of a 1:2000 dilution of the HBAg antiserum-tube.

IV. DETERMINATION OF CONJUGATED PROTEIN

The quantity of antiserum protein attached to a solid support may be determined by one of several methods. The simplest is to observe the disappearance of absorbance at 280 nm from solution during the course of coupling (8,49). However, covalently attached protein frequently yields erroneously high values because many of the carriers adsorb or include proteins that are not removed unless the conjugate is subsequently washed in buffered salt solutions. When polymers devoid of nitrogen are employed, the quantity of coupled protein may be determined from the total nitrogen content of the conjugate (7,8). In the case of nitrogen-containing polymers, the protein content is determined by amino acid analysis of an acid hydrolyzate of the conjugate (18,19,21); or by determining constituent amino acid, e.g. arginine (54), or histidine (25,26) content of an acid hydrolyzate of the conjugate.

The following method for the determining of the histidine content of covalently attached protein conjugates has been found to be of general utility with many polymers, including agarose and organofunctional glass beads (25,26). Histidine is released by hydrolysis of the conjugate under 6 N HCl for 18 to 24 hours. The liberated amino acid is then determined colorimetrically by its ability to form a biz-azo derivative with diazonium-

1-H-tetrazole (<u>55</u>). This method could not be used to determine the protein content of conjugates formed by reaction with a diazotized arylamino carrier, because the imidazole group necessary for this reaction is susceptible to diazonium coupling with the carrier.

A. Method

The conjugate (1 gram - wet weight of agarose and 100 to 200 mg glass adducts prepared as described earlier) is overlayed with 2 to 3 ml 6 N HCl in a 16 x 150 mm PYREX$^{(R)}$ brand screwcap serological tube. The tube is capped and the contents are hydrolyzed 18 to 24 hours at 110°C. The carrier particles (charred polysaccharide breakdown products with agarose adducts) are removed by filtering through a fine 15-ml sintered glass funnel. The filtrate is collected in a distillation flask (50 to 125 ml), the side arm of which is connected to a vacuum source as shown in Fig. 2a. Most organic polymers (particularly polysaccharides) release acid-soluble materials into the hydrolyzate, causing considerable bumping and foaming in the subsequent volume reduction. The oversize flask permits control of the foaming and prevents loss of material during concentration under reduced pressure.

Table 4

Protein Content of Representative Immunoadsorbents[a]

Immunoadsorbent	% w/w Moisture	mg solid / ml	Nanomole histidine / mg solid wet	Nanomole histidine / mg solid dry	% w/w protein wet	% w/w protein dry	Coupling Efficiency %
native rabbit serum	----	----	---	278[b]	---	---	---
anti-T_3-agarose	80.3	----	4.15	21.1	1.49	7.40	74
anti-digoxin-agarose	84.6	----	2.1	13.7	0.70	5.00	50
normal rabbit serum-agarose	85.5	----	3.5	24.2	1.28	8.80	88
ITC-glass-anti-digoxin	----	14.2	---	23.8	---	8.6	86
ITC-glass-normal rabbit serum	----	13.3	---	25.0	---	9.0	90

a. Determined from the histidine content as described in Sec. IV (27,28).

b. Determined on a weighed quantity of lyophilized serum.

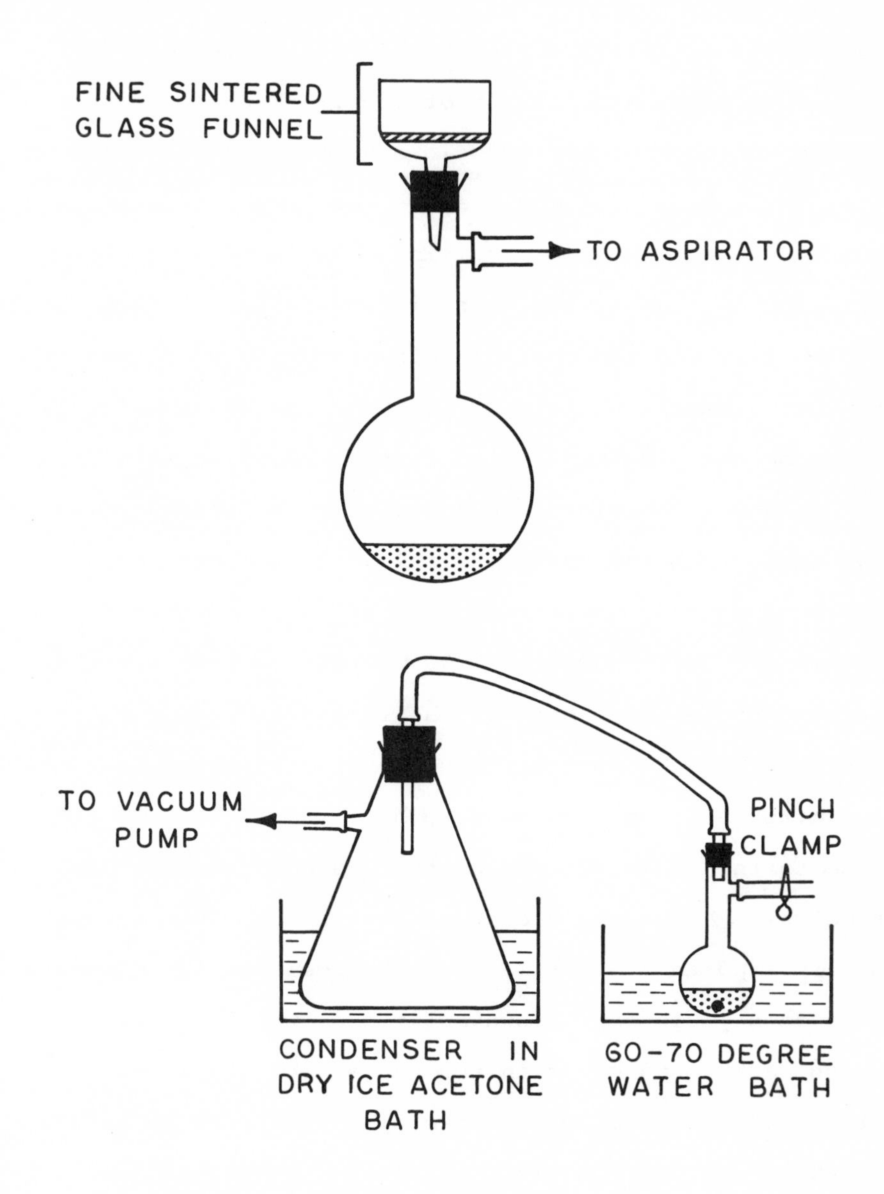

Fig. 2: Apparatus for reducing conjugate hydrolyzates to dryness.

The side arm rubber connection is clamped off with a Day pinch clamp, and the top of the flask is connected to a condensation trap immersed in a dry ice-acetone bath as outlined in Fig. 2b. The trap is then connected to a vacuum pump while a glass boiling chip is added to the distillation flask. The flask is immersed in a 60 to 70°C. water bath and the contents are dried under reduced pressure. Excessive foaming is prevented by momentarily releasing the side arm pinch clamp as required until maximum vacuum is obtained. By attaching the flasks to the condensate trap through a manifold, the authors have concentrated as many as four samples at one time. It is recommended that a 500- or 1000-ml filter flask serve as the condensate trap, because the large volumes of fluid that are removed generally plug most coldfingers.

After all the liquid is removed, 5 ml of water are added to the flask and the contents are allowed to dry as before. After the second concentration, the residue is dissolved in 4 ml of water. The residues from agarose conjugates yield dark brown solutions, which contribute a high background color in the subsequent histidine determination. They can be easily decolorized by suspending 40 mg charcoal (Norit A.N.F.X.)/g conjugate (wet weight) originally hydrolyzed in the solution for three minutes and filtering to remove the charcoal. Other polysaccharide carriers (e.g. cellulose and dextran) must also be decolorized in this manner. Since there is no interference with the histidine determination, the hydrolyzate from the glass conjugates may be used directly.

An aliquot (generally 0.1 to 0.5 ml) of the dissolved residue is diluted to 4.0 ml with 0.8 M $NaHCO_3$, pH 8.8. Then add one ml of diazonium-1-H-tetrazole (DHT) prepared as follows: 1.1 g 5-amino-1-H-tetrazole are dissolved in 23 ml 1.7 N HCl. Ten ml of an aqueous solution containing 0.7 g $NaNO_2$ is added dropwise with stirring to the tetrazole solution which had been chilled by immersion in an ice-water bath. Following addition of the $NaNO_2$ solution, the pH of the solution is adjusted to 5.0 with 5 N KOH, keeping the solution in ice-water bath. During this step, the solution turns from colorless to a deep yellow. The solution is then diluted fourfold with water for use in the histidine determination. The solution should be prepared fresh daily and kept stored on ice during the day.

After 30 minutes, the absorbance at 480 nm is measured and the quantity of histidine is calculated from a standard curve run with histidine ($E_{1\,cm}^{480} = 1.4 \times 10^5$). The protein content of the conjugate is determined from the quantity of histidine liberated by the native antiserum when hydrolyzed under the same conditions.

No background (except the color interference noted earlier for polysaccharide conjugates) is contributed by agarose or glass carriers when hydrolyzed alone. When the carriers are hydrolyzed in the presence of a known quantity of histidine, the histidine recovery is quantitative.

V. METHODS OF LABELING

A. Radiolabeled Antigens and Antibodies

Most RIA procedures utilized ^{125}I-labeled antigens or antibodies. The isotope is introduced into proteins and polypeptides by electrophillic substitution at the activated benzene ring provided by the pendant tyrosine residues. Substitution takes place at the 3 position with the formation of a 3-iodotyrosyl residue or at the 3 and 5 positions with the formation of a 3,5 diodotyrosyl residue. Generally, the I electrophile is generated from the Na ^{125}I by oxidation either by using chloramine T (56,57) according to some modification of the method of Hunter and Greenwood, or by using Na ^{125}I as an electron donor in a coupled enzymatic reduction of H_2O_2 by lactoperodixase (58,59).

The method best suited for labeling varies with the antigen and is beyond the scope of this chapter. The reader should consult the references already cited and those listed in Table 1 for specific examples.

Many of the RIA procedures for hapten antigens (e.g. digoxin, digitoxin, cortisol and other sterodis) employ ^3H-labeled antigens, most of which can be obtained from commercial radiochemical firms (26,37,38). A method described for introducing ^{3}H into proteins and polypeptides (60) is based on the reductive alkylation of primary amines originally described by Means and Feeney (61).

B. Enzyme-Labeled Immunogens

In an attempt to circumvent the problems of handling radioisotopes, some investigators have devised methods for attaching enzymes with high turnover numbers (and for which simple colorimetric determinations are available) to the immunogen to be detected. Two enzymes frequently employed for this purpose are alkaline phosphatase and horseradish peroxidase (4,44). Generally, the enzyme is conjugated to the antigen or antibody with a bifunctional reagent such as glutaraldehyde or tolylene 2, 4 disocyanate (4,29,44).

When employed in solid-phase immunoassays, the conjugated enzymatic activity may be determined by assaying either the supernatant (i.e. disappearance of activity upon antibody binding) or the solid phase (i.e. appearance of activity as the label is bound) by methods for immobilized enzymes outlined elsewhere in this book. Again, the methods for attaching enzymes to immunogens vary with both moieties and are beyond the scope of this chapter. The reader is encouraged to consult the references presented in Table 1 for more details.

An interesting competitive binding homogeneous enzyme-linked system has been described by Rubenstein, et al (62). This method was developed for detecting morphine and related drugs. It consists of attaching the antigen to an enzyme near the active site. Antibody binding to the labeled antigen results in inhibition of conjugated enzymatic activity. When unlabeled antigen competes with the enzyme-labeled antigen, displacement may be monitored by reversal of inhibition.

VI. IMMUNOASSAY SYSTEMS

Requirements for an immunoadsorbent (with regard to titer, sensitivity, and specificity of the antibody) are essentially the same as for the antiserum from which it is made. In addition, the carrier itself should not bind significant quantities of antigen. Nonspecific antigen adsorption may be checked by coupling serum from host animals (i.e. guinea pig, rabbit, goat, etc.) which have not been immunized under the same conditions as employed for antiserum coupling (25,26). The normal serum control-carrier adduct is titered with labeled antigen over the same range as for the immunoadsorbent. Any background associated with the control must be subtracted from that obtained with the immunoadsorbent. Ideally, the background should be less than 2% in which case it can be neglected. A list of commercial solid-phase RIA kits is given in Table 4.

Table 5

Commercially Available Solid-Phase RIA Kits

Carrier	Trade Name	Attached Immunogen	Method of Attachment	Method of Assay
* dextran (Sephadex)	Phadebas Insulin Kit	insulin antibody	covalent	competitive binding RIA
"	Phadebas IgE Kit	antibody to the IgE subfraction of human - globulin	covalent	competitive binding RIA
"	Phadebas B_{12} Kit	intrinsic factor	covalent	competitive binding radioassay
# polypropylene tubes	AUSRIA	antibody to HBAg-hepatitis associated antigen	adsorbed	sandwich RIA
"		antibody to human growth hormone	adsorbed	competitive binding RIA
+ organofunctional controlled-pore glass	IMMO PHASE digoxin kit	antibody to digoxin	covalent	competitive binding RIA

c. The test is not an immunoassay. Intrinsic factor is a glycoprotein which has a high affinity for B_{12}. The principles of the method are the same as outlined for competitive binding RIA. *Vendor - Pharmacia, Piscataway, N.J.; #Vendor - Abbott Labs., North Chicago, Ill.; +Vendor - Corning Glass Works, Medfield, Mass.

The use of solid-phase antisera in RIA is illustrated by three systems: The detection of triiodothyronine (T_3) by competitive binding RIA employing an antiserum covalently attached to CNBr-activated agarose (25); the detection of digoxin by competitive binding RIA employing an antiserum covalently attached to CNBr-activated agarose and to the isothiocyanate of controlled-pore glass beads (26); and by the detection of hepatitis associated antigen or Australian antigen (HBAg) by direct (sandwich) RIA using antisera adsorbed on plastic tubes (3).

A. Triiodothyronine RIA

The production of specific antisera to l-triiodothyronine (T_3) and a homogeneous RIA for this hormone was first described by Gharib et al (34) and Brown et al (35). A more rapid homogeneous RIA has recently been described by Sekadde et al (36). The hapten is rendered antigenic by coupling to a human or bovine serum albumin employing a water-soluble carbodiimide (34,35). Antisera to this conjugate were then raised in rabbits. The reader is referred to the original references for details.

T_3 detection in serum is of clinical significance in cases of suspected thyroid dysfunction. Aside from titer, the principal requirement of the antiserum is that it be able to differentiate T_3 from l-thyroxine (T_4). This requirement is fairly rigid, because T_4 is the predominant circulating thyroid hormone (T_3 = 50 to 220 ng %;

T_4 = 4.5 to 13.2 ug %) (63). Recently, Siegel et al (25) described a solid-phase T_3 RIA employing an antiserum covalently coupled to CNBr-activated agarose by the method outlined in Section III-C-1.

Materials needed include assay buffer - 0.01 M phosphate buffer, 0.14 M NaCl, 0.1 % w/v NaN_3, 2.0% w/v normal rabbit serum, pH 7.4. Endogenous T_3 and T_4 are removed from the buffered serum by stirring 1 liter with 50 grams of Sephadex A-25 (Cl^- from, 200-400 mesh) overnight. The resin is removed by filtration. Also needed are ^{125}I-tagged T_3 (available from Abbott, New England Nuclear, and others); platform or rotary mixer; and a clinical centrifuge.

T_3 antiserum may be raised in rabbits or some other suitable animal by injecting a T_3-albumin conjugate by methods described elsewhere (25). Initial bleedings may be taken about one month after immunization and tested for titer, displacement and selectivity of T_3 over T_4 in a suitable homogeneous RIA (25,34-36). Host animals whose sera possess satisfactory properties may be sacrificed by cardiac puncture and the sera pooled. The pool should be characterized as above for the individual bleedings in a homogeneous RIA. The pool may be stored frozen or lyophilized. The antisera may be used without further purification in coupling to agarose.

In the following examples, the antiserum that was coupled to agarose was obtained from rabbits. In this case, the animals were sacrificed six weeks after immunization and the antiserum pool lyophilized. Employing an 18-hour incubation period, the pool was titered vs T_3 ^{125}I at 4^oC in a variation of the method described by Gharib et al (25,34). The results of the homogeneous titration are presented in Fig. 3, curve 1. The region of maximum sensitivity occurred at dilutions from 1:5000 to 1:1000 (final dilution), as evidenced by the rapid rise in the titration curve. The titer is defined as that antiserum dilution which binds 50% of the available T_3 ^{125}I. For this antiserum pool, the titer was reached at a final dilution of 1:2700, corresponding to 26 ug serum protein (Table 6).

Agarose-Anti-T_3 and agarose-NRS are prepared by coupling both T_3 antiserum and normal rabbit serum (NRS) to CNBr-activated Bio-Gel A, 0.5 m, by the method outlined in Section III-C-1. The properties of the conjugates used for illustration properties are presented in Table 5. The protein content based on the dry weight of the gel was 74% for the immunoadsorbent and 88% for the NRS conjugate. Maximum coupling in either case would have yielded conjugates containing 10% w/w. Thus, the coupling efficiency (protein coupled/theory x 100) was 74% for the immunoadsorbent, and 88% for the NRS adduct.

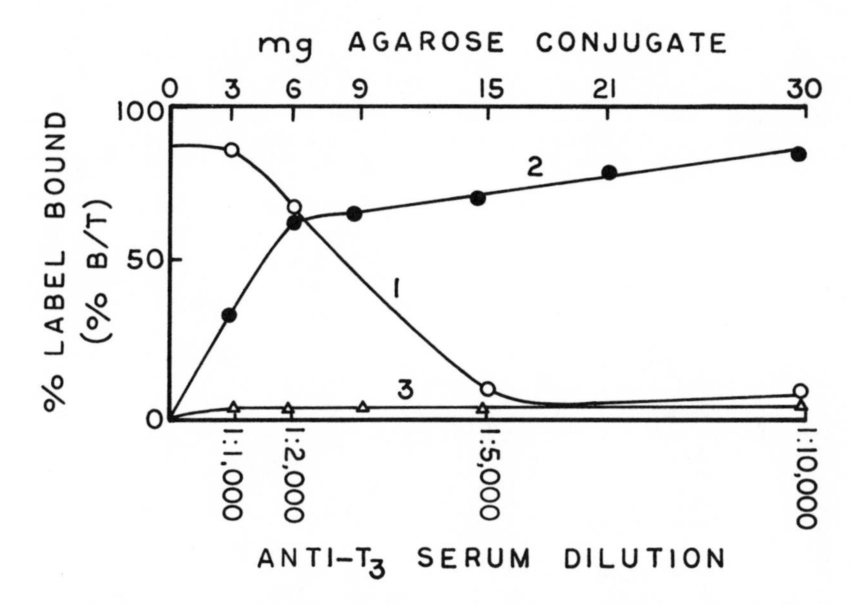

Fig. 3: Titration curve of soluble and immobilized T_3 antiserum. Curve 1 (○) = titration curve of native T_3 antiserum. Dilution factors are presented on the lower abscissa. Curve 2 (●) = anti-T_3-agarose. The weights of the immunoadsorbent are on the upper abscissa. Curve 3 (△) = NRS-agarose (upper abscissa). The ordinate is presented as the percentage total added $T_3{}^{125}I$ bound to the antibody (% B/T). Each tube contained 0.2 ng $T_3{}^{125}I$ which gave an initial total count of about 10,000 cpm. Reproduced from Ref. 25.

1. Titration

Aliquots of the moist conjugate were weighed into small plastic tubes, each equipped with a coarse filter and a closed plastic tip (Whale Scientific) at the bottom. The immunoadsorbent is then overlayed with 1.8 ml of the assay buffer. A quantity of $T_3{}^{125}I$ sufficient to give 10,000 cpm (dilutions made in assay buffer) is added, and the tubes are capped and mixed continuously for two hours at room temperature. The tubes are then inverted, the are removed, and the supernatant is centrifuged through the tube filter into disposable test tubes.

Table 6

Titer of Representative Immunoadsorbents

Immunoadsorbent	Dilution or mg immunoadsorbent to each 50% B/T	Protein Content %	Protein Equivalent at titer (50% B/T)	% Titer Retention
native anti-T_3	1:2700	----	26[a]	100
Anti-T_3-agarose	3.6[b]	1.28	53	49.2
native anti-digoxin	1:35,000	----	2.0[a]	100
ITC-glass-anti-digoxin	0.05[c]	8.6	4.3	47
Agarose-anti-digoxin	3.8[b]	0.7	27	7.4

a. The native antisera are employed as lyophillized powders and reconstituted with H_2O prior to use. It was assumed that the protein content of rabbit serum was 70 mg/ml.

b. moist weight

c. dry weight

The radioactivity remaining in the filtrate (free T_3) is measured in a gamma ray crystal scintillation spectrophotometer, and the percentage of antibody-bound T_3 is calculated by difference. The system is checked for nonspecific adsorption of T_3 ^{125}I by titering the NRS-agarose over the same range as the immunoadsorbent. In addition, three total count tubes (i.e. the incubation mixture minues either conjugate) should be run to obtain the total T_3 ^{125}I available to the immunoadsorbent and the NRS conjugate.

Figure 3, Curve 2 illustrates the titration curve obtained with the T_3 immunoadsorbent described by Siegel et al (25). In this case, the titer is defined as the weight of immunoadsorbent necessary to bind 50% of the available isotope. This point was reached at 3.6 mg (wet weight) immunoadsorbent, which corresponds to 53 ug serum protein (from the protein content in Table 4). Thus, the retention of activity was about 50% upon coupling. These results are summarized in Table 6.

Curve 3 of Fig. 3 shows the titration curve obtained with the NRS-agarose control. As can be seen, there is essentially no background adsorption of isotope through the range tested. Thus, binding of T_3 ^{125}I by the T_3 immunoadsorbent is solely a function of attached antibody, and it may be concluded that neither the carrier nor any component of the serum from nonimmunized rabbits coupled to the carrier binds any T_3.

2. Displacement

Displacement reactions of T_3 ^{125}I by T_3 or T_4 are performed as for the titration except that the quantity of immunoadsorbent is fixed (generally at a level which binds a 50 to 60% of the available T_3 ^{125}I in the absence of any T_3 or T_4). Serial dilutions of T_3 and T_4 stocks are made in the assay buffer. T_3 ^{125}I is then added to each of the dilutions to a final concentration of about 10,000 cpm/ml. The immunoadsorbent may then be weighed into the assay tubes or dispensed from a stirring stock suspension (made in the assay buffer). The volume is adjusted to 1.5 ml with the buffer and 1.0 ml of the serial T_3 or T_4 dilutions are added. The tubes are then capped and mixed continuously for two hours at room temperature. The remaining steps are as described previously for titration. The initial binding capacity (i.e. binding of T_3 ^{125}I in the absence of unlabeled antigen, or B_o/T) is obtained by treating the immunoadsorbent with 1.0 ml of a solution containing only the isotope. The displacement of T_3 ^{125}I with increasing T_3 (or T_4) concentration is then plotted on a semi-logarithmic scale (% B/T vs log antigen).

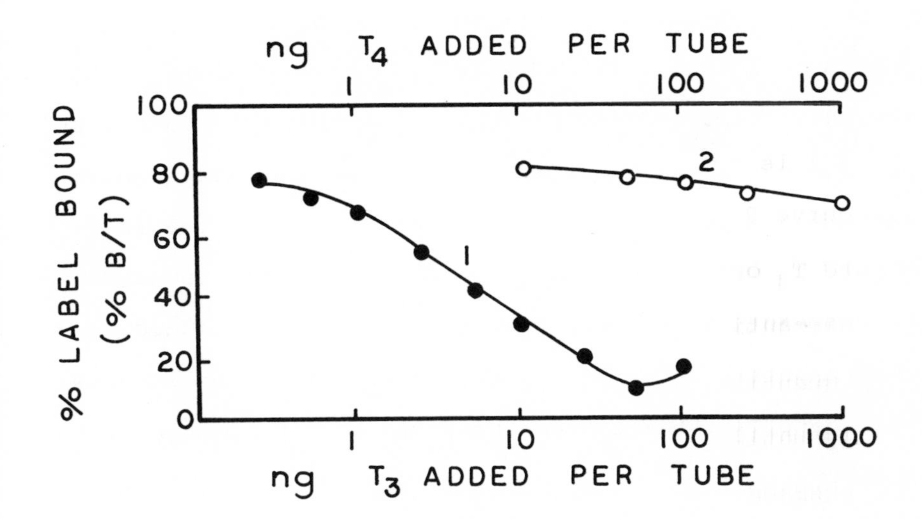

Fig. 4: Displacement of T_3 ^{125}I by T_3 and T_4 for anti-T_3-agarose. Curve 1 (●) = lower abscissa-displacement by T_3. Curve 2 (○) = upper abscissa-displacement by T_4. The quantity of immunoadsorbent employed in each case was 30 mg (wet weight). Each tube contained 0.2 ng T_3 ^{125}I which gave an initial total count of about 10,000 cpm. Reproduced from Ref. 25.

Curve 1 of Fig. 4 shows the displacement of T_3 ^{125}I by T_3 for the agarose-anti-T_3 previously described (25). In this case, maximum sensitivity occurred in the region 1 to 10 ng T_3. Curve 2 shows the displacement of labeled T_3 by T_4, and it is evident that no significant displacement occurred up to 1 ug T_4, or 1000 times the effective T_3 concentration (Curve 1). Further, the immunoadsorbent failed to bind any T_4 ^{125}I. In fact, the immunoadsorbent demonstrated improved selectivity over the antiserum from which it was synthesized. The displacement of T_3 ^{125}I from the soluble antiserum in the homogenious RIA described by Siegel et al. (25) is shown in Fig. 5.

Curve 1 is the displacement function obtained with T_3 and Curve 2 that obtained with T_4. Maximum sensitivity toward T_3 occurred in the region 1 to 10 ng as with the agarose-anti-T_3. However, T_4 began to displace significant quantities of $T_3\ ^{125}I$ at about 50 ng, increasing slowly until at 1 ug about 55% of the $T_3\ ^{125}I$ bound in the absence of unlabeled antigen had been displaced. The reasons for the improved selectivity of the immunoadsorbent over the native antiserum remain obscure at this time.

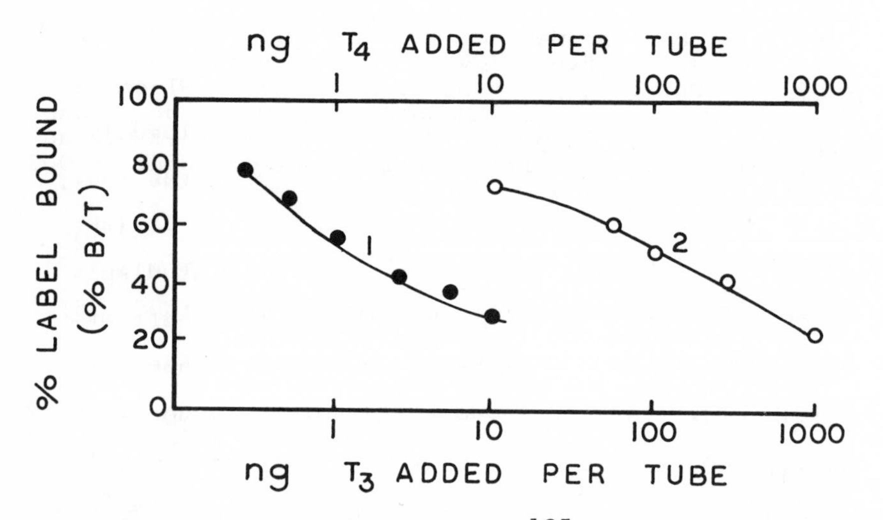

Fig. 5: Displacement of $T_3\ ^{125}I$ by T_3 and T_4 in the Homogenious System. Curve 1 (●) = lower abscissa-displacement by T_3. Curve 2 (○) = upper abscissa-displacement by T_3. The antiserum dilution was 1:1000 in both cases. Each tube contained 0.2 ng $T_3\ ^{125}I$ which gave an initial total count of about 10,000 cpm. Reproduced from Reg. 25.

The kinetics of T_3 ^{125}I binding by agarose-anti-T_3 are illustrated in Fig. 6. In this case, the quantity of immunoadsorbent employed (20 mg - wet) was capable of binding 70% of the available isotope in the standard two-hour incubation. However, the results presented in Fig. 6 show that 60% of that quantity is bound in the first 5 to 15 minutes, indicating that the incubation time could be reduced further.

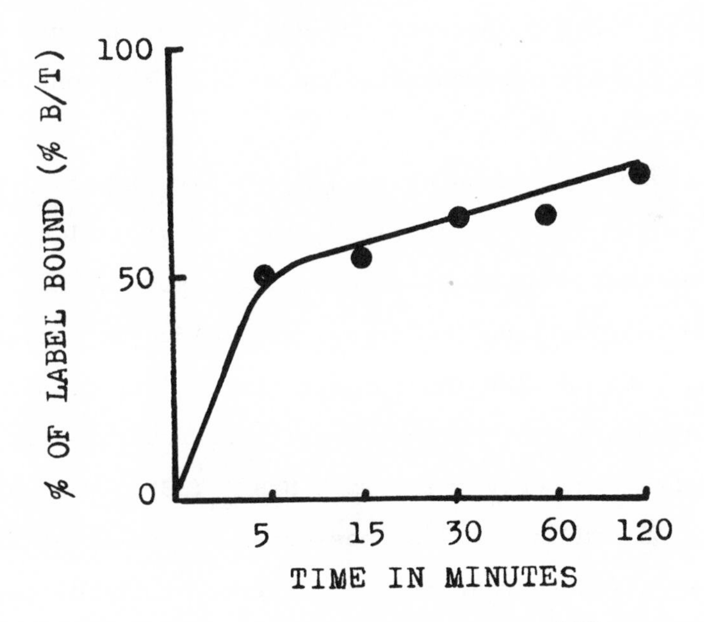

Fig. 6: Kinetics of Binding of T_3 ^{125}I by anti-T - agarose. Each tube contained 0.2 ng T_3 ^{125}I which gave and initial total count of about 10,000 cpm. Reproduced from Ref. 25.

B. Digoxin RIA

Digoxin and digitoxin are cardiac glycosides derived from digitalis. They are employed in the long-term treatment of chronic heart patients. They are metabolized and excreted slowly and can accumulate to toxic levels in the circulatory system with protracted use. Thus the determination of circulating levels is important.

The production of digoxin-specific antibodies and homogenious RIA for this drug was first described by Butler et al (37). The glycoside was rendered immunogenic by conjugating it to a foreign albumin using a water-soluble carbodiimide (presumably to form an ester between the albumin -COOH groups and the -OH pendant groups provided by the trisaccharide, the steroid aglycone, or both).

Subsequently, Smith et al (38,39) have improve the immunogenicity by first oxidizing the cis vicinal hydroxyls on the terminal glucose of digitoxose with periodate and coupling the resulting dialdehyde derivative with foreign albumins through its pendant primary amines to form a Schiff's base, which is subsequently reduced to a secondary amine with $NaBH_4$ (38). The antiserum was raised in rabbits, possessed an excellent titer, and was specific in that it could differentiate digoxin from digitoxin, delanoside and other cardiac glycosides (34). Although the antiserum exhibited greater sensitivity toward digoxin than digitoxin, the latter could still be detected by using higher concentrations of the antiserum (45).

The initial RIA procedure (37,38) employed ^{3}H digoxin as the marker antigen. Subsequently, methods employing ^{125}I-labeled cardiac glycoside have been described. This necessitates the introduction of an activated benzene ring into the cardiac glycoside. Briefly, this is accomplished by first succinylating the hydroxyls on the digoxigenin or digitoxigenin (Steroid stripped of digitoxose) or on digitoxose with succinic anhydride. A tyrosine ester is then coupled with a water-soluble carbodiimide to yield the corresponding tyrosyl amide, which then may be iodinated by one of the procedures described previously.

Recently, Line et al (26) have described a solid-phase digoxin RIA employing an antiserum covalently coupled to CNBr-activated agarose and to a controlled-pore glass isothiocyanate by the methods outlined in Section III-C.

Materials needed are assay buffer (0.019 M sodium barbital - 0.003 M sodium acetate - 0.15 M NaCl, pH 7.4.), normal human serum (available from Metrix, and Dade, among others), 12 - (^{3}H) - digoxin (5 Ci/mMol) - (New England Nuclear), Digoxin (Sigman Chemical, or the U. S. Pharmacoepial Convention, Inc.), and a platform or rotary mixer as outlined for the T_3 RIA.

Digoxin antiserum may be raised in rabbits by injecting a digoxin-bovine albumin conjugate synthesized as described by Smith, et al (38). The immunization protocol is that described in the same paper (38). A high-titer antiserum capable of differentiating digoxin from digitoxin may be formed in about 12 weeks. Sample bleedings may be characterized by the homogeneous RIA described by Butler et al (37) or by Smith et al (39). Host animals whose antisera possess satisfactory titer and specificity may be sacrificed by cardiac puncture, and the antisera pooled and lyophilized without loss in immunological activity (26). The dried pool may be employed in the coupling reactions without further purification.

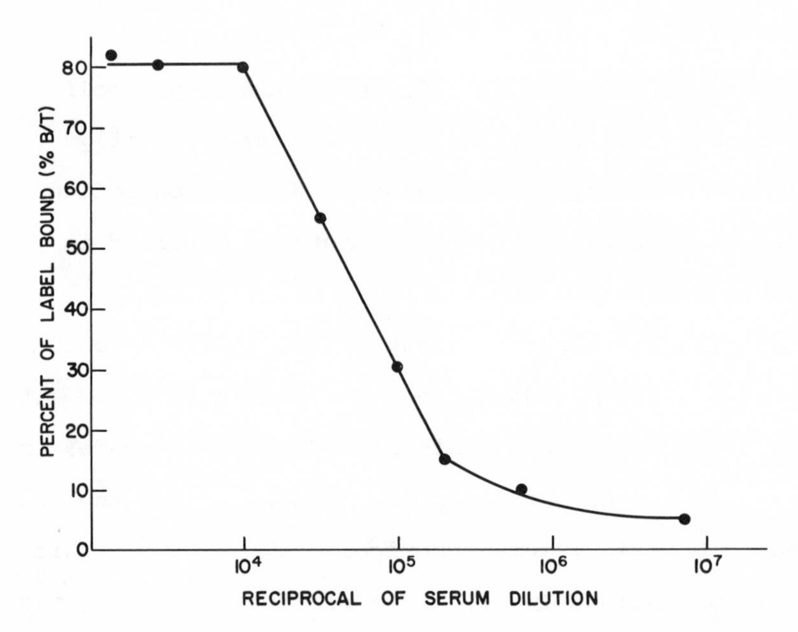

Fig. 7: Titration Curve for Native Anti-digoxin Serum. The percent of the total ^{3}H digoxin bound to the antibody is presented on the ordinate (% B/T), and the final dilutions of digoxin antiserum are on the abscissa. The titer of this antiserum pool was 1:35,000. Each tube contained 3.1 ng ^{3}H digoxin/ml, which gave an initial count of 16,000 cpm/ml. Reproduced from Ref. 26.

Figure 7 shows the titration curve obtained with the soluble antiserum pool prior to coupling obtained by Line et al (26) using the homogenious RIA procedure described by Smith (39). The region of maximum sensitivity occurred between 1:10,000 and 1:100,000 final dilution. The titer (again defined as the quantity of antiserum which bound 50% of the available ^{3}H digoxin) occurred at a 1:35,000 final dilution, corresponding to a 2.0 ug serum protein (Fig. 7, Table 6).

Agarose-anti-digoxin and agarose-NRS are prepared by coupling digoxin antiserum and NRS to CNBr-activated Bio-Gel A, 0.5 m, by the method outlined in Section III-C-1. The properties of the conjugates used for illustration are presented in Table 4. The protein content of the dry gel was 50% for the immunoadsorbent. As for the T_3 conjugate, maximum coupling would have yielded a conjugate containing 10% w/w serum protein. Thus, the coupling efficiency was 50% for the immunoadsorbent.

ITC-glass-anti-digoxin and ITC glass-NRS are prepared by coupling the antiserum and NRS to the isothiocyanate formed from alkylaminofunctional controlled-pore glass by the methods described in Section III-C-2. The properties of these conjugates are also outlined in Table 4. The protein content was 8.6% for the immunoadsorbent and 9.0% for the NRS adduct. Thus, the coupling efficiencies were about 90% since the systems were loaded to contain 10%.

In titrating the immunoadsorbents with ^3H Digoxin, aliquots of the immunoadsorbents may be dispensed into the assay tubes directly by weighing or by dispensing from a stirring stock suspension made in the assay buffer. The volume is made to 2.0 ml with the assay buffer and 0.5 ml normal human serum is added to each tube. The titration and displacement reactions are performed in the presence of excess serum to normalize any nonspecific binding by serum components such as albumin, etc., when the determination is applied to serum digoxin levels (37,38). The ^3H digoxin is diluted in the assay buffer so that the addition of 50 ul to the incubation mixture will yield a final concentration of about 3.0 ng/ml (equivalent to about 16,000 cpm/ml, when counted as described later). The tubes are mixed for 30 minutes at room temperature, after which antibody bound digoxin is separated from digoxin by centrifugation in a table-top centrifuge. One ml of the supernatant is delivered into 10 ml Instagel scintillation fluid contained in a liquid scintillation vial. The vial is capped, and the contents are well mixed. The quantity of free ^3H digoxin is determined by counting the sample for two minutes in a liquid scintillation counter. Antibody-bound ^3H digoxin is calculated by difference. The respective NRS conjugates are titered over the same range as the immunoadsorbent to check for nonspecific binding.

Figure 8 shows the titration of the ITC-glass-anti-digoxin previously described (26) (Curve 1) as well as that obtained with the ITC-glass-NRS control (Curve 2). Fifty percent of the available ^{3}H digoxin was bound with 0.05 mg of the immunoadsorbent, corresponding to 4.3 ug attached serum protein (Tables 5, 6). The retention of immunological activity was 47% upon coupling to ITC-glass. As Curve 2 shows, no background ^{3}H digoxin binding occurred with the NRS adduct. Thus, digoxin binding was solely a function of attached antibody.

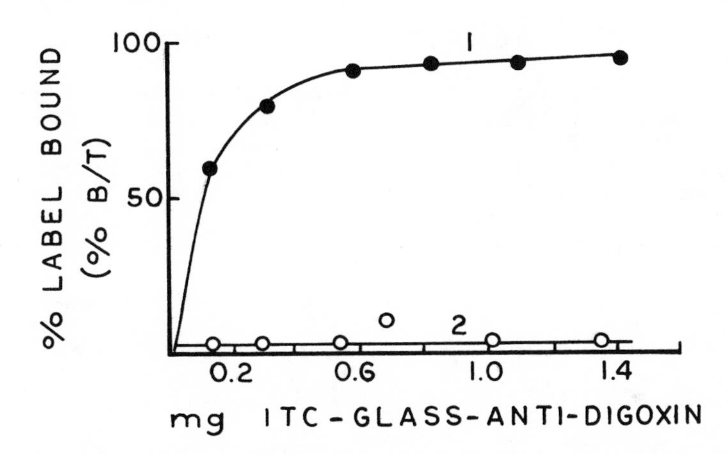

Fig. 8: Titration of ITC-glass-anti-digoxin with ^{3}H Digoxin. Curve 1 (●) = ITC-glass-anti-digoxin. Curve 2 (○) = ITC-glass-NRS. Each tube contained 3.1 ng ^{3}H digoxin/ml which gave an initial count of 16,000 cpm/ml. Reproduced from Ref. 26.

Figure 9 shows the titration curve obtained with agarose-anti-digoxin (Curve 1) and its NRS control (Curve 2). The titer occurred at 3.8 mg (wet weight) immunoadsorbent, corresponding to 27 ug serum protein or to a titer retention of 7.4% (Tables 5, 6). Again, there was no background adsorption associated with the NRS adduct (Curve 2).

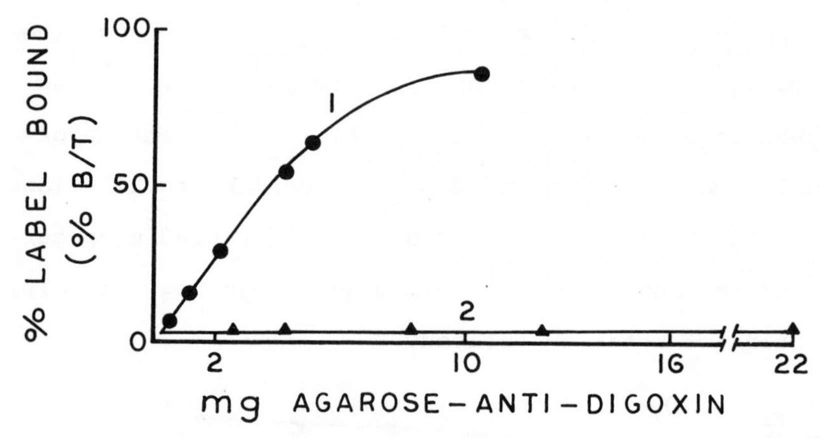

Fig. 9: Titration of Agarose-anti-digoxin with ^{3}H-digoxin. Curve 1 (●) = agarose-anti-digoxin. Curve 2 (▲) = agarose-NRS. Each tube contained 3.1 ng ^{3}H digoxin/ml which gave an initial count of 16,000 cpm/ml. Reproduced from Ref. 26.

Displacement reactions of ^{3}H digoxin by digoxin are performed as described for the titration except that the quantity of immunoadsorbent is fixed (again at a quantity which binds 50-60% of the available isotope in the absence of unlabeled antigen). Serial dilutions of digoxin stock are made in the assay buffer and the ^{3}H digoxin concentration is adjusted to about 7.5 ng/ml in each of the dilutions. The immunoadsorbent is overlayed with 1.0 ml assay buffer followed by 0.5 ml normal human serum. Then 1 ml of each of the serial dilutions containing isotope is added, and the tubes are continually mixed for 30 minutes at room temperature. At the end of this period the immunoadsorbent is removed by centrifugation. The quantity of ^{3}H digoxin remaining in the supernatant is then determined as described for titration. Displacement results may then be plotted on a semi-logarithmic scale as described for T_3.

Figure 10 shows the displacement of ^{3}H digoxin by increasing digoxin concentration from the soluble antiserum in Curve 3, to the agarose anti-digoxin in Curve 2 and the ITC-glass anti-digoxin in Curve 1. The results presented in the figure illustrate the displacement over the range 1 to 5 ng digoxin/ml. They are presented on a linear scale to illustrate a difference between the displacement function yielded by the soluble antiserum (Curve 3) on one hand and the two immunoadsorbents (Curves 1 and 2) on the other. Displacement of ^{3}H digoxin with increasing digoxin is linear in the case of the agarose-anti-digoxin and the ITC-glass-anti-digoxin, but not with the native antiserum in the corresponding homogenious RIA. This may be due to the fact that higher B_o/T values (60 to 70%) were obtained with the immunoadsorbents than with the native antiserum (about 40%). In any event, the range of the assay in this case (26) was increased and the relationship between radioactivity displacement with increasing digoxin concentration simplified. Over a broader digoxin range, a semi-logarithmic plot (% B/T vs log digoxin) is linear in the region of maximum sensitivity for both the soluble antiserum and both immunoadsorbents.

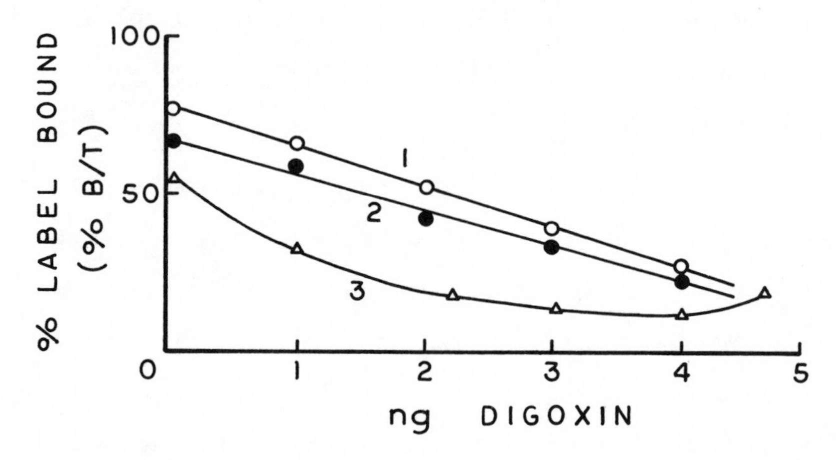

Fig. 10: Displacement of ^{3}H Digoxin by Digoxin. Curve 1 (○) = ITC-glass-anti-digoxin. Curve 2 (●) = agarose-anti-digoxin. Curve 3 (△) = native anti-digoxin. Each tube contained 3.1 ng ^{3}H digoxin/ml which gave an initial count of 16,000 cpm/ml. Reproduced from Ref. 26.

C. Determination of Australian Antigen by Direct RIA

Hepatitis-associated antigen or Australian antigen is a virus associated with infectious hepatitis. The virus is most frequently transmitted by direct inoculation. In the past, transfused patients showed high incidences of the disease. This is probably due to the fact that donors came from a population with a significant percentage of chronic alcoholism or narcotic addiction. Blood banks are now required to screen their donors for the presence of HBAg. The diagnostic techniques for this virus are of recent vintage and include agar-gel diffusion (41), counter-immunoelectrophoresis (42), and competitive binding RIA (43). In 1972, Ling and Overby (3) developed a direct solid-phase RIA employing HBAg antisera adsorbed to polypropylene tubes as outlined in Section III-C-3. This determination is the one most frequently used to detect the virus and is available as a diagnostic kit (AUSRIA-Abbott). The test materials consist of an antiserum-coated polypropylene tube, purified ^{125}I-labeled anti-HBA immune globuline, standard normal human plasma, and HBAg-negative human plasma.

The antiserum is prepared by injecting guinea pigs with purified HBAg-positive donor serum and is purified by density gradient centrifugation. The antiserum described by Ling and Overby had an agar-gel diffusion titer of 1:512 vs the ad subtype of the antigen and 1:256 vs the ay subtype.

The antibody is purified from the antiserum by first complexing the antibody with purified HBAg. The insoluble complex formed is then washed thoroughly to remove other serum contaminants. It is then dissociated by incubation in acetate-formate buffer, pH 2.2. Separation of antigen from purified antibody is accomplished by density gradient centrifugation, after which it is neutralized. The purified antibody is then labeled with Na^{125}I by the chloramine T method of Hunter and Greenwood (56,57).

1. Method

One-tenth ml of the test sample is placed on the bottom of an assay tube. The tube is capped and allowed to stand at room temperature for 16 hours. It should be noted that a three-hour incubation is possible using a modification of the procedure described in Section III, C, 3. At the end of this time, the contents of the tubes are removed by aspiration and the tubes are washed with 1 to 5 ml aliquots of 0.01 M tris-HCl, pH 7.1. Then, 0.1 ml of the ^{125}I anti-HBAg immune globulin is added to each tube and is incubated for 90 minutes. The tube contents are removed by aspiration, each tube is washed, and the quantity of ^{125}I-labeled antibody incorporated into the sandwich is determined using a gamma ray spectrophotometer. Negative controls are run with standard HBAg negative control plasma described above.

The original procedure (3) recommends running 10 negative controls per determination. Sera are considered positive for HBAg if they had count rates higher than the average of the negative controls plus 5 standard deviations. A typical standard curve is shown in Fig. 11. In this case, purified HBAg was added to the negative control plasma to yield the concentrations on the abscissa.

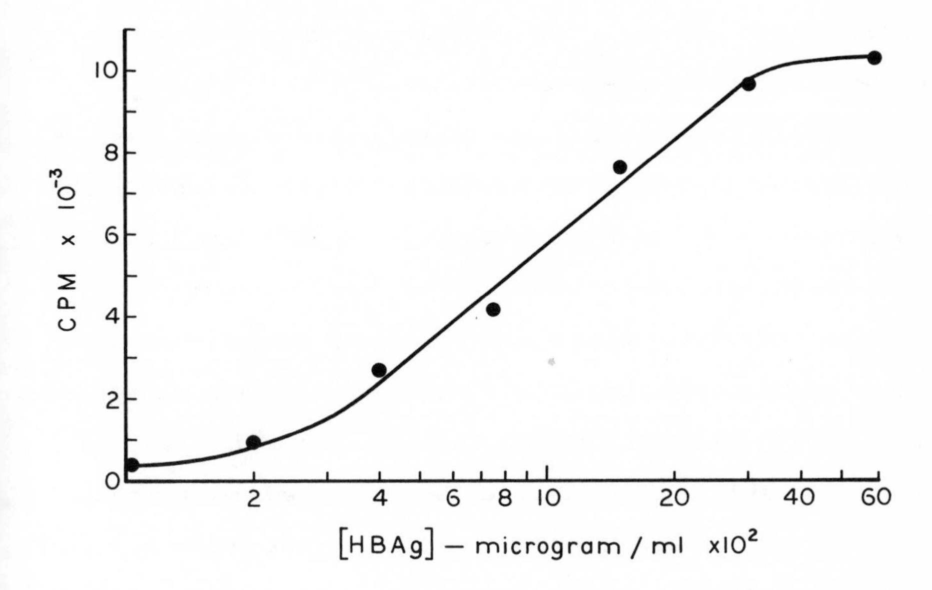

Fig. 11: Titration Curve of Anti-HBAg adsorbed on Polypropylene Tubes with Purified HBAg. The data was plotted from Table 1.

VII. CONCLUSION

The three methods described here are representative of solid-phase RIA techniques. Although several literature procedures employ antisera fixed by adsorption (1,2), competitive binding analysis was illustrated using the example of antisera covalently attached to macroporous carriers (25,26). The sandwich technique was illustrated with antisera adsorbed on plastic tubes (3), but could be employed with covalent immunoadsorbents (21,32).

Enzyme-linked immunogens have been employed in place of radiolabeling in both competitive binding (7,29) and direct immunoassays (4). In these cases, only the mode of detection changes. The majority of enzyme-linked methods employ solid-phase antisera.

The advantages of solid-phase antisera which have been discussed are mainly advantages of convenience. The potential use of an immunoadsorbent is predicated on a well-characterized antiserum. The difficult work of raising and characterizing an antiserum (with respect to tier, specificity, etc.) in a homogenious immunoassay must precede attachment to a carrier. In addition, the carrier itself and the carrier and attached serum from unimmunized animals must be tested to assure minimal background binding of labeled immunogen in every system. Covalent attachment of any protein to a carrier proceeds best from relatively concentrated protein solutions. In conducting feasibility studies, it is always best to load the system for relatively high percentages of attached protein at first and then work back to that initial load that yields the best coupling efficiency and activity retention. Thus, it becomes essential that the antiserum to be attached be available in ample supply. If it is not, the authors recommend that optimum conditions for attachment be ascertained with a relatively cheap commercial antiserum, e.g. anti-bovine albumin, anti IgG, etc., before application to expensive or difficult-to-obtain antisera.

If it is necessary to employ dilute antisera, one may still operate with concentrated protein solutions by diluting the antiserum with serum from unimmunized host animals.

Given the proper conditions, however, immunoadsorbents provide useful adjuncts in the performance of sensitized immunoassays. Their use permits a universal experimental design in performing these assays. Further, most of them are amenable to regeneration and reuse, which contributes to antibody conservation. Finally, many can be used in column procedures which may prove useful when some of the processes described are automated.

The brief review of solid phase LIA presented in this article is far from complete. However the authors overlooked two contributions, which will be of considerable interest to the reader and which therefore will be summarized here.

Goodfiend, et al (64) and Updike, et al (65) described solid-phase RIA procedures employing antisera entrapped in polyacrylamide gels. The RIA is carried out by incubating labelled and unlabelled antigens with the gel-entrapped antiserum contained in small disposable columns. After incubation, the gel is washed by elution of the column, and either phase may be counted (^{125}I-labelled antigens were employed). Updike, et al (65) employed columns containing dry gel-entrapped antibody, and it was possible, in many cases, to incorporate labelled hormone into the gel during the drying process. The authors (65) discuss the applicability of entrapped antibody systems to automated RIA.

Bolton and Hunter (66) covalently attached antisera to several steroid, peptide, and protein hormones to agarose, microcrystalline cellulose and Sephadex gels. They then compared the various solid-phase antisera with each other and with the soluble antisera to determine the effect of the matrix on titer retention and sensitivity to displacement of labelled antigen in competitive binding RIA.

REFERENCES

(1). K. Catt, and G. Tregar, Science, 58, 1570 (1967).
(2). K. Catt, U. S. Patent 3,652,671 (1972).
(3). C. Ling, and L. Overby, J. Immunology, 109, 834, (1972).
(4). E. Engvall, and P. Perlmann, J. Immunology, 109, 129, (1972).
(5). A. Hagiwara, Nature (Lond.) 202, 1019, (1964).
(6). M. Mitz, and L. Summaria, Nature (Lond.), 189, 576, (1961).
(7). A. Schuurs, and K. VanWeemen, U. S. Patent 3,654,090, (1972).
(8). J. Torresani, J. Immunological Methods, 3, 35, (1973).
(9). D. Campbell, E. Luescher, and L. Lerman, Proc. Natl. Acad. Sci. (U.S.), 37, 575, (1951).
(10). H. Isliker, Adv. Prot. Chem., 12, 387, (1957).
(11). L. Kent, and J. Slade, Biochem. Jour., 77, 12, (1960).
(12). A. Jagendorf, A. Patchnornik, and M. Sela, Biochim. et. Biophys. Acta, 78, 516, (1963).
(13). H. Weetall, and N. Weliky, Nature (Lond.), 204, 896, (1964).
(14). H. Weetall and N. Weliky, Biochim. et. Biophys. Acta, 107, 150, (1965).

(15). L. Wide, and J. Porath, Biochim. et. Biophys. Acta, 130, 257 (1966).
(16). L. Wide, R. Axen, and J. Porath, J. Immunochem., 4, 381, (1967).
(17). R. Axen, and J. Porath, Nature (Lond.), 210, 367, (1966).
(18). R. Axen, J. Porath, and S. Ernback, Nature (Lond.) 214, 1302, (1967).
(19). J. Porath, R. Axen, and S. Ernback, Nature (Lond.) 215, 1491, (1967).
(20). R. Axen, J. Porath, and L. Wide, U. S. Patent 3,555,143, (1971).
(21). R. Axen, J. Porath, and S. Ernback, U. S. Patent 3,645,852, (1972).
(22). P. Cuatrecasas, J. Agr. Food Chem., 19, 600, (1971).
(23). P. Cuatrecasas, and C. Anfinsen, Methods in Enzymology, XXII, 345, (1971).
(24). P. Cuatrecasas, Nature (Lond.), 228, 1327, (1970).
(25). S. Siegel, W. Line, N. Yang, A. Kwong, and C. Frank, J. Clin. End. Metab. 37, 526, (1973).
(26). W. Line, S. Siegel, A. Kwong, C. Frank, and R. Ernst, Clin. Chem. 19, 1361, (1973).
(27). R. Messing, and H. Weetall, U. S. Patent 3,519,538 (1970).
(28). H. Weetall, U. S. Patent 3,652,761 (1972).
(29). E. Ishikawa, J. Biochem. (Tokyo), 73, 1319, (1973).
(30). I. Silman, and E. Katchalski, Ann. Rev. Biochem. 35, pt. II, 873, (1966).
(31). N. Weliky, and H. Weetall, Immunochem., 2, 293, (1965).
(32). A. Gurvich, and G. Drizlikh, Nature (Lond.), 203, 648, (1964).
(33). R. Yalow, and S. Berson, J. Clin. Invest., 39, 1157, (1960).
(34). H. Gharib, W. Mayberry, and R. Ryan, J. Clin. End. Metab., 31, 709, (1970).
(35). B. Brown, R. Elkins, S. Ellis, and S. Reith, Nature (Lond.) 226, 359 (1970).
(36). C. Sekadde, W. Slaunwhite, and T. Aceto, Clin. Chem., 19, 1016 (1973).
(37). W. Butler, and P. Chen, Proc. Nat'l. Acad. Sci. (U.S.) 57, 71, (1966).
(38). T. Smith, W. Butler, and E. Haber, Biochem., 9, 331, (1970).
(39). T. Smith, V. Butler, and E. Haber, New England J. Med., 281, 1212, (1969).
(40). G. Oliver, B. Parker, D. Brasfield, and C. Parker, J. Clin. Invest. 47, 1035 (1968).
(41). B. Blumberg, A. Sutnick and W. London, Bull. N.Y. Acad. Med., 44, 1566, (1968).
(42). D. Gocke, and C. Howe, J. Immunol., 104, 1031, (1970).

(43). J. Walsh, R. Yalow, and S. Berson, J. Infect. Dis., 121, 550, (1970).
(44). P. Nakane, and B. Pierce, J. Cell. Biol., 33, 307, (1967).
(45). G. Grodsky, and P. Forsham, J. Clin. Invest., 40, 799, (1960).
(46). S. Genuth, L. Frohman, and H. Lebovitz, J. Clin. Endocr., 25, 1043, (1965).
(47). Y. Levin, M. Pecht, L. Goldstein, and E. Katchalski, Biochem., 3, 1905, (1964).
(48). L. Goldstein, Biochem. et Biophys, Acta, 315, 1, (1973).
(49). J. Inman, and H. Dinitzis, Biochem., 8, 4074, (1969)
(50). J. Turkova, O. Hubalkova, M. Krivakova, and J. Coupek, Biochim. et Biophys. Acta 322, 1, (1973).
(51). R. Axen, and S. Ernback, Eur. J. Biochem., 18, 351, (1971).
(52). L. Ahgren, L. Kagedal, and S. Akerstrom, Act. Chem. Scand., 26, 285, (1972).
(53). O. Zaborsky, "Immobilized Enzymes", CRC Press, Cleveland, O., (1973).
(54). W. Line, A. Kwong, and H. Weetall, Biochim. et Biophys. Acta 242, 194, (1971).
(55). H. Horinishi, Y. Hachimori, and K. Shibata, Biochim. et Biophys. Acta, 86, 477, (1964).
(56). W. Hunter, and F. Greenwood, Nature (Lond.) 194, 495, (1962).
(57). F. Greenwood, W. Hunter, and J. Glover, Biochem. J., 89, 114, (1963).
(58). Y. Miyachi, J. Vaitukaitis, E. Nieschlag, and M. Lipsett, J. Clin. End., 34, 23, (1972).
(59). J. Marchalonis, Biochem., J. 113, 299, (1969).
(60). R. Price, and G. Means, Jour. Biol. Chem. 246, 831, (1971).
(61). G. Means and R. Feeney, Biochem., 7, 2191 (1968).
(62). D. Rubenstein, R. Schneider, and E. Ullman, Biochem. Biophys. Res. Comm., 47, 846, (1972).
(63). D. Fisher, J. Pediatrics, 82, 1, (1973).
(64). Goodfriend, T., Ball, D., and Updike, S., Immunochem., 6, 484, (1969).
(65). Updike, S., Simmons, J., Grant, D., Magnuson, J., and Goodfriend, T., Clin. Chem., 19, 1339, (1973).
(66). Bolton, A., and Hunter, W., Biochim. et Biophys. Acta, 329, 318, (1973).

Chapter 10

SOLID-PHASE SEQUENCING OF PEPTIDES AND PROTEINS

Richard A. Laursen

Department of Chemistry
Boston University
Boston, Massachusetts 02215

I. INTRODUCTION

The Edman degration (1), in which amino acids are cleaved stepwise from the N-terminus of a peptide of protein, has proved to be the single most useful chemical tool for protein sequence analysis. Since its introduction nearly 25 years ago, this procedure has undergone almost continual modification (2-4), culminating in the development of two automated versions (5,6).

The first step (coupling) of the Edman degradation (Fig. 1) involves reacting the peptide or protein N-terminal with phenylisothiocyanate at alkaline pH. After removing excess reagents, the resulting phenylthiocarbamyl peptide is treated with acid, which causes cyclization and splitting off of the first amino acid as a thiazolinone (cleavage step). After separation from the peptide, the thiazolinone is converted in aqueous acid to a phenylthiohydantoin, which can be identified by one of a number of procedures that will be discussed later. The residual peptide is then subjected to additional cycles of the Edman degradation.

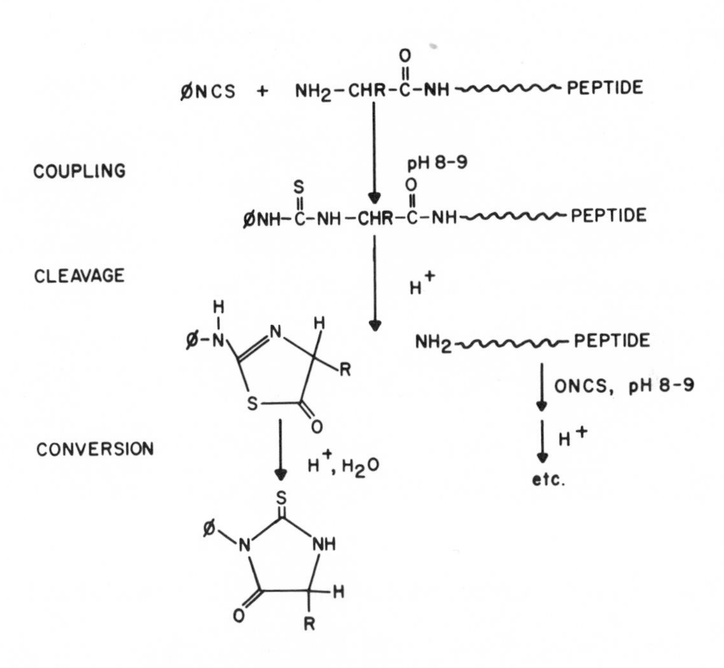

Fig. 1: Edman degradation.

In principle, one could perform an indefinite number of degradative cycles, thus sequencing an entire protein. In practice, this is not possible because of incomplete reactions, side reactions, and mechanical losses of materials.

In 1967, Edman and Begg (5) described the automatic protein 'sequenator' that minimized many of these problems and permitted degradations through as many as 80 cycles.

In the sequenator, the protein is deposited in a spinning cup as a thin film. Reagents and solvents flow into the cup, over the protein, and out the top of the cup to provide efficient extraction of excess reagents and of the thiazolinones. This efficient extraction becomes a problem, however, when sequencing peptides, because the peptides themselves are often extracted from the cup. Following the success of Merrifield (7) in solid-phase peptide synthesis, the solid-phase version of the Edman degradation was devised (8-10) for peptides that are not adequately handled by the protein sequenator. In this modification, peptides are anchored at or near the C-terminus to an insoluble resin, and degradation is carried out as usual. Excess reagents and the liberated thiazolinone are removed from the peptide simply by washing the resin. Extraction losses are eliminated because the peptide is bound to the resin. The solid-phase procedure has also been automated (6).

For the reader's convenience, the following abbreviations are used in this chapter: 1,2-Dichloroethane, DCE; Dimethylformamide, DMF; Phenylthiohydantoin, PTH; Phenylisothiocyanate, PITC; Triethylenetetramine, TETA; and Trifluoroacetic acid, TFA.

II. METHODS

A. Reagents

Sequencing reagents from a variety of suppliers have

been used with no noticeable difference in results. Unless otherwise state, the lowest-priced analytical-grade reagent is used without further purification. Following is a list of reagents and some suppliers whose products have given satisfactory results: Methanol (Baker and Adamson, absolute; Fisher, E. Merck); 1,2-Dichloroethane (Matheson, Coleman and Bell; Fisher; E. Merck); Trifluoroacetic acid (Eastman; E.Merck): Methyl Isothiocyanate (Eastman); _t_-Butyloxycarbonyl azide (Pierce); Phenyl isothiocyanate (Eastman); Acetonitrile (Matheson, Coleman and Bell; Fisher; E. Merck); Pyridine (Fisher; E. Merck); N-Methylmorpholine (Pierce, Sequenal grade); N-Dimethylaminopropyl-N'-ethyl carbodiimide hydrochloride (Pierce; Aldrich); Triethylamine (Pierce, Sequenal grade); and Hydrazine (Eastman).

_p_-Phenylene diisothiocyanate (Eastman) is used directly if obtained as pale yellow crystals. If the reagent is orange, it is recrystallized from acetic acid. Dimethylformamide (Fisher; E. Merck) is distilled at 50°C/20mm Hg and stored over Linde Type 3A molecular sieves in Hypo-Vials (Pierce).

The sequencing buffer is a 3:2 mixture of pyridine and N-methylmorpholinium trifluoroacetate (pH 8.1). The latter is prepared by diluting a mixture of 28 ml of N-methylmorpholine and 3.75 ml of trifluoroacetic acid with water to 200 ml. The resulting solution is 1 _M_ in N-methylmorpholine and 0.25 _M_ in the conjugate acid.

Radioactive phenyl isothiocyanate (^{3}H-phenyl isothiocyanate) is purchased (Malinkrodt) or is synthesized (11) from tritiated aniline (New England Nuclear) diluted with unlabelled phenyl isothiocyanate and distilled at $90^{o}C$/10 mm Hg. The specific activity is determined by converting a sample to leucine phenylthiohydantoin, which is recrystallized and counted. Specific activities of about 60 dpm/nanomole are used in most experiments. A 5% solution of phenyl isothiocyanate in dry acetonitrile is used for degradations.

Carbonyldiimidazole is synthesized by the method of Paul and Anderson (12) or is purchased (Pierce). Stock solutions of carbonyldiimidazole (usually 80 mg/ml) in anhydrous DMF are kept in Hypo-Vials (Pierce). The titer of the solution can be checked periodically by adding aliquots to excess aniline, diluting the mixture with water, and weighing the diphenylurea formed. Protected from moisture, solutions are stable for several months.

Polystyrene beads (BioBeads SX-1, minus 400 mesh) are purchased from BioRad Laboratories. Before use, the beads are washed several times with benzene, chloroform, dioxane and, finally, methanol to remove low-molecular-weight polymers.

Glass beads (200-235 mesh, Microbeads Division of Cataphote Corporation, Jackson, Miss.) are washed with petroleum ether and then heated in concentrated HCl for about two hours to dissolve metal particles. The beads are washed thoroughly with water on a fritted glass filter and are then dried at 100°C.

B. Synthesis of Peptide Supports

1. Chloromethylpolystyrene (6)

A stirred suspension of 27 g of dioxane-washed-styrene-1% divinylbenzene copolymer (minus 400 mesh; Bio-Rad Laboratories Biobeads SX1) in 160 g of chloromethyl methyl ether is cooled to 0°C in an ice-salt bath. (Note: chloromethyl methyl ether is a potent carcinogen and should be handled only in an efficient hood.) A solution of 3.0 ml of anhydrous $SnCl_4$ in 40 g of chloromethyl methyl ether is added dropwise over a period of 15 minutes more, and the resin is filtered on a fritted glass filter. The resin is washed with 400 ml of dioxane-water (3:1), then alternately with dioxane and water until the washings give no precipitate with $AgNO_3$, and finally with methanol. The resin is dried under vacuum at 100°C; yield, 25.5 g.

The product is analyzed for chlorine by heating about 150 mg of the resin in 2 ml of pyridine on the steam bath for 15 minutes, diluting the mixture with 50 ml of water, acidifying with concentrated HNO_3, and titrating chloride ions potentiometrically. Found: 1.23 mequiv/g.

2. Ethylenediamine Resin

A 5.0-g portion of chloromethylpolystyrene is added to 15 ml of ethylenediamine (Fisher). The mixture is stirred at room temperature for 15 minutes and then heated in a steam bath for 30 minutes. The resin is filtered off and washed thoroughly with water. It is heated for 15 minutes in hot water containing about 5% triethylamine before being washed with water and, finally, methanol. The resin is dried under vacuum at 60°C overnight.

3. Triethylenetetramine (TETA) Sequencing Resin (13)

Ten grams of chloromethyl polystyrene is stirred with 125 ml of triethylenetetramine (Fisher) for 30 minutes room temperature. The mixture is then heated in a steam bath for 1.5 hours. The resin is filtered, washed thoroughly with methanol, stirred with 20 ml of triethyl-amine, and filtered. After washing with copious amounts of methanol, water, and again with methanol, the resin is dried under vacuum at 60°C overnight; yield, 11.5 g of white resin. The resin is stored in small batches under nitrogen in a freezer.

4. Aminopolystyrene

Keeping the temperature below 2°C, a 50-g sample of polystyrene (BioBeads SX1, minus 400 mesh) is added slowly with stirring over a period of about 15 minutes to 600 ml of fuming nitric acid (90%) that was previously cooled to 0°C. Stirring is continued for one hour at 0°C. The mixture is poured onto about 2000 ml of ice, and the resin is separated on a fritted glass filter and alternately washed several times with dioxane and water and finally with methanol. After drying, the nitro resin weighed 59 g.

A stirred suspension of 56 g of the nitro resin with 1000 ml of dimethylformamide is heated to 75°C, and a solution of 420 g of $SnCl_2 \cdot 2H_2O$ in 350 ml of warm dimethylformamide is added slowly (exothermic reaction). The temperature is raised and maintained at 140 to 150°C for 15 minutes, after which the mixture is cooled and 350 ml of concentrated HCl is added. Heating is continued at about 100°C for one hour. The resin is separated on a fritted glass filter and is washed thoroughly with water and HCl to remove tin salts. Neutralization is effected by washing the resin with pyridine-water (1:1) followed by dimethylformamide-triethylamine (3:1) until no chloride ions are detected in the washings. After final washes with water and methanol, the yellow-brown product is dried at 60°C under vacuum; yield 44 g. The resin is stored in a refrigerator.

5. Aminopropyl Glass

Porous glass beads (BioRad Laboratories BioGlass 200, 500, 1000, etc.) are used as received without further drying. The beads are placed in a flask with twice their volume of a 4% solution of γ-aminopropyltriethoxysilane (Pierce) in dry acetone. After the flask is evacuated in a vacuum desiccator for a few minutes to remove air from the pores, it is stoppered, and heated at 50°C for 24 hours. During heating, the flask is shaken occasionally. The resin is filtered off and washed several times alternately with water and methanol and is finally dried under vacuum. Different batches typically contain 40 to 100 micromoles of amino groups per gram of resin.

6. Isothiocyanato Glass (14)

A sample of aminopropyl glass is added in small portions to a solution of *p*-phenylene diisothiocyanate (25 moles per mole of amino groups) in dimethylformamide (2-3 volumes per volume of glass) over an hour with slight stirring (vigorous stirring will grind up the glass). The mixture is allowed to stand for two more hours at room temperature and is washed on a fritted glass filter with dimethylformamide and methanol. The product is dried under vacuum and stored in a freezer.

C. Estimation of Amino Groups on Amino Glass Supports

This assay depends on alkylation of the amino groups with trinitrobenzenesulfonic acid and hydrolysis of the trinitrophenylamino resin with KOH, which not only cleaves the Si-O bonds with the glass, but also converts the trinitrophenyl group to picric acid. Under controlled conditions, the procedure is reproducible to within a few percent, although absolute quantitation may be somewhat in error because of incomplete conversion to picric acid.

A 50-mg sample of amino glass is mixed with 15 mg of trinitrobenzenesulfonic acid in 1 ml of 0.1 *M* sodium borate (pH 9) by rotation in a small test tube for 1.5 hours. The resin is then washed three times by centrifugation with 8-ml portions of water. To the moist resin is added one ml of 2 M KOH. The tube is kept at 45°C for 60 minutes and is shaken occasionally. The mixture is then diluted with an appropriate amount of water and the absorbance at 358 nm is measured. The number of amino groups is calculated using a molar extinction coefficient for picric acid of 14,100 at alkaline pH. Typical loadings are 20 to 150 micromoles of amino groups per gram of amino glass.

D. Resin Washing and Manipulation

Because of the small quantities of peptide generally available for sequence work, it is important that manipulations be kept to a minimum to prevent losses. Although a resin washing device has been described (6), it is more convenient to carry out all attachment procedures in common 13 x 100-mm test tubes. After peptide attachment, the resin is washed simply by centrifugation and pipetting off the supernatant solvent.

E. Attachment of Peptides to Resins

1. N-terminal Blocking (6)

A sample (up to about 1 micromole) of salt-free peptide dissolved in triethylammonium bicarbonate buffer or dilute triethylamine is placed in a 13 x 100-mm test tube and lyophilized or evaporated in a vacuum desiccator. To the residue is added 0.5 ml of dimethylformamide, 0.05 ml of N-methylmorpholine, and 0.075 ml of *t*-butyloxycarbonyl azide. The mixture is stirred using a small magnetic stirring bar until solution is complete. In cases where peptide does not dissolve readily, up to 0.1 ml of water may be added. The tube is covered with Parafilm and heated at 50°C for at least 5 hours, after which the solvent and excess reagents are removed by evaporation in a vacuum desiccator using a good vacuum pump. Evacuation of the desiccator should be cautious at first to prevent bumping. When the tube is dry, 0.5 ml more of dimethylformamide is added and evaporated. Drying under vacuum is continued for 10 hours or more at room temperature.

2. Attachment of Peptides Using Carbonyldiimidazole (6)

To a dry sample (up to 500 nmole) of BOC-peptide in a 13 x 100-mm test tube is added a solution of 8 mg of carbonyldiimidazole in 0.2 ml of dry dimethylformamide. The tube is quickly flushed with dry nitrogen and carefully sealed with Parafilm. The mixture is stirred at room temperature for 30 minutes and 2 μl of water are added to destroy excess carbonyldiimidazole. After 10 minutes, the mixture is transferred to a second test tube containing 50 mg of TETA resin pre-swollen in 0.2 ml of dimethylformamide. The first tube is washed with 0.1 ml of dimethylformamide. The resin mixture is stirred at room temperature for at least two hours. Excess amino groups are then blocked as described next.

3. Attachment of Peptides by the Carbodiimide Method (15)

The peptide (50-250 nmole) is treated with BOC-azide as described earlier to mask the amino function, and is then dissolved in 0.2 ml of dimethylformamide containg 1 to 2 μl of N-ethylmorpholine. N-ethyl-N'-dimethylaminopropyl carbodiimide hydrochloride (5 mg) is added, and the reaction mixture is allowed to stand at 40°C. After 90 minutes (or after 30 minutes at 50°C), 50 mg of TETA-sequencing resin is added and the suspension is maintained at 40°C for at least three hours. Excess resin amino groups are blocked as described below.

4. Attachment of Peptides Using p-Phenylenediisothiocyanate (16)

The peptide sample (50 to 250 nmole) is evaporated to dryness in a 13 x 100-mm test tube, is redissolved in 0.5 ml of 5% triethylamine, and is evaporated again to remove traces of ammonia. The residue is then dissolved in 0.05 ml of N-methylmorpholine-water (1:1, adjusted to pH 9.5 with CF_3COOH). To this is added 1.0 mg (5 μmole) of *p*-phenylene diisothiocyanate in 0.1 ml of dimethylformamide. The solution is heated at 45°C for 30 minutes and is transferred with a Pasteur pipette to a second 13 x 100-mm test tube containing 35 mg of peptide resin (aminopolystyrene) previously swollen in 0.25 ml of dimethylformamide. The first tube is washed with 0.1 ml of dimethylformamide which is also added to the resin.

The resin mixture is then treated with methyl isothiocyanate to block resin amino groups as described in the next section.

5. Hydrazinolysis of arginine peptides (16)

A peptide sample in a 13 x 100-mm test tube is heated in an oil bath at 105°C for 30 minutes in 0.50 ml of 20% hydrazine. Vapors are condensed by circulating water through thin-walled rubber tubing wrapped around the upper portion of the tube. The solution is then evaporated on a rotary evaporator. Traces of hydrazine are removed by coevaporation twice with 0.5 ml of 5% aqueous N-methylmorpholine and finally by evacuation in a desiccator over P_2O_5 using an efficient vacuum pump. The residue is then attached to the peptide resin by the dissothiocyanate method described earlier.

6. Attachment of Peptides Using the Homoserine Lactone Method

A solution of the C-terminal homoserine peptide (usually 20 to 200 nanomoles) is evaporated to dryness in a small vial or test tube, and the residue is redissolved in 1.0 ml of anhydrous trifluoroacetic acid. The solution is kept at room temperature for one hour, and the solvent is evaporated over KOH in a vacuum desiccator. The peptide sample is dissolved in 0.1 to 0.15 ml of dimethylformamide, and the solution is added to a 13 x 100-mm test tube containing 40 mg of TETA sequencing resin previously swolled in 0.2 ml of dimethylformamide. The first tube is washed with 0.1 ml of dimethylformamide, which is added to the resin mixture. (In cases where the peptide is poorly soluble in dimethylformamide, up to 20% of the total dimethylformamide volume of water can be added to promote solution.) An 0.05-ml aliquot of triethylamine is added to the resin mixture, which is then stirred at 45°C for two hours. Excess amino groups on the resin are blocked as described in the following section.

7. Blocking of excess resin amino groups

To the peptide attachment mixture, containing resin (up to 50 mg), peptide, and activating agent, is added 0.1 ml of $CH_3NCS-CH_3CN$ (1:1,v/v) and 0.1 ml of N-methylmorpholine-dimethylformamide (1:1, v/v). The mixture is stirred at room temperature for 75 minutes, and the resin is washed by centrifugation three times with 8-ml volumes of methanol, the solvent being removed with a Pasteur pipette after each washing. The damp resin is then placed in the original test tube in a vacuum desiccator at water aspirator pressure until dry (30 to 60 minutes). At this point, the resin is either stored in a refrigerator or mixed with glass beads and packed into a reaction column for sequencing.

8. Attachment of Cytochrome c to Isothiocyanato Glass (17)

A 5-mg (400 nanomole) sample of *Candida krusei* cytochrome *c* is dissolved in 0.3 ml of water and 0.3 ml of pyridine is added. The red solution is shaken with 200 mg of isothiocyanato glass for two minutes, during which time the red color disappears from solution. Ethanolamine (0.1 ml) is then allowed to react with the resin to block excess NCS groups. The loaded resin is filtered, washed with methanol, and dried under vacuum.

F. Packing of Reaction Columns

The reaction column consists of an approximately 240-mm-long glass tube with a 2.5-mm internal diameter. A 120-mm section of the column is thermostatted by circulating warm water (6). This is illustrated in Fig. 4.

The column is packed by attaching the bottom fitting and then loosely clamping the column in a vertical position. Using a small funnel, the tube is filled with glass beads to the level of the thermostatted region. A mixture of resin (50 mg maximum) and glass beads (about 900-1000 mg) is then added to fill the heated section of the column. The funnel and tube which originally held the resin are 'rinsed' with two small portions of glass beads, and the column is filled to about 1 cm from the top with beads. During the pouring procedure, the column is packed by tapping it on the benchtop. It is very important that the column be uniformly packed and that no bands of resin form between layers of glass beads, otherwise resin swelling in certain solvents will block the column. To insure uniformity, the top of the packed column is covered with Parafilm and the column is inverted several times with tapping to mix the glass and resin beads. Finally, when no discrete resin bands are visible, the column is packed down and filled to the top with glass beads. The top fitting is then screwed on and the column is placed in the sequencer.

G. Conversion of Thiazolinones to Phenylthiohydantoins

Thiazolinones are collected in the sequencer in 18 x 100-mm test tubes. Each tube contains about 1.5 ml of trifluoroacetic acid and about 3 ml of methanol, which is evaporated in a water bath at $40^{\circ}C$ under a stream of nitrogen. To each tube is added 0.2 ml of 20% trifluoroacetic acid in water. After the tubes are heated at $80^{\circ}C$ for 10 minutes, they are cooled and the contents diluted with about 1 ml of methanol. Each solution is applied to a column (0.5 x 1.0 cm) of Dowex-50-X2 packed in a plugged Pasteur pipette. The resin is washed further with 3 x 1 ml of methanol, and the washings are collected in 10-ml centrifuge tubes and evaporated under nitrogen in a $40^{\circ}C$ bath.

H. Thin-Layer Chromatography of Phenylthiohydantoins

The PTH residues are dissolved in 50 l of acetonitrile and part of the solution (usually 30%) is spotted with appropriate standards on silica gel-coated aluminum plates containing a fluorescent indicator (E. Merck). The plate is eluted with one of three solvents: (I) chloroform-ethanol (98:2); (II) chloroform-methanol (9:1); or (III) chloroform-methanol (7:3). Solvent I elutes, in decreasing order of elution, the PTHs of Pro, Leu, Ile, Val, α-Thr, Phe, Met, Ala, Trp, Gly, PTC-Lys, and Tyr. The PTHs of aspartic and glutamic acid methyl esters, formed on the Dowex-50 column, elute near Ala. All the other derivatives remain at the origin and can be eluted with solvent II which produces, in decreasing order of elution, Thr, MTC-Lys, Hse, Ser, Gln, Asn, Glu, Asp, CM-Cys, and CysOH. For analysis of Arg and His, the Dowex-50 columns are washed with 5 ml of 6 N HCl and the washings are collected and evaporated. These derivatives can be resolved in Solvent III, although they are sometimes obscured by other uv-absorbing impurities leached off the resin. For quantitation, in cases where ^{3}H-PITC is used, a grid is drawn on the plate using the reference PTHs as markers, the plate is cut into pieces, and the spots counted in a toluene-ethanol scintillation solution (18).

III. DISCUSSION

A. General Considerations

The solid-phase Edman degradation is illustrated in Fig. 2. The N-protected peptide is first attached to a resin-bearing amino group using a peptide bond forming agent. After removal of the amino protecting group, the peptide is degraded as usual, the excess reagents and thiazolinone being removed by filtration. As discussed in

Section III.E.4, peptides can also be attached by activation of lysine ε-amino groups. Certain attachment methods do not require prior blocking of peptide amino groups.

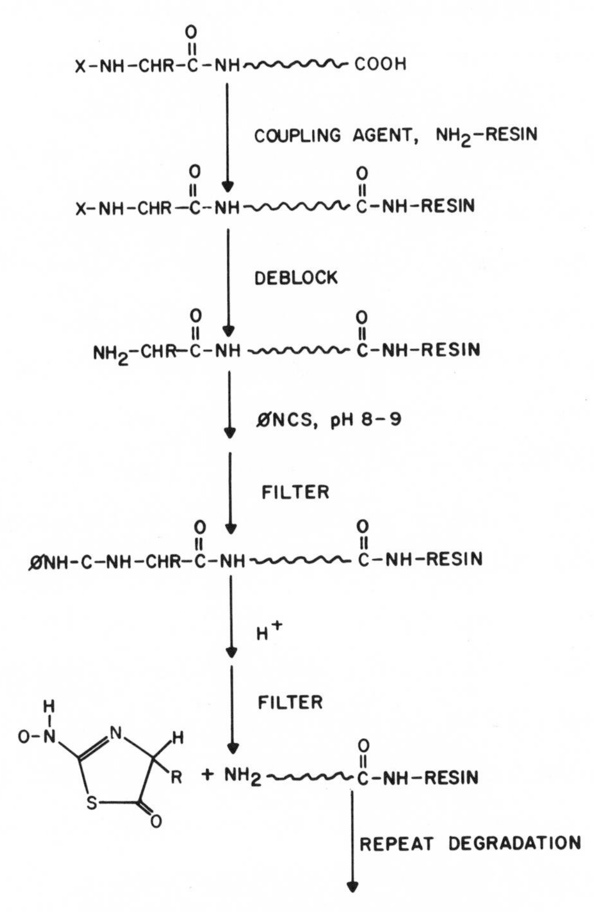

Fig. 2: Solid-phase Edman degradation.

Protection of the N-terminal amino acid is essential when the peptide is to be attached by carboxyl activation, otherwise blocking of the amino group may occur. This is particularly true when carbonyldiimidazole (12) is used, since this reagent rapidly forms ureas with amines. With carbodiimides, polymerization may occur, although Schellenberger et al (19) have reported success in attaching and sequencing a peptide without amino blocking. It may be that the excess resin amino groups compete more favorably than peptide aminos for the carbonyl groups.

We have found *t*-butyloxycarbonyl azide (BOC azide) to be most useful for blocking amino groups because its volatility permits easy removal of excess reagent before activation of the peptide. The BOC group is also easily removed from the peptide by trifluoroacetic acid under the same conditions used in the Edman degradation. Blocking is usually carried out in dimethylformamide in the presence of a tertiary amine such as N-methylmorpholine. In some cases, adding 10 to 20% water is necessary to promote solution of the peptide in the organic solvent. If lysine, whose side chain amino has a higher pK_a than an α-amino group, is present in the peptide, it is advisable to use triethylamine (pK_a 10.9) instead of N-methylmorpholine (pK_a 7.4). The resulting BOC-peptides are usually soluble in dimethylformamide.

B. Construction of the Sequencer

The construction of the solid-phase sequencer has been described in detail elsewhere (6) and will be discussed here only in general. Recent improvements include an improved programmer and a dual column system that allows simultaneous degradation of two peptides.

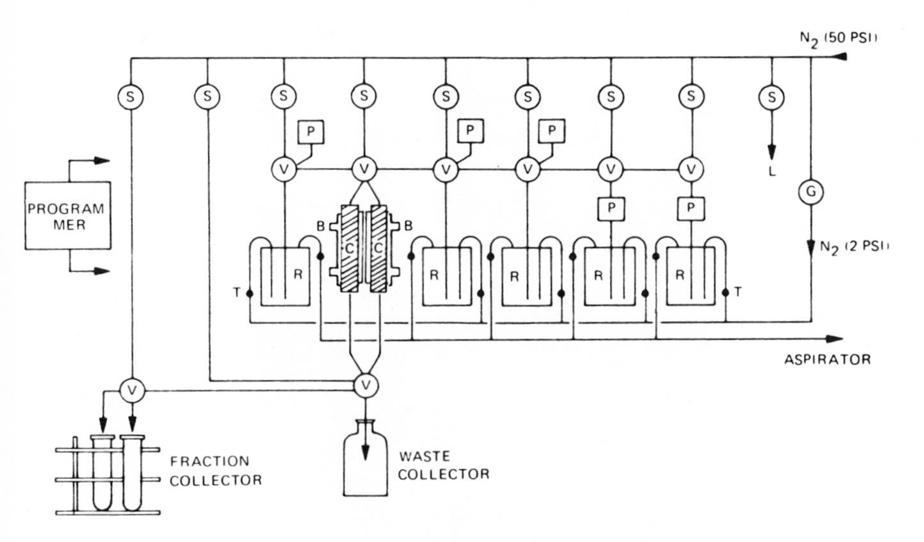

Fig. 3: Schematic diagram of the Sequemat Model 12 peptide sequencer; B, water bath; C, reaction column, G reducing valve; L, Auxiliary nitrogen line; P, pumps; R, reagent and solvent reservoirs; S, solenoid valves; T, stopcocks; V, reagent and solvent valves.

The construction of the Sequemat Model 12 (Sequemat, Inc.; Watertown, Mass.) is shown schematically in Fig. 3. The reagents (phenyl isothiocyanate, trifluoroacetic acid and buffer) and solvents (methanol and 1,2-dichloroethane) are pumped through a series of valves into reaction columns, so arranged that cleavage with TFA in one column can occur simultaneously with PITC coupling in the other. Effluent from the columns is directed either to a waste bottle or to the fraction collector. A photography of the Sequemat (Fig. 4) shows the physical arrangement of components. The reaction columns are maintained at a temperature of about 45°C by circulating water.

Fig. 4: Photograph of the Sequemat sequencer.

C. Programming

In the original sequencer (6), the valves and pumps were controlled by a disk programmer. In more recent models, this has been replaced by a 12-channel punched type programmer (Industrial Timer Co., Model 212), which is more flexible and permits easy change-over between single- and dual-column operation. Two sample programs are shown in Figs. 5 and 6. The length of the program is governed principally by the reaction times in the PITC-coupling and TFA-cleavage steps, which are about 15 minutes each. Approximately 15 minutes must be added to each time to allow for filling the column and the void volume between the valves and the column. In single column systems, cycle times of 90 minutes (16 cycles per day) have been achieved using the pumping rates indicated in Fig. 6. Shorter times should be possible by increasing the pump rates to reduce the time needed to fill the columns.

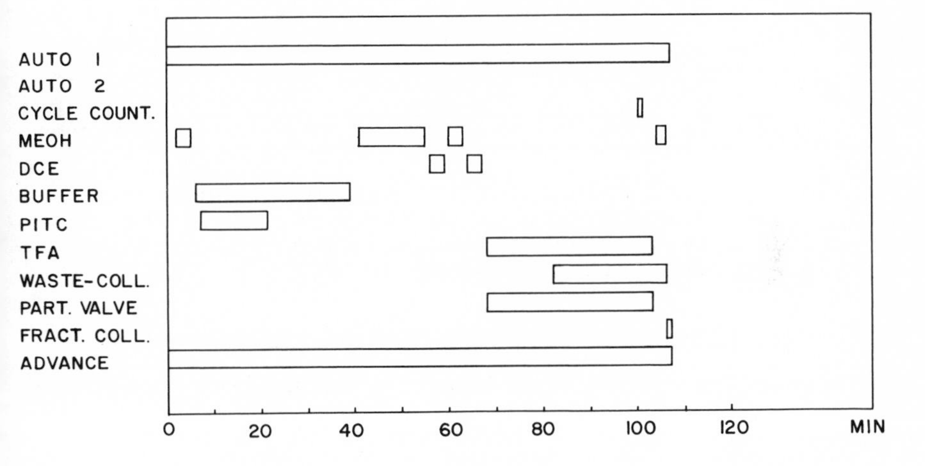

Fig. 5: Single column program for the Sequemat Sequencer. Pumping rates: MeOH and DCE, 1 ml/min; buffer and TFA, 0.057 ml/min; PITC 0.028 ml/min. Operations are controlled by a 12-channel tape programmer.

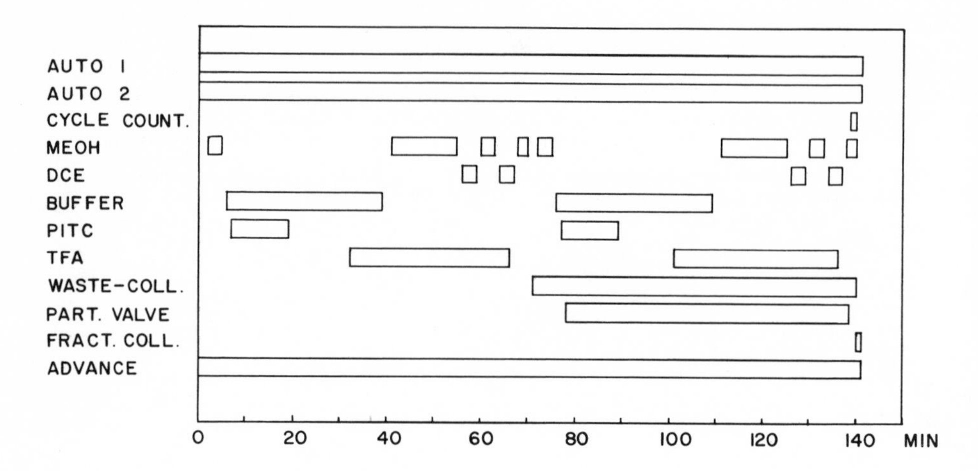

Fig. 6: Dual column for the Sequemat sequencer. Pumping rates are the same as in Figure 5. With the dual column system, cleavage in one column and coupling in the other occur simultaneously.

D. Synthesis and Properties of Resins

Sequencing resins must have two essential properties: accessibility of peptides and reagents to the reactive functional groups, and chemical and mechanical stability to conditions of the Edman degradation. Unfortunately, some of these properties are mutually incompatible, so a compromise must be made.

Most of the resins described for peptide sequencing have been derived from cross-linked polystyrene, largely because it is easy to obtain and modify. Historically, the first resin used for sequencing was the ethylene diamine resin 1 (Fig. 7) to which peptides could be attached rather efficiently (7). It has been found, however, that degradation of peptides occurs only with appreciable 'overlap' or carry-over of amino acids from one cycle to the next (6). This is due to incomplete reaction in the trifluoroacetic acid cleavage step, because of poor penetration of the trifluoroacetic acid into the hydrophobic resin. This is reflected in the swelling properties of resin 1, which swells readily in a dimethylformamide and pyridine used in peptide attachment and the coupling step of the Edman degradation, but hardly at all in trifluoroacetic acid (6). Addition of aryl amino groups, as in resin 2, gives a resin that swells nearly five-fold in trifluoroacetic acid, presumably because protonation of the aromatic amino group results in delocalization of the charge into the ring and a concommitant increase in polarity of the resin.

During development of the diisothiocyanate peptide attachment procedure (16), it was found that the simple aminopolystyrene 3 was just as satisfactory as resin 2 for attachment. Since 3 is much easier to make than 2, it is the resin of choice in this case. It is not, however, satisfactory for attachment *via* carboxyl activation methods, presumably because of the relatively poor nucleophilicity of aryl amines.

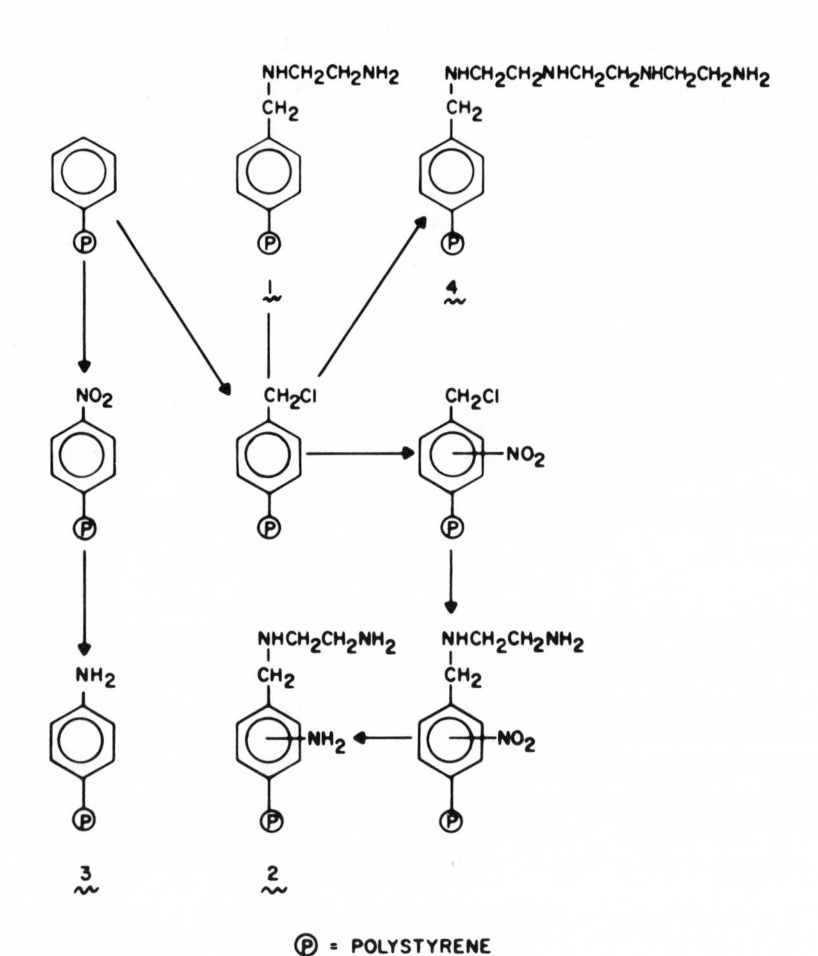

Fig. 7: Synthesis of sequencing resins from cross-linked polystyrene.

Continued evaluation has resulted in triethylenetetramine resin 4 (13), which has all the desirable properties of 2, but is easier to prepare. In this case, the additional aliphatic amino groups located away from the polystyrene backbone are more accessible to reagents;

in contrast with resin 1, overlap is not observed. This resin is quite good for attachment _via_ carboxyl activation, but is unsatisfactory for diisothiocyanate coupling. For unknown reasons, triethylenetetramine resin 4 has a tendency to become ineffective for attaching peptides after a few months. However, its stability is improved by storage in small batches under nitrogen in a freezer.

Other factors which affect reactivity of resins are degree of substitution and cross-linkage. Most of the resins used for sequencing have been derived from polystyrene crosslinked with 1% of divinylbenzene, more highly cross-linked resins being less permeable to reagents. Mross (20) has successfully used derivatives of 2% cross-linked polystyrene. To facilitate resin washing, very fine (minus 400 mesh) particles have been used, although 200-400 mesh resins are probably also satisfactory. Resins 1, 2 and 4 have been synthesized from polystyrene that is only about 15% chloromethylated, since further substitution results in increased cross-linkage by methylene bridges (21).

One of the problems with polystyrene-derived resins is that they swell appreciably in certain solvents (pyridine, trifluoroacetic acid) but not in others (methanol). Because of this, the resins cannot be used directly in reaction columns without the danger of solvent channeling or column blockage. This difficulty has been overcome (6) by diluting the resin with a 20-fold excess of glass beads, which allows swelling to occur in the interstices between the beads. It would be desirable, however, to have an incompressible resin with reactive functional groups on the surface. Possibilities include pellicular (22) and macroporous resins (23), Kel-F-polystyrene graft copolymers (24) and silylated porous glass beads (See III. G); some of these are under investigation.

E. Peptide Attachment Procedures

1. Carbonyldiimidazole

The first reagent to be used for attaching peptides to resins (6,7) was carbonyldiimidazole, which was chosen because of its reported (12) high yields and absence of side reactions. During activation by this reagent, the N-blocked peptide is converted to an imidazolide which in turn reacts with nucleophiles (Fig. 8). Since carbonyldiimidazole reacts readily with water and other nucleophiles, reactions must be carried out in anhydrous solvents, such as dimethylformamide which has been freed from water and dimethylamine by distillation from P_2O_5. In practice, a 50- to 100-fold (usually about 75 mg/ml) excess of reagent is used to scavenge traces of water. The excess must be removed before adding the resin to the activation mixture, because carbonyldiimidazole will react with resin amino groups to form ureas, thus reducing the number of attachment sites. This is done conveniently by adding a small amount of water (approximately one mg per mg of carbonyldiimidazole) at the end of the activation period. This destroys the excess reagent, without affecting the imidazolide, which is stable, for short periods, in contact with water (12).

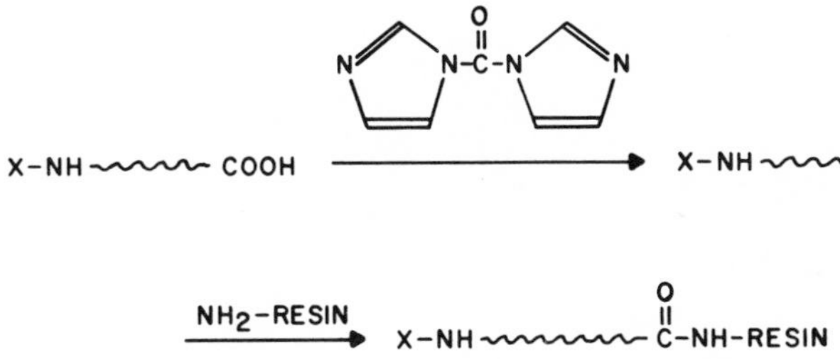

X-NH~~~~COOH → (carbonyldiimidazole) → X-NH~~~~C(=O)-N(imidazole)

$\xrightarrow{NH_2\text{-RESIN}}$ X-NH~~~~C(=O)-NH-RESIN

Fig. 8: Carbodiimidazole attachment procedure.

Attachment yields by this procedure are fairly good (80 to 100%). However, there is a major problem in that the side chain carboxyls of aspartic and glutamic acid also become activated. In the latter case, the side chain becomes attached to the resin, and a phenylthiohydantoin is not detected for Glu. In the former case, Edman degradation stops completely at aspartic acid, presumably because of cyclic imide formation (Fig. 9), which prevents further degradation. Because of these shortcomings, other attachment procedures have been sought.

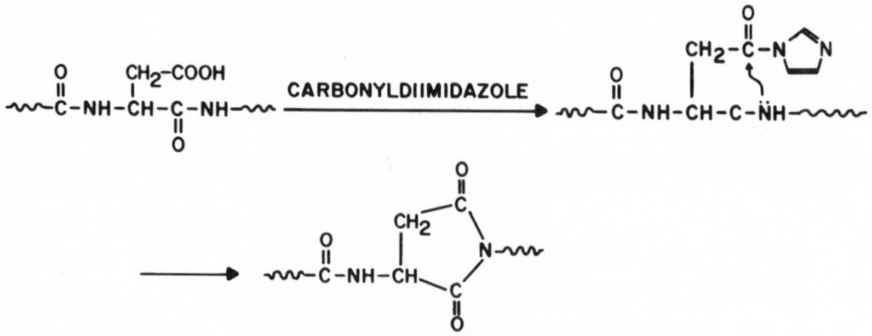

Fig. 9: Carbonyldiimidazole promoted cyclic imide formation at aspartic acid.

2. Carbodiimides

Aliphatic carbodiimides are among the most widely used amide bond-forming agents in peptide chemistry, and they have also been used, though not extensively, for attaching peptides to resins (8,19,20,25,26). The water-soluble carbodiimides, in particular, are useful because they permit attachment in partially aqueous solution. Unfortunately, like carbonyldiimidazole, they activate side chain carboxyls, although cyclic imide formation does not seem to be such a problem in this case.

Mross (20,25) has described one approach to circumventing the Asp-Glu problem. Tryptic peptides are coupled by means of a carbodiimide with glycineamide, which converts all the carboxyls to amides (Fig. 10). Then the peptide is digested again with trypsin to specifically liberate the C-terminal carboxyl, which is subsequently linked to the resin. The problem here is that a number of steps are required and the peptide must be repurified, often with loss of material.

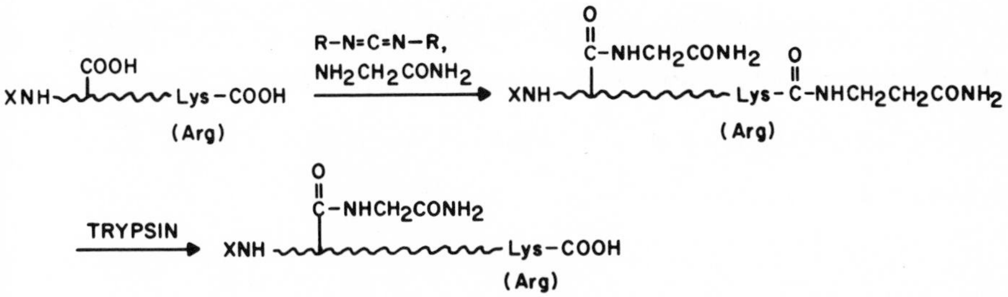

$$XNH\text{~}\underset{\displaystyle |}{\overset{COOH}{}}\text{~~~}Lys-COOH\ (Arg) \xrightarrow[NH_2CH_2CONH_2]{R-N{=}C{=}N-R,} XNH\text{~}\overset{\overset{O}{\|}}{C}-NHCH_2CONH_2\text{~~~}Lys-\overset{\overset{O}{\|}}{C}-NHCH_2CH_2CONH_2\ (Arg)$$

$$\xrightarrow{TRYPSIN} XNH\text{~}\overset{\overset{O}{\|}}{C}-NHCH_2CONH_2\text{~~~}Lys-COOH\ (Arg)$$

Fig. 10: Selective protection of side chain carboxyl groups in tryptic peptides according to Mross (20)

Recently, Previero et al (15) described a procedure which allows simultaneous activation of the C-terminal and protection of the side-chain carboxyls (Fig. 11). This process makes use of some well-known (27) reactions of carbodiimides. The peptide is first activated in a solvent such as dimethylcarbodiimide using a water-soluble carbodiimide, N-dimethylaminpropyl-N'-ethyl carbodiimide HCl being the most effect, in the absence of a nucleophile. Under these conditions, the O-acyl intermediate of the side chain carboxyl rearranges to the inactive N-acyl form, while the activated C-terminal undergoes azlactone formation. Since the azlacton is still an activated form of the carboxyl, it can then react selectively with the resin amino groups. Although this procedure would seem to be applicable to all peptides, results have been variable. While some peptides (e.g., glucagon, insulin A-chain) (15) are attached in good yield, others are coupled in very poor yield, or not at all (R. Laursen, unpublished). Furthermore, cyclic imide formation sometimes occurs when Asp-Gly bonds are present. Presumably, the lack of steric hindrance provided by glycine permits imide formation to compete with N-acylurea formation.

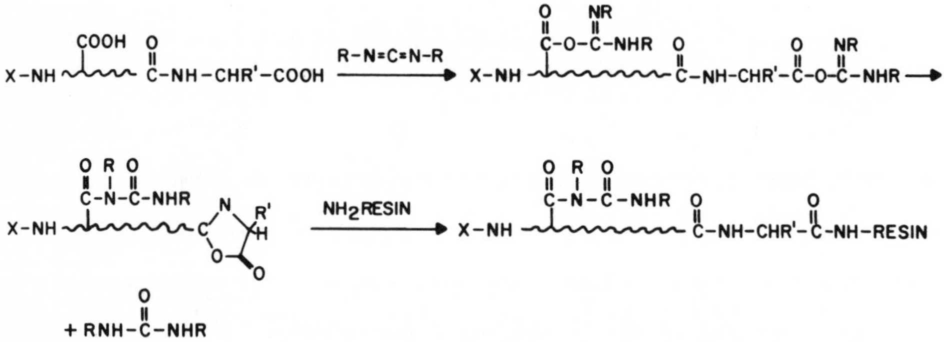

X-NH~(COOH)~C(=O)-NH-CHR'-COOH —(R-N=C=N-R)→ X-NH~(C(=O)-O-C(=NR)-NHR)~C(=O)-NH-CHR'-C(=O)-O-C(=NR)-NHR →

X-NH~(C(=O)-N(R)-C(=O)-NHR)~(oxazolone, R', H) —(NH_2RESIN)→ X-NH~(C(=O)-N(R)-C(=O)-NHR)~C(=O)-NH-CHR'-C(=O)-NH-RESIN

+ RNH-C(=O)-NHR

Fig. 11: Selective blocking of side chain carboxyl groups and activation of the C-terminal carboxyl according to Previero, et al (15).

Two factors may account for the generally low (about 50%) coupling yields with carbodiimides. First, traces of amines, such as dimethylamine in dimethylformamide, may react with the activated peptide before the resin has a chance to react. (Carbodiimides, unlike carbonyldiimidazole, do not scavenge nucleophiles.) Second, N-acylurea formation at the C-terminus may compete more favorably with azlactone formation in certain cases, leading to unreactive peptide.

3. Homoserine Lactone

Cleavage of peptides at methionine with cyanogen bromide (28) produces peptides containing C-terminal homoserine. Recently, it has been found (13) that these peptides can be conveniently attached to amine resins without protecting side chain carboxyl and amino groups. The C-terminal carboxyl is activated by conversion to the lactone in dry trifluoroacetic acid (29). Aminolysis of the lactone by the amine resin results in attachment of the peptide to the resin (Fig. 12). This has proved to be one of the most convenient and reliable attachment procedures in terms of time required and yield (> 80%). Since cyanogen bromide peptides are often fairly large, it is frequently possible to sequence as many as 20 or 30 amino acids in a single sequencer run.

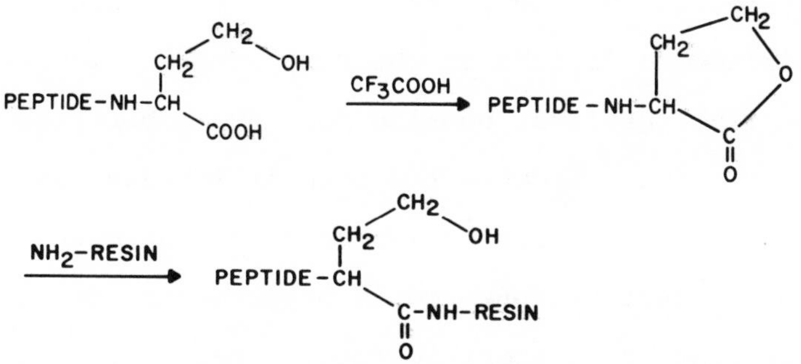

CH_2 — CH_2 — OH
PEPTIDE-NH-CH — COOH
$\xrightarrow{CF_3COOH}$
PEPTIDE-NH-CH — CH_2 — CH_2 — O — C(=O) (lactone ring)
$\xrightarrow{NH_2\text{-RESIN}}$
PEPTIDE-CH — CH_2 — CH_2 — OH
C(=O)-NH-RESIN

Fig. 12: Homoserine lactone attachment procedure.

In using this procedure, it is important that the reaction mixture of peptide and resin be basic, since the resin amino groups must be in the neutral form. This is accomplished by evaporating excess trifluoroacetic acid under vacuum and adding excess triethylamine. The usual solvent is dimethylformamide, containing 10 to 20% water, if necessary. The resin should be added before the triethylamine to minimize basic hydrolysis of the lactone and to maximize the possibility of aminolysis by the resin.

4. p-Phenylene Diisothiocyanate

The diisothiocyanate procedure (16) was devised as a means of circumventing the carboxyl problem, and is most useful for tryptic peptides containing C-terminal lysine. The peptide amino groups are activated with p-phenylene diisothiocyanate and coupled with an amino resin (Fig. 13). In this process, both the lysine ε-amino groups and the α-amino group of the N-terminal become attached. However, treatment of the resin with acid under the conditions of the Edman degradation results in cleavage of the first peptide bond and liberation of a new amino group. Because the phenylthiohydantoin (or thiazolinone) of the first amino acid remains bound to the resin, it must be detected by other means, or by difference from the composition of the peptide.

This procedure can be carried out in dimethylformamide or pyridine solvent and in the presence of appreciable amounts of salts (e.g., tertiary amine salts), provided a pH of about 9.5 is maintained. It is important to remove primary and secondary amino which can react with the coupling agent. The yields are generally greater than 80%, and the entire procedure can be accomplished in 3 to 4 hours. Peptides containing aminoethyl cysteine have been attached with equal success (M. Horn, unpublished).

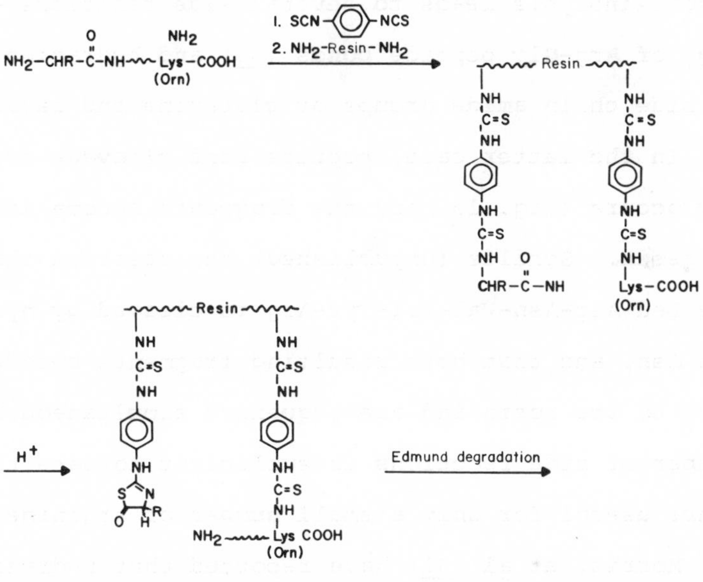

Fig. 13: Diisothiocyanate attachment procedure.

Arginine peptides can also be attached (16) by the diisothiocyanate procedure, if the arginine is first deguanidated with hydrazine (Fig. 14) using the hydrazinolysis conditions of Shemyakin et al (20% hydrazine, 105°C, 30 minutes) (30). Before activation, it is essential that all traces of hydrazine be removed by codistillation with triethylamine, because the hydrazine will react with the diisothiocyanate.

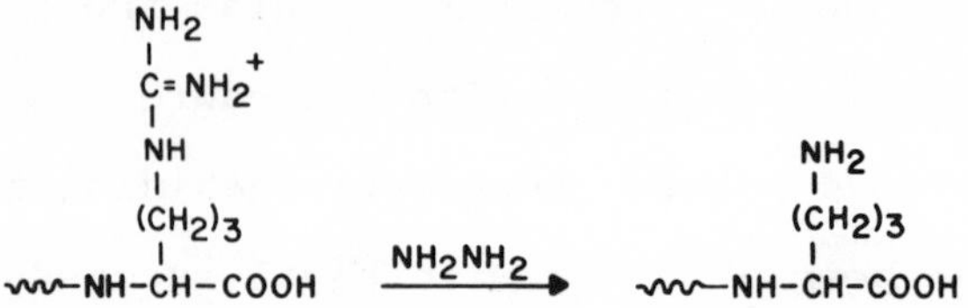

$$\sim\text{NH-CH(COOH)-(CH}_2)_3\text{-NH-C(NH}_2)\text{=NH}_2^+ \xrightarrow{NH_2NH_2} \sim\text{NH-CH(COOH)-(CH}_2)_3\text{-NH}_2$$

Fig. 14: Hydrozinolysis of arginine peptides.

Hydrazinolysis leads to several side reactions - cleavage of Arg-Gly peptide bonds (30) and hydrazinolysis of the side chain amide groups of glutamine and asparagine. In the latter case, peptide bond cleavage frequently occurs (Fig. 15) and the fragments become attached to the resin. Schiltz (unpublished) has observed that the peptide Leu-Asp-Asn-Val-Val-Tyr-Arg is cleaved by hydrazine at Asn, and that both resulting fragments become attached to the resin and are sequenced simultaneously. The number of side reactions is sufficient to make this procedure useful for only a small number of arginine peptides. Morris, et al (31) have reported that hydrazinolysis in 50% hydrazine at 75°C for 15 minutes gives fewer side reactions, but these conditions have not yet been tried in connection with the solid-phase Edman degradation.

F. Factors Affecting the Edman Degradation

In order for the Edman degradation to be carried through many cycles, reactions must be quantitative at each step. In actual practice, however, the Edman degradation is never quantitative because of two effects -- carry-over (or overlap), and blockage. Carry-over results from incomplete coupling or cleavage, and is characterized by the appearance of a particular amino acid in subsequent cycles. Even one percent of carry-over can be a serious problem in sequencing large peptides, since the effect is additive. For example, after 20 cycles, one would see nearly 20% of the preceding amino acid. Carry-over is relatively easily remedied by increasing the reaction times or the temperature. This effect is almost never seen in solid-phase degradation of peptides, but has been observed with proteins, particularly when the resin is heavily loaded (17). Apparently, the amino-terminal groups become buried and inaccessible to reagents. In this case, overlap can be reduced by working at higher temperatures (R. Laursen, unpublished).

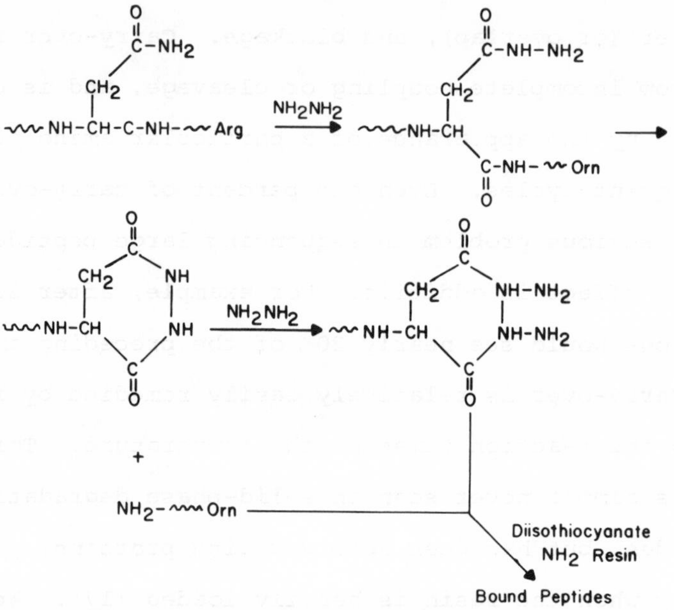

Fig. 15: Peptide bond cleavage at asparagine by hydrazine.

The cause of peptide blockage is not known. It is manifested by a 5 to 10% decrease in the amount of PTH liberated in each cycle, and is seen to about the same extent in both solid- and liquid-phase (4,32,33) versions of the Edman degradation. It is not due simply to loss of peptide material, since it has been demonstrated (7) that the peptide remains attached to the resin after degradation. The most plausible explanation is desulfurization (Fig. 16) of the phenylthiocarbamyl peptide to a phenylcarbamyl peptide which is incapable of cyclization and cleavage under normal conditions. Ilse and Edman (34) have suggested that oxygen or oxidizing agents are responsible for desulfurization, although it has been our experience that neither scrupulous removal of oxygen nor addition of anti-oxidants to reagents is of any benefit. An alternate explanation is hydrolytic desulfurization by water in the buffers, perhaps catalyzed by components of the buffer.

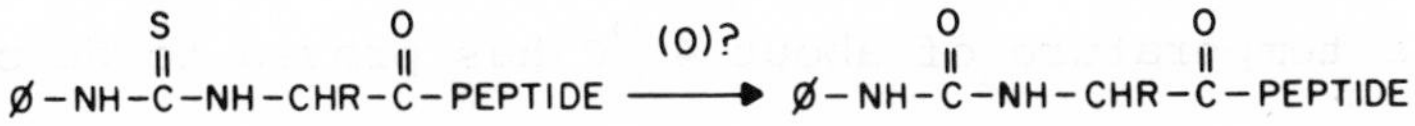

$$\phi\text{-NH-}\overset{\overset{\displaystyle S}{\|}}{C}\text{-NH-CHR-}\overset{\overset{\displaystyle O}{\|}}{C}\text{-PEPTIDE} \xrightarrow{(O)?} \phi\text{-NH-}\overset{\overset{\displaystyle O}{\|}}{C}\text{-NH-CHR-}\overset{\overset{\displaystyle O}{\|}}{C}\text{-PEPTIDE}$$

Fig. 16: Desulfurization of phenylthiocarbamyl peptides.

Edman (5) and others (4,33) have emphasized the need for using highly purified reagents in the liquid-phase sequencer, with particular care being taken to remove aldehydes, and it seems well-established that ordinary reagents do not work well in liquid-phase sequencers. By contrast, reagent purity does not seem to be an important factor with solid-phase degradations, and one can use ordinary reagent grade chemicals without further purification. Perhaps a reason for this is the resin itself. The resin contains up to a 1000-fold excess of amino groups over peptide amino groups. Ordinarily, these have been converted to methylthiocarbamyl groups by reaction with methylisothiocyanate. Therefore any reagent impurity that could react with a peptide pehnylthiocarbamyl group should be scavenged by the resin.

Temperature also affects the efficiency of the Edman degradation. If, under a given set of reaction times, the temperature is too low, the reactions do not go to completion, and carry-over is observed. At high temperatures, carry-over is eliminated, and peptide blockage increases, resulting in a decrease in the amount of degradable peptide. For the solid-phase Edman degradation, a temperature of about $45^{\circ}C$ has proved to be optimal (6).

The pH of buffers used in the degradation does not seem to be very critical. Edman (5) and others (4,33) have used Quadrol and dimethylallylamine buffers at about pH 9 to 9.5 in the liquid sequenators, while we have used pH 8 N-methylmorpholine buffers in the solid-phase sequencer. Since the pK_2 of peptide amino groups is about 7.5, high pH buffers are unnecessary, except to modify the -amino group of lysine (pH 10.4). In the diisothiocyanate and homoserine coupling methods, the lysines are modified before sequencing begins and special procedures are unnecessary.

The PITC used in degradations is stored as a 5% solution in acetonitrile, which is essentially inert and is soluble in the aqueous sequencing buffer. The approximately 60% of pyridine in the buffer tends to swell the resin and allow penetration of the PITC.

Washing of the resin after the PITC and TFA steps is accomplished with two solvents, methanol and 1,2-dichloroethane. After PITC coupling, it is essential to remove all traces of reactants and by-products that might interfere with PTH analysis. The column is washed alternately with methanol and DCE to remove both polar and hydrophobic impurities. DCE swells the resin and methanol shrinks it, so the effect is like that of squeezing a sponge. However, it may be that DCE is not actually needed for this purpose, because satisfactory sequencing runs have been performed even when the DCE reservoir accidentally ran dry. DCE, which is the last solvent to be used before the TFA cleavage step, also serves to dry out the resin, thus minimizing the possibility of acid-catalyzed methanolysis of peptide bonds. Random acid-catalyzed cleavage of peptide bonds leads to the liberation of new amino groups and is the main cause of a gradual increase in background as degradation proceeds. The larger the peptide, the more pronounced the effect.

After the TFA cleavage step, the resin is washed with methanol rather than DCE. When DCE was used, it was found that degradation yields dropped appreciably when serine was reached. An explanation for this is that the serine hydroxyl becomes trifluoroacetylated by TFA. As soon as the N-terminal amino is deprotonated in the

PITC coupling step, an O N acyl migration can take place resulting in blocking of the N-terminal (Fig. 17). When methanol is used for washing, deacylation of serine by transesterification can take place.

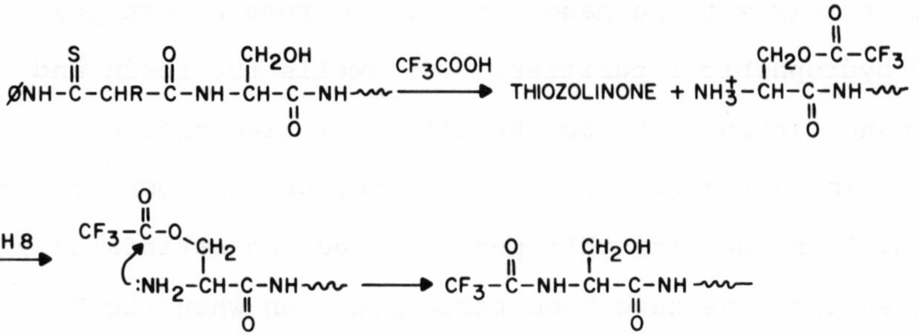

$$\text{ØNH-C(=S)-CHR-C(=O)-NH-CH(CH}_2\text{OH)-C(=O)-NH}\sim \xrightarrow{CF_3COOH} \text{THIOZOLINONE} + \text{NH}_3^+\text{-CH(CH}_2\text{O-C(=O)-CF}_3\text{)-C(=O)-NH}\sim$$

$$\xrightarrow{pH\ 8} \text{:NH}_2\text{-CH(CH}_2\text{-O-C(=O)-CF}_3\text{)-C(=O)-NH}\sim \longrightarrow \text{CF}_3\text{-C(=O)-NH-CH(CH}_2\text{OH)-C(=O)-NH}\sim$$

Fig. 17: Proposed mechanism of blockage of N-terminal serine peptides.

G. Degradation of Proteins

Until recently, solid-phase degradation of proteins was not possible because of the difficulty of attaching proteins to polystyrene-derived resins. These resins are not porous enough to allow attachment of proteins except on the surface. This has changed, however, as a result of the work of Machleidt and coworkers (14,17) who attached proteins to aminopropyl glass resins which have a very high effective surface area. Aminopropyl glass is prepared by reaction of controlled-pore glass (Corning, Bio-Rad) with trimethoxyaminopropylsilane (Fig. 18). The amino groups are then activated by reaction with *p*-phenylene diisothiocyanate (27,17) to give an isothiocyanate resin, and proteins (or peptides) are coupled by their lysine ε-amino groups (14,16,17). This process is analogous to that of Dowling and Stark (36), who used polystyrene isothiocyanate resins for degradation of peptides. Isothiocyanato glass can also be prepared by the reacting aminopropyl glass with thiophosgene (37), although the noxious properties of thiophosgene make this route inconvenient.

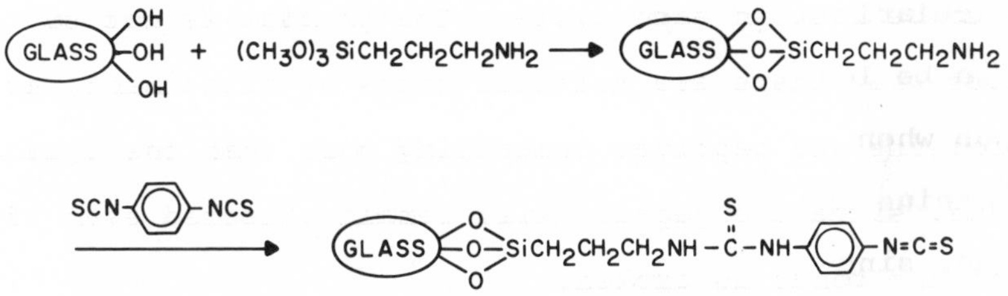

GLASS(OH)$_3$ + $(CH_3O)_3SiCH_2CH_2CH_2NH_2$ ⟶ GLASS(O)$_3$SiCH$_2$CH$_2$CH$_2$NH$_2$

$\xrightarrow{SCN-C_6H_4-NCS}$ GLASS(O)$_3$SiCH$_2$CH$_2$CH$_2$NH–C(=S)–NH–C$_6$H$_4$–N=C=S

Fig. 18: Synthesis of isothiocyanato glass.

Another important virtue of resins derived from porous glass is their mechanical stability. They do not swell and shrink as polystyrene resins do, so there is no danger of high backpressures in the sequencer (14) and it is not necessary to mix them with glass beads (6). Although it is possible to attach large amounts of protein to NCS glass resins, 1 to 2 nmole/mg of resin seems to be optimal (17), since greater amounts lead to overlap. Apparently, when the protein concentration is too high, unfavorable sites become occupied and Edman degradation is too high, unfavorable sites become occupied and Edman degradation reactions do not go to completion.

Using a 75-Å mean-pore-diameter resin, Machleidt et al (17) attached and degraded through 35 cycles the cytochrome c from Candida krusei. The same investigators were also able to attach a much larger protein, 3-phosphoglycerate kinase (MW 47,000) to 240-Å pore glass.

The glass-derived resins are somewhat less satisfactory for peptides. The Si-O bonds which hold the ligand to the glass are not completely stable to acid (particularly aqueous acid) and small amounts of peptide tend to be lost in each cycle. The problem is not so serious when there are multiple sites of attachment, as in proteins and peptides containing more than one lysine residue, since the peptide will remain attached even if one of the bonds is broken.

H. Work-up and Analysis of Phenylthiohydantoins

During Edman degradation of peptides, amino acids are liberated as thiazolinones (34), and must be converted to PTHs. In the solid-phase version, thiazolinones are collected in solution with TFA and methanol (which rapidly forms water and methyl trifluoroacetate). The solvents are evaporated under a stream of nitrogen and the residue is dissolved in 0.2 ml of 20% TFA in water heated at 80°C for 10 minutes (Edman and Begg (5) suggest 1 N HCl at 80°C for 10 minutes) to effect conversion to the PTHs.

At this point, the PTH must be extracted from the aqueous solution to separate it from salts that might interfere with analysis. Although most investigators use ethyl acetate extraction (5,33), we have found it convenient to pass the solution through a small column of Dowex-50 (H^+), which removes all cationic species such as buffer salts (and also PTH-His and PTH-Arg) (18). The advantage here is that difficultly extracted derivatives such as PTH-CysOH and PTH-Ser are recovered quantitatively, because they pass straight through the column. Also manipulations are simplified. The Dowex-50 column is eluted with methanol, and we have found that PTH-Asp and PTH-Glu are partly converted to the methyl esters, which chromatograph near alanine. This side reaction often provides a cross-check in identifying PTH-Glu and PTH-Asp. PTH-His and PTH-Arg can be eluted and identified separately from the Dowex-50 columns with 6 N HCl. The HCl also elutes the cationic impurities, which sometimes interfere with analysis.

A variety of procedures are available for PTH analysis and will be discussed here only briefly. No one method is entirely satisfactory. The first method to be used with an automatic sequencer was thin layer chromatography (TLC) (5), which has the virtue of being fast, inexpensive, and capable of resolving all amino acids. A variety of solvent systems and supports have been described (5,18,38-41), some of which are useful for nanomole amounts of PTHs (42,43). In addition, PTHs react with ninhydrin to give different colors, which is often helpful in identification (39). The major drawback to TLC is that it is not quantitative, so one is forced to make subjective evaluations based on spot intensity. One can quantitate the process by eluting the spots and measuring concentrations in a spectrophotometer, but this is very time-consuming.

We have found it possible to semi-quantitate TLC analysis by carrying out degradations with radioactive PITC (18) and chromatographing the radioactive PTHs on aluminum-backed silica gel plates. After the spots have been located under a uv lamp, the plate is marked into a grid using unlabeled PTHs for references and is cut into pieces with scissors. In this way one can confirm that a uv-absorbing spot is a PTH by measurement of its radioactivity. The use of radioactivity also permits sequence analysis down to about 5 nanomole or lower. The disadvantages of this procedure are that counting the plates is somewhat tedious and time-consuming, and one must use radioactive PITC in the sequencer, which requires observance of certain precautions.

Gas liquid chromatography (GLC) is probably the most widely used procedure for PTH analysis. The advantages are that GLC is sensitive, fairly fast, and semi-quantitative (33,44). On the other hand, this method of analysis requires a fairly sophisticated gas chromatograph, which must be carefully calibrated. To improve volatility the PTHs must be silylated, which adds another step. PTH-His, PTH-Arg, PTH-CysOH and PTH-CMCys are not volatile and must be analyzed in other ways, and PTH-Lys, PTH-Asn and PTH-Gln are difficult to quantitate (33).

Mass spectrometry has also been used for PTH analysis (45,46), but the equipment is too expensive and tempermental for routine work in most laboratories. It has, however, proved useful in conjunction with isotope dilution techniques for sequence analysis of peptide mixtures (47).

PTHs can be hydrolyzed back to their parent amino acids and identified by amino acid analysis. Smithies et al (48) have used hydrolysis in 57% HI at 127°C for 20 hours, and in 0.2 M NaOH + 0.1 M $Na_2S_2O_4$ at 127°C for 3.5 hours for hydrolysis of PTHs. The main disadvantage of this procedure is that an amino acid analyzer has to be devoted almost exclusively to PTH analysis and cannot be used for anything else. Furthermore, a number of PTHs undergo side reactions in HI. For example, serine, carboxymethylcysteine, cysteine and tryptophan are all converted, at least partly, to alanine.

Liquid chromatography may provide the ultimate means for PTH analysis. Although column chromatographic separations of PTHs were described by Sjoquist (49) over 15 years ago, only the recent development of high-performance liquid chromatography, employing small columns and high pressures (1000 - 3000 psi), has made this technique practical. Graffeo et al (50) have eluted silica gel columns with mixtures of heptane, chloroform and methanol, whereas Zimmerman et al (51) have employed reversed phase supports (Corasil C_{18}) and elution with aqueous solvents. Although this work is still in the developmental stage, several virtues of the technique are apparent. Liquid chromatography is fast, requiring only 30 to 45 minutes for analysis, and is sensitive to the subnanomole level. Modification of PTHs before analysis is not required, and decomposition does not occur because the columns are eluted at room temperature. Furthermore, all PTHs (even PTH-His, PTH-Arg and PTH-CysOH) can be eluted (B. Karger, personal communication). Liquid chromatographs are comparable in price to gas chromatographs.

I. Other Versions of Solid-Phase Edman Degradation

The first to attempt Edman degradation on a solid support was Stark (36,52,53) who used an insoluble Edman reagent, NCS-polystyrene, prepared as shown in Fig. 19. In this case, peptides are first attached by their N-terminal amino groups to the polymer. On treatment with trifluoroacetic acid, the peptide minus the N-terminal amino acid is cleaved off and is then subjected to another round of degradation (Fig. 20). One advantage of this process is that problems with overlap are not encountered.

Once the peptide is attached to the resin, the only way it can be removed is by degradation, so there is no possibility of contamination with undegraded peptide.

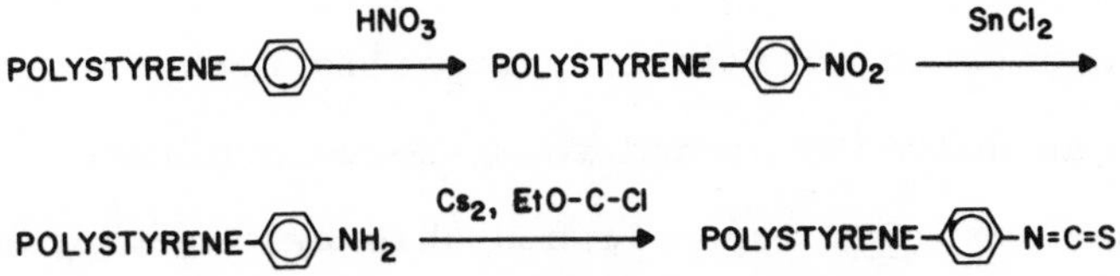

POLYSTYRENE—⬡ —(HNO_3)→ POLYSTYRENE—⬡—NO_2 —($SnCl_2$)→

POLYSTYRENE—⬡—NH_2 —(Cs_2, EtO-C-Cl)→ POLYSTYRENE—⬡—N=C=S

Fig. 19: Synthesis of NCS-polystyrene.

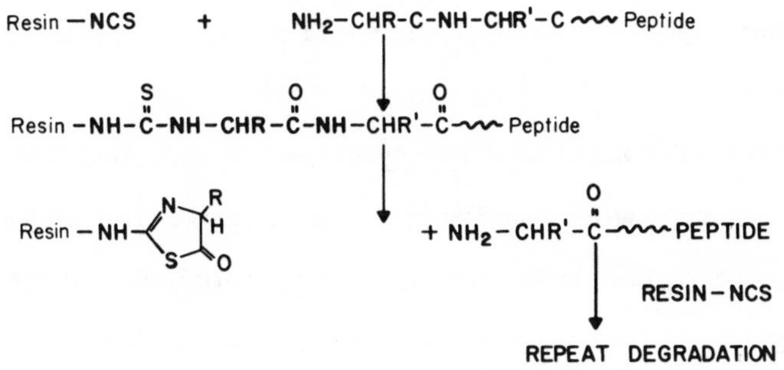

Resin —NCS + NH_2—CHR-C-NH-CHR'-C ~~ Peptide

Resin —NH-C(=S)-NH-CHR-C(=O)-NH-CHR'-C(=O)~~ Peptide

Resin —NH—(thiazolinone, R, H, S, O) + NH_2-CHR'-C(=O)~~PEPTIDE

RESIN-NCS

REPEAT DEGRADATION

Fig. 20: Degradation of peptides on NCS-polystyrene according to Dowling and Stark (37).

Unfortunately, several disadvantages limit the utility of the Stark method. Most important is that the peptide must be reattached to the resin after each round; frequently, the attachment yields are low (75% average), so material is lost. Material is also consumed by the analytical procedure which involves hydrolysis and amino acid analysis of a portion of the peptide. The Stark degradation is useful for peptides containing 10 or fewer residues (36). Because the peptide must be reattached after each cycle and samples must be removed for analysis, this procedure is not readily amenable to automation.

The NCS-polystyrene resin is very hydrophobic and must be converted to a more hydrophilic form for use in the aqueous solvents needed to dissolve peptides. Dowling and Stark (36) accomplished this by coupling about 60% of the NCS groups to glucosaminol, thereby introducing a nymber of polar hydroxyl groups into the resin. To promote maximal swelling and penetration of the peptides, they used a resin derived from 0.25% cross-linked polystyrene.

Mross and Doolittle (20,25) have described the degradation of polymer-bound peptides using thioacetylthioglycolic acid, instead of phenyl isothiocyanate. This reagent converts peptide amino groups into thioacetamides which undergo cyclization in acid to thiazolinones (Fig. 21). The methylthiazolinones, which cannot rearrange to thiohydantoins, are readily hydrolyzed in aqueous acid back to the parent amino acids, which are identified by amino acid analysis.

Edman degradation has been carried out on non-cross-linked aminopolystyrene resins which allows attachment of peptides and degradation in solution, where reactions are considerably faster than in the solid phase (54). The peptide-resin complexes are purified after each step by precipitation in water.

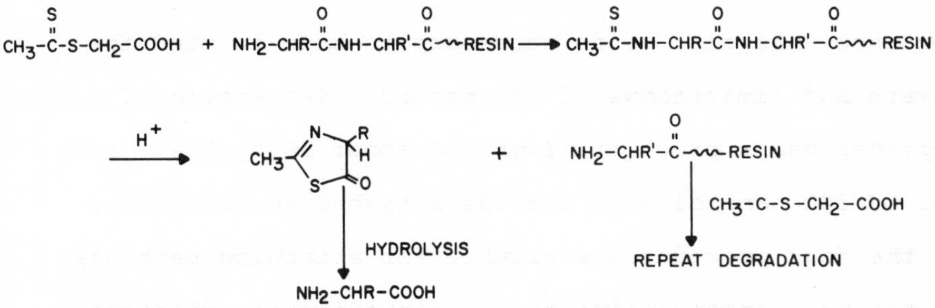

$$CH_3-\overset{S}{\overset{\|}{C}}-S-CH_2-COOH + NH_2-CHR-\overset{O}{\overset{\|}{C}}-NH-CHR'-\overset{O}{\overset{\|}{C}}\sim RESIN \longrightarrow CH_3-\overset{S}{\overset{\|}{C}}-NH-CHR-\overset{O}{\overset{\|}{C}}-NH-CHR'-\overset{O}{\overset{\|}{C}}\sim RESIN$$

$\xrightarrow{H^+}$ (thiazolinone: CH_3, N, R, H, S, O) + $NH_2-CHR'-\overset{O}{\overset{\|}{C}}\sim RESIN$

HYDROLYSIS → $NH_2-CHR-COOH$

$CH_3-C-S-CH_2-COOH$ → REPEAT DEGRADATION

Fig. 21: Peptide degradation with thioacetylthioglycolic acid.

Birr et al (55) have explored the possibility of Edman degradation of peptides immobilized by adsorption to aluminum oxide. N-Protected peptides are coupled to aminomethyl groups of azoarylsulfonic acid dyes after activation with ethyl chloroformate. The peptide-dye complex is then adsorbed onto acidic aluminum oxide and Edman degradation is carried out in a small reactor that allows for washing of the alumina. To prevent the dye complex from being washed off the support, the polarity of some of the reagents, for example TFA, must be reduced by dilution with non-polar organic solvents. An advantage of the adsorption procedure is that modification of the carboxyl groups can be done in homogenous solution where reactions are faster and easier to control than on resins.

J. Sequencing Strategies

When using the solid-phase Edman degradation for sequence analysis, it is necessary to keep in mind the powers and limitations of the method. Sequencing of peptides has a practical limit of about 30 cycles, given 200 to 300 nanomoles of peptide attached to the resin. Of the four procedures available for attaching peptides to resins, carbonyldiimidazole, carbodiimide, diisothiocyanate, and homoserine lactone, the latter two are by far the most useful in terms of attachment efficiency and lack of side reactions. Ideally, therefore, one should try to obtain protein fragments which terminate in lysine or homoserine and contain 30 or fewer amino acids.

For large proteins, simple fragmentation with enzymes such as trypsin or chymotrypsin is impractical because a large number of rather small peptides is formed. Isolation of peptides from such complex mixtures is very difficult and invariably some are lost. Furthermore, the alignment of the peptides in the polypeptide chain becomes more difficult, the larger the number of peptides. For this reason, it is preferable to use limited cleavage and obtain a small number of large peptides. The most commonly used procedures are cleavage at methionine with BrCN or at arginine with trypsin after first blocking the lysines with a citraconyl or other removable group.

How such procedures can be used with solid-phase Edman degradation is perhaps best described with a hypothetical example.

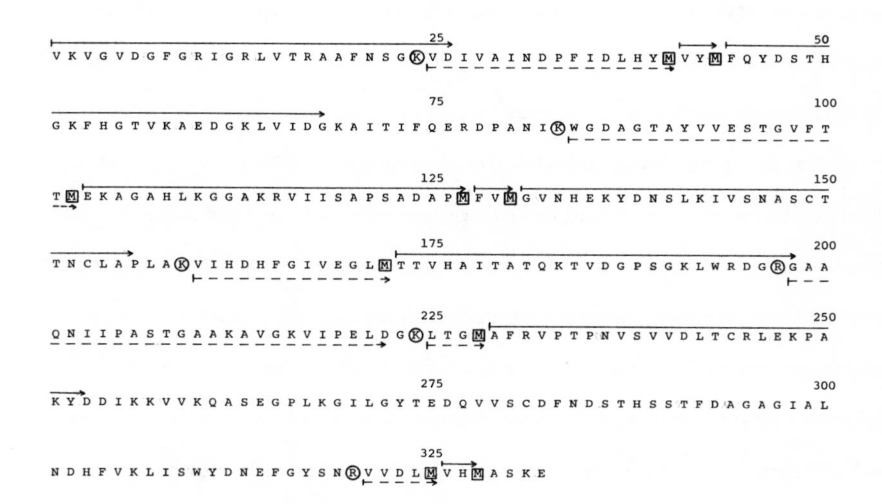

Fig. 22: Amino acid sequence of porcine glyceraldehyde-3-phosphate dehydrogenase. Solid lines indicate the portion of the BrCN peptide that would be determined on sequencing the intact peptide. Dashed lines indicate the portion that would be sequenced after cleavage of the BrCN peptide with trypsin and attachment of the fragment by homoserine lactone.

Porcine glyceraldehyde-3-phosphate dehydrogenase contains 332 amino acid residues, including 10 Arg, 26 Lys, and 9 Met residues (56) (Fig. 22). Cleavage with cyanogen bromide would give 10 peptides containing from 3 to 97 residues. Five of the peptides contain fewer than 25 residues and could be sequenced directly after attachment by the homoserine lactone procedure. For the larger peptides, for example, the N-terminal peptide which contains 40 residues, one could first sequence the first 25 or so residues. In a second experiment, one could digest a sample of the peptide with trypsin and then perform a homoserine coupling with the mixture. Only the C-terminal fragment containing homoserine would become attached to the resin, and in this way the remaining 15 residues could be sequenced (Fig. 23). Using this strategy, it should be possible in principle to obtain about 80% of the protein sequence from the cyanogen bromide peptides alone, without having to isolate any subfragments. Very long peptides, sich as the 97-residue peptide in glyceraldehyde-3-phosphate dehydrogenase, may not attach efficiently to the TETA resin by homoserine coupling. In this case, one might try attaching the peptide by the lysine ε-amino groups to the isothiocyanato glass described by Machleidt et al (17).

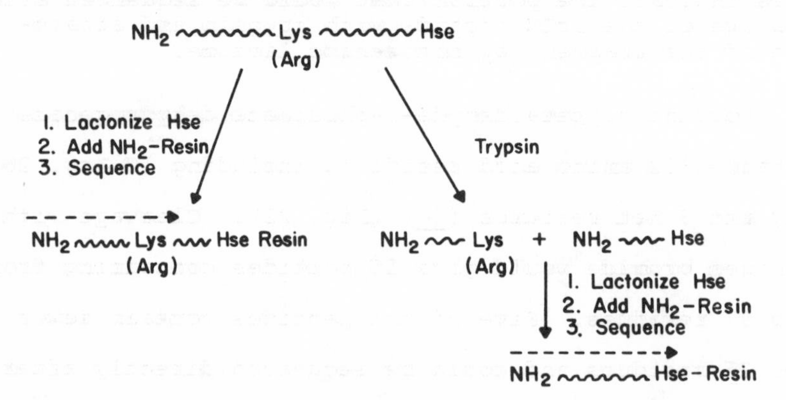

Fig. 23: Strategy for sequencing long BrCN peptides containing lysine and arginine (See text).

An alternative strategy for long tryptic peptides that contain methionine, like the tryptic peptide comprising residues 25-52 in Fig. 22, is to degrade the peptide as far as possible in the usual way. In a second experiment, a portion of the peptide can be digested with cyanogen bromide and the mixture attached to aminopolystyrene using the diisothiocyanate method. In this case, only the C-terminal peptide can be attached and sequenced (Fig. 24).

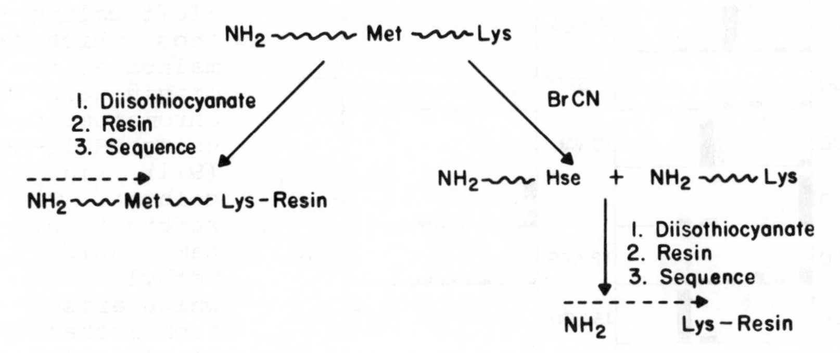

Fig. 24: Strategy for sequencing long tryptic peptides containing methionine.

If a protein is first treated with citraconic or maleic anhydride and then cleaved with trypsin, only peptides containing C-terminal arginine will be obtained. Those peptides which also contain lysine can be attached by the diisothiocyanate method and then sequenced as far as the last lysine in the peptide, at which point it will become detached from the resin.

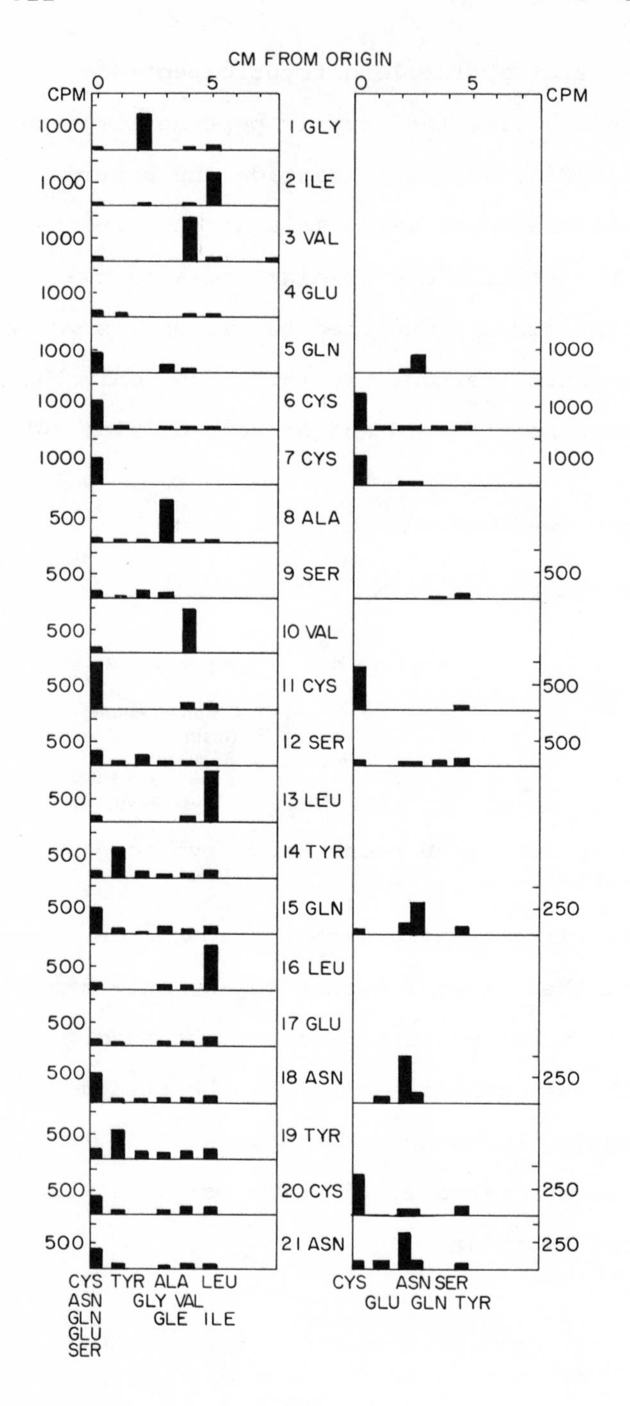

Fig. 25: Degradation of 240 nmole of oxidized insulin A-chain with ^{3}H-PITC. 10 to 20% of the PTH sample was used for analysis of each cycle. Samples were chromatographed first using $CHCl_3$-EtOH (98:2) as solvents (left column); those which remained at the origin were then chromatographed using $CHCl_3$-MeOH (9:1) (right column). Gle refers to glutamic acid methyl ester, which arises from methanolysis of glutamine. Reprinted from Ref. (6), p. 100 by courtesy of Springer-Verlag.

K. Examples

Although peptides from a wide variety of proteins have been sequenced, only a few examples will be discussed here. Figure 25 shows the results of degrading a 240-nmole sample of oxidized insulin A-chain that has been attached to a resin by the carbonyldiimidazole method. Blanks occur at cycles 4 and 17, because the glutamic acid residues at these positions remain attached by their side chains to the resin. PTH-Ser is largely decomposed during the PTH work up procedure and does not give rise to strongly radioactive spots. A plot of spot radioactivity versus cycle number (Fig. 26) indicates an average yield per cycle of 93%. In general, yields average between 90 and 94%. An increased background due to random hydrolysis of peptide bonds and generation of new amino terminals becomes apparent in later cycles.

The first proteins to be sequenced primarily by means of solid-phase Edman degradation were the proteins L7 and L12 from *E. coli* ribosomes (56), which are identical except that the N-terminal serine of L7 is blocked with an acetyl group (Fig. 27). All of the tryptic peptides except T6, T13 and T1Phe were sequenced by solid-phase Edman degradation after attachment to aminopolystyrene by the diisothiocyanate procedure. The 30 residue peptide T1Phe, which is very hydrophobic, could not be coupled to the resin because of its insolubility in the coupling solvent. However, another large peptide, T1Met, was degraded successfully through 24 cycles (Table 1). It should be noted that all three of the methionine residues in the protein are located in this peptide; thus, cyanogen bromide cleavage would not have been very useful in this case. Most of the peptides were sequenced using about 50 to 70 nmole of material. Only 15 nmole was needed for the octapeptide T14. Chymotryptic cleavage provided overlapping peptides.

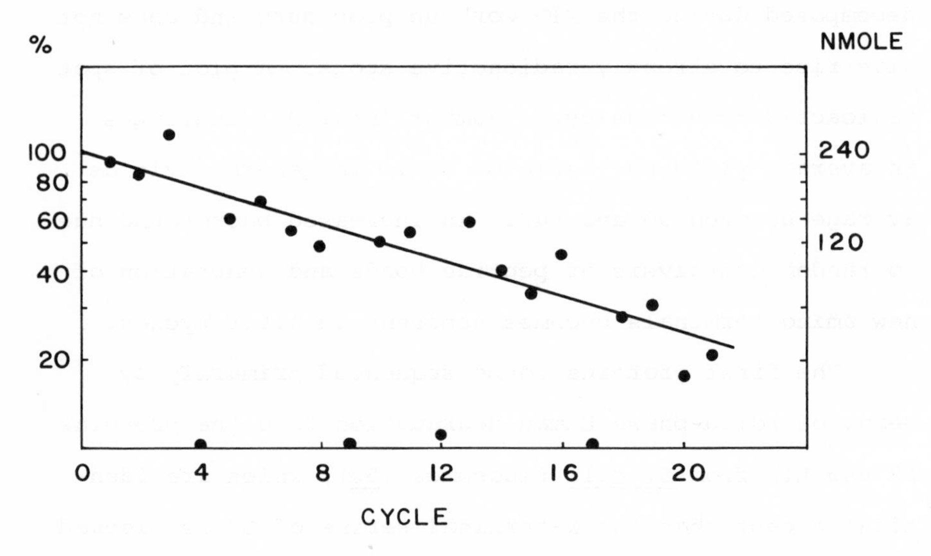

Fig. 26: PTH yields obtained during degradation of oxidized insulin A-chain. An average yield of 93% per cycle can be calculated from the straight line.

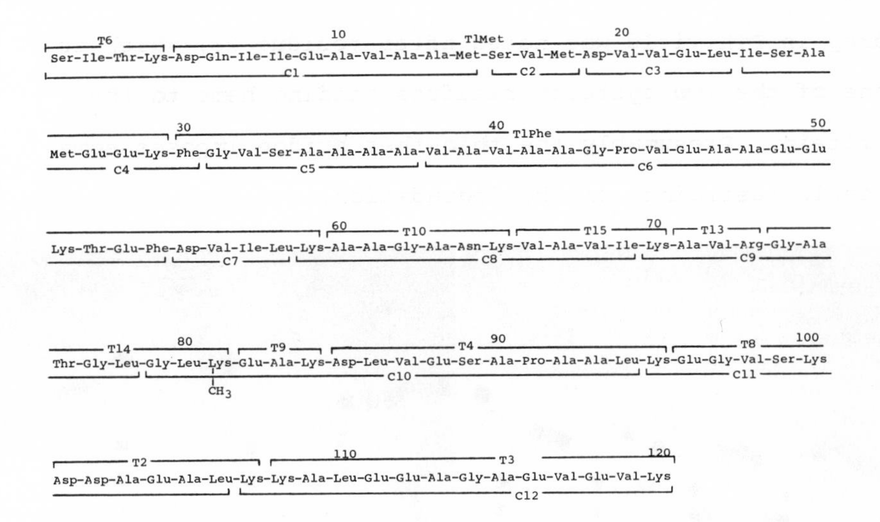

Fig. 27: Amino acid sequence of the ribosomal proteins L7 and L12; T, tryptic peptides; C, chymotryptic peptides.

Figure 28 shows a photograph of a thin-layer chromatogram of PTHs cleaved from a cyanogen bromide peptide obtained from the myoglobin of the mollusc *Busycon canaliculatum* (A. Bonner, unpublished). The peptide was attached by the homoserin lactone method. The sequence is Asp-Thr-Thr-Gly-Val-Gly-Lys-Ala-His-Gly-Val-Ala-Val-Phe-Lys-Gly-Leu-Gly-Ser-Hse.

The last example is the degradation of *Candida Krusei* cytochrome *c* attached to isothiocyanato glass (17) (Fig. 29). Approximately 400 nmole of protein was attached to the glass and degraded through 35 cycles. Gas-liquid chromatography was used for PTH analysis. The average yield per cycle was about 95%. However, an appreciable drop in PTH yield was noted after residue 23, which is one of the two cysteine residues binding heme to the protein. Presumably the thioether bridge imposes some steric restraints on the degradation.

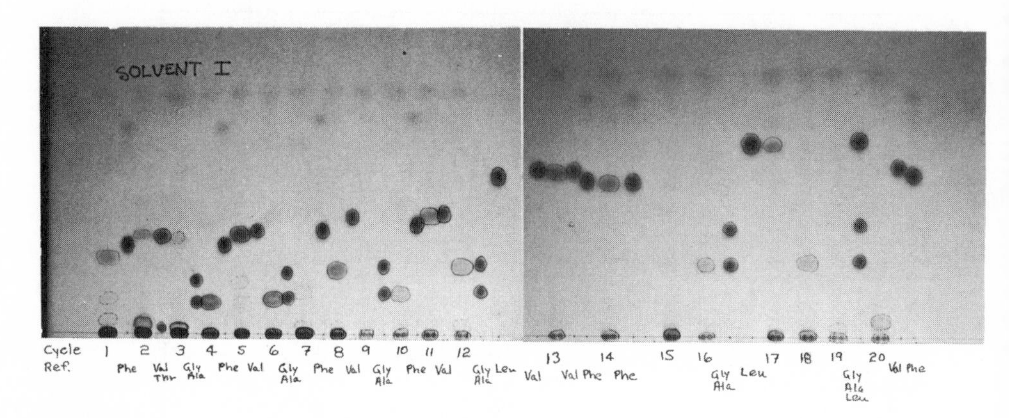

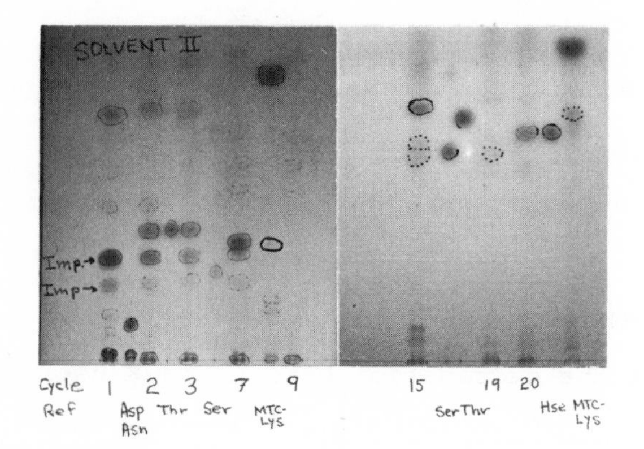

Fig. 28: Photograph of a thin-layer chromatogram showing PTHs obtained by degradation of a cyanogen bromide peptide attached by the homoserine lactone procedure to TETA sequencing resin; PTHs which did not elute in solvent I ($CHCl_3$-EtOH, 98:2) were run in solvent II ($CHCl_3$-MeOH, 9:1). Selected PTH references were run beside the unknowns. Residue 9 is His, whose PTH remained absorbed on the ion exchange column (see Section III.H).

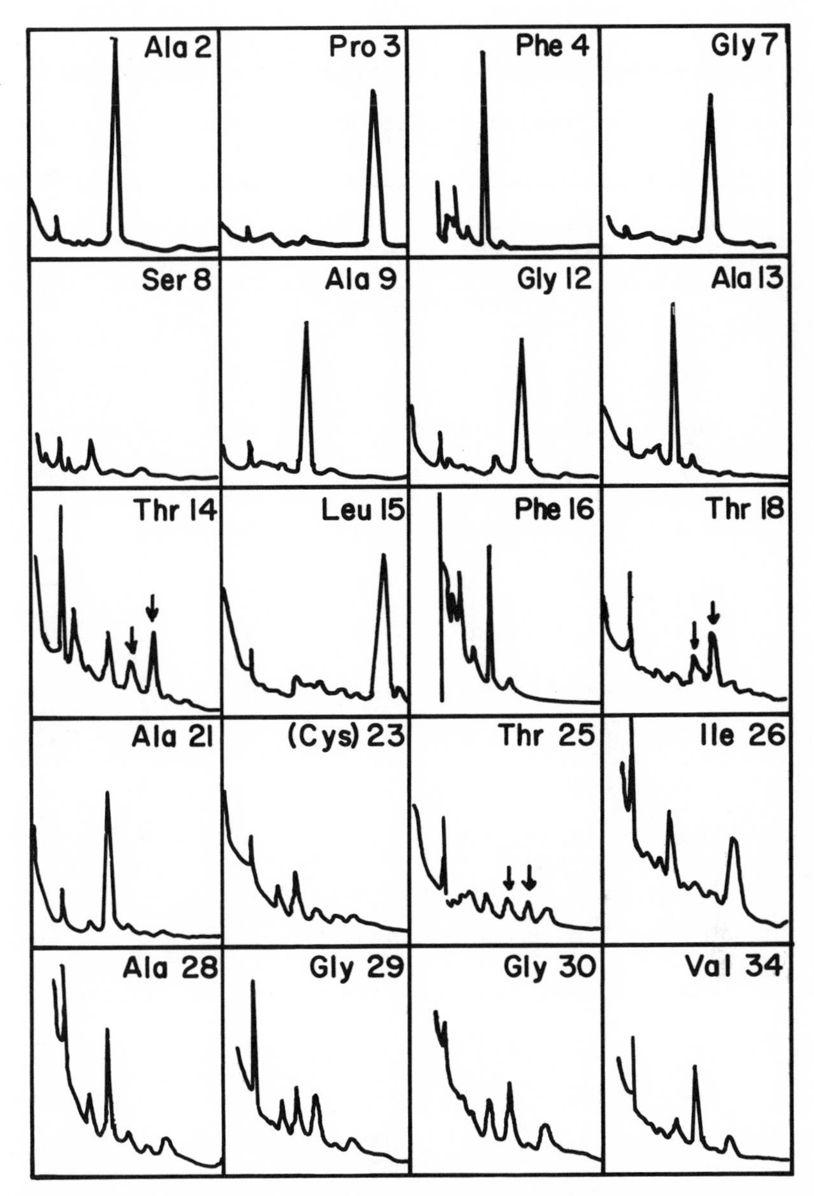

Fig. 29: Gas chromatograms of PTHs from the degradation of cytochrome c from *Candida krusei*. Sensitivities are increased 2-fold for steps 7, 8, 12, 14, 15 and 21, 4-fold for step 18 and steps 23 - 34 when compared to step 1. Arrows indicate the two peaks obtained for PTH-Thr. Reprinted from Ref. (17), p. 218, by courtesy of North Holland Publishing Company.

Table 1

Solid-Phase Edman Degradation of 130 nmol of L7-TIMet

Cycle	Radioactivity of Glx Asx Ser	Ala Gle	Met	Val	Ile	Leu	Amino
	counts/min						
Solvent: I							
1	0	0	0	0	0	0	---
2	748	43	0	0	0	0	---
3	0	0	0	27	384	121	Ile
4	0	0	0	35	317	70	Ile
5	667	247	0	11	9	7	---
6	87	560	0	0	0	10	Ala
7	46	27	64	435	28	4	Val
8	62	594	20	32	0	2	Ala
9	0	400	0	5	0	7	Ala
10	64	25	269	31	0	7	Met
11	169	0	0	0	0	5	---
12	0	0	35	257	0	7	Val
13	179	48	157	0	0	19	Met
14	313	110	28	0	0	22	---
15	18	6	26	282	20	0	Val
16	0	0	9	231	33	0	Val
17	177	99	0	25	10	0	---
18	57	9	0	0	81	175	Leu
19	9	0	0	23	178	75	Ile

Table 1 (continued)

Cycle	Glx Asx Ser	Ala Gle	Met	Val	Ile	Leu	Amino
	counts/min						
20	154	6	0	0	23	0	---
21	0	228	0	0	0	0	Ala
22	---	---	---	---	---	---	---
23	82	10	5	0	0	0	---
24	86	0	0	0	0	0	---

Cycle	Radioactivity of					Amino Acid
	Asp	Glu	Asn	Gln	Ser	
	counts/min					
Solvent: II						
2	18	362	71	716	3	Gln
5	25	219	0	0	0	Glu
11	0	0	0	2	20	Ser
14	224	0	0	0	4	Asp
17	0	131	0	3	0	Glu
20	0	7	0	14	71	Ser
23	0	95	0	0	7	Glu
24	0	114	0	0	0	Glu

Sequence: (Asp)-Gln-Ile-Ile-Glu-Ala-Val-Ala-Ala-Met-Ser-Val-Met-Asp-Val-Val-Glu-Leu-Ile-Ser-Ala-(Met)-Glu-Glu-(Lys).

(See next page for legend)

Table 1: The peptide was attached to 40 mg of sequencing resin by the diisothiocyanate method. PTHs were chromatographed on silica gel plates first in Solvent I (see text); derivatives which remained at the origin were then separated in Solvent II. About 30% of each PTH sample was used for analysis in each solvent. After chromatography the plate was cut into pieces and counted. Background radioactivity has been substracted from the values quoted. The assigned residues in brackets were deduced by the difference by the difference between the sequence and amino acid composition of the peptide, or from overlapping peptides. Residue 22 (Met) was lost during analysis. Gle refers to glutamic acid methyl ester. Reprinted from Ref. (57), p. 144, by courtesy of Springer-Verlag.

L. Potential and Limitations

Using solid-phase Edman degradation it is now possible to routinely sequence peptides through 20 or 30 residues if: 1) the peptide has a free N-terminal; 2) the peptide can be attached covalently to a resin; and 3) there is sufficient material. As has been discussed, a variety of procedures exist for coupling peptides and even proteins to resins. As new resins and coupling procedures are developed, it should eventually be possible to obtain sequence data on virtually any polypeptide.

Needed not only for this version of the Edman degradation, but also for others, are improved analytical procedures that will allow quick and reliable identification of all PTHs at the nanomole level. Perhaps high-performance liquid chromatography will fill this need. Presently, between 150 and 300 nmole of attached peptide is required for 30-cycle degradations on a routine basis. For shorter peptides, much less material is needed.

The results of Machleidt et al (14,17) demonstrate that solid-phase sequencing of proteins is possible, and there is no reason a priori why degradations through 40 or 50 residues should not be possible. The limiting factors here, as with liquid-phase Edman degradation, are an increase in background after many cycles and the blocking side reaction that inactivates 5 to 10% of the N-terminal groups during each cycle. If the blocking reaction could be reduced by just 3%, much longer sequences could be determined.

With sufficiently sensitive procedures for PTH analysis, it should be possible to use the Edman degradation to analyze the purity of peptides and proteins. Frequently, N-terminal analysis of a protein, for example, shows a single amino acid. This does not rule out the possible presence of proteins with the same N-terminal amino acid. A 4- to 5- cycle run on the sequencer would, in most cases, determine whether one or more proteins is present.

Another application is in monitoring solid-phase peptide synthesis. The appearance of a particular amino acid ahead of sequence (prelap) indicates a 'failure sequence'. Furthermore, the effect is additive-- the farther the peptide is degraded, the more 'prelap' will be observed. Synthetic peptides have in fact been degraded successfully on Merrifield synthesis resins (C. Birr, unpublished).

It is possible to sequence as many as 24 residues per day using a dual column system like the Sequemat. The instrument is no more difficult to operate and maintain than an average amino acid analyzer. Since specially purified reagents are not required, operating cost is low. A further development, which seems feasible, is automatic conversion of thiazolinones to PTHs, which would further reduce the time necessary for sequence analysis of peptides and proteins.

ACKNOWLEDGEMENTS

Much of the work described here was supported by grants from the National Institutes of Health and the National Science Foundation. I also wish to acknowledge the receipt of a NIH Career Development Award and an Alfred P. Sloan Fellowship.

REFERENCES

(1). P. Edman, Acta Chem. Scand., 4, 277 (1950).
(2). W. Schroeder, Methods Enzymol., 25, 298 (1972).
(3). A. M. Weiner, T. Platt and K. Weber, J. Biol. Chem., 247, 3242 (1972).
(4). H. D. Niall and J. T. Potts, in Peptides: Chemistry and Biochemistry Proceedings of the First American Peptide Symposium, Aug. 1968 (B. Weinstein and S. Lande, Eds.), Marcel Dekker, New York, 1970, p. 215.
(5). P. Edman and G. Begg, Eur. J. Biochem., 1, 80 (1967).
(6). R. A. Laursen, Eur. J. Biochem., 20, 89 (1971).
(7). R. B. Merrifield, Biochemistry, 3, 1385 (1964).
(8). R. A. Laursen, J. Amer. Chem. Soc., 88, 5344 (1966).
(9). A. Schellenberger, A. Jeschkeit, R. Henkel and H. Lehmann, Z. Chem., 7, 192 (1967).
(10). A. Dijkstra, H. A. Billiet, A. H. Van Doninck, H. Van Velthuyzen, L. Maat and H. C. Beyerman, Rec. Trav. Chim., 86, 65 (1967).
(11). F. B. Dains, R. C. R. Brewster, C. P. Olander, Org. Synth., 1, 447 (1941).

(12). R. Paul and G. W. Anderson, J. Amer. Chem. Soc., 82, 4596 (1960).
(13). M. J. Horn and R. A. Laursen, FEBS Letters, 36, 285,(1973).
(14). E. Wachter, W. Machleidt, H. Hofner and J. Otto, FEBS Letters, 35, 97 (1973).
(15). A. Previero, J. Derancourt, M.-A. Coletti-Previero and R. A. Laursen, FEBS Letters, 33, 135 (1973).
(16). R. A. Laursen, M. J. Horn and A. G. Bonner, FEBS Letters 21, 67 (1972).
(17). W. Machleidt, E. Wachter, M. Scheulen and J. Otto, FEBS Letters, 37, 217 (1973).
(18). R. A. Laursen, Biochem. Biophys. Res. Comm. 37, 663 (1969).
(19). A. Schellenberger, H. Graubaum, C. Meck and G. Sternkop, Z. Chem., 12, 62 (1972).
(20). G. A. Mross, Ph. D. Thesis, University of California, San Diego, 1971.
(21). J. A. Patterson, in Biochemical Aspects of Reactions on Solid Supports (G. R. Stark, Ed.), Academic Press, New York, 1971, p. 189.
(22). R. P. W. Scott, K. K. Chan, P. Kucera and S. Zolty, J. Chrom. Sci., 9, 577 (1971).
(23). J. R. Millar, D. G. Smith, W. E. Marr and T. R. E. Kressman, J. Chem. Soc., 218 (1963).
(24). G. W. Tregear, K. Catt and H. D. Niall, Proc. Roy. Aust. Chem. Inst., 345 (1967).
(25). G. A. Mross and R. F. Doolittle, Federation Proc., 30, 1241 (1971).
(26). J. J. Maher, M. E. Furey and L. J. Greenberg, Tetrahedron Lett., 29 (1971).
(27). M. Smith, J. C. Moffatt and H. G. Khorana, J. Amer. Chem. Soc., 80, 6204 (1958).
(28). E. Gross, Methods Enzymol., 11, 238 (1967).
(29). R. P. Ambler, Biochem. J., 96, 32P (1965).
(30). M. M. Shemyakin, E. I. Vinogradova, Yu. A. Ovchinnikov, A. A. Kiryushkin, M. Yu. Feigina, N. A. Aldanova, Yu. B. Alakhov, V. M. Lipkin, B. V. Rosinov and L. A. Fonina, Tetrahedron, 25, 5785 (1969).
(31). H. R. Morris, R. J. Dickinson and D. H. Williams, Biochem. Biophys. Res. Comm., 51, 247 (1973).
(32). H. D. Niall, H. T. Keutmann, D. H. Copp and J. T. Potts, Proc. Nat. Acad. Sci (USA), 64, 771 (1969).
(33). M. A. Hermodson, L. H. Ericsson, K. Titani, H. Neurath and K. A. Walsh, Biochemistry, 11, 4493 (1972).
(34). D. Ilse and P. Edman, Aust. J. Chem., 16, 411 (1963).
(35). P. J. Robinson, P. Dunnill and M. D. Lilly, Biochim. Biophys. Acta, 242, 659 (1971).
(36). L. M. Dowling and G. R. Starck, Biochemistry, 8, 4728 (1969).
(37). Corning Glass Works, Biomaterial Supports Technical Bulletin, Corning Biological Products Group, Medfield, Mass., 1973.

(38). T. Inagami, Anal. Biochem., 52, 318 (1973).
(39). A. S. Inglis and P. W. Nicholls, J. Chromatog., 79, 344 (1973).
(40). J. M. Boigne, N. Boigne and J. Rosa, J. Chromatog 47, 228 (1970).
(41). J. O. Jeppsson and J. Sjoquist, Anal. Biochem., 18, 264 (1967).
(42). M. Cohen Solal and J. L. Bernard, J. Chromatog., 80, 140 (1973).
(43). M. R. Summers, G. W. Smythers and S. Oroszlan, Anal. Biochem., 53, 624 (1973).
(44). J. J. Pisano, T. J. Bronzert and H. B. Brewer, Anal. Biochem., 45, 43 (1972).
(45). T. Fairwell and R. E. Lovins, Biochem. Biophys. Res. Comm., 43, 1280 (1971).
(46). H. Tschesche and E. Wachter, Eur. J. Biochem., 16, 187 (1970).
(47). T. Fairwell, W. T. Barnes, F. F. Richards and R. E. Lovins, Biochemistry, 9, 2260 (1970).
(48). O. Smithies, D. Gibson, E. M. Fanning, R. M. Goodfliesh, J. G. Gilman and D. L. Ballantyne, Biochemistry, 10, 4912 (1971).
(49). J. Sjoquist, Arkiv. Kem., 11, 151 (1957).
(50). A. P. Graffeo, A. Haag and B. L. Karger, Anal. Letters, 6, 505 (1973).
(51). C. L. Zimmerman, J. J. Pisano and E. Appella, Biochem. Biophys. Res. Comm., 55, 1220 (1973).
(52). G. R. Stark, Federation Proc., 24, 225 (1965).
(53). G. R. Stark, in Biochemical Aspects of Reactions on Solid Supports (G. R. Stark, Ed.), Academic Press, New York, p. 171 (1971).
(54). J. J. Maher, M. E. Furey and L. J. Greenberg, Tetrahedron Letters, 29 (1971).
(55). C. Birr, C. Just and T. Wieland, Ann. Chem., 736, 88 (1970).
(56). J. I. Harris and R. N. Perham, Nature, 219, 1025 (1968).

AUTHOR INDEX

"The number in parantheses after the page number represents the reference number in the bibliography at the end of the chapter."

A

B

D

E

F

G

L

M

N

O

P

T

Y

Z

SUBJECT INDEX

D

E

F

G

P

R

S

T

U